AF783087

Photonic Packaging Sourcebook

Ulrich H.P. Fischer-Hirchert

Photonic Packaging Sourcebook

Fiber-Chip Coupling for Optical Components, Basic Calculations, Modules

Springer

Ulrich H.P. Fischer-Hirchert
Photonic Communications Lab
Harz University of Applied Sciences
Wernigerode
Germany

ISBN 978-3-642-25375-1 ISBN 978-3-642-25376-8 (eBook)
DOI 10.1007/978-3-642-25376-8

Library of Congress Control Number: 2015936158

Springer Heidelberg New York Dordrecht London
© Springer-Verlag Berlin Heidelberg 2015
This work is subject to copyright. All rights are reserved by the Publisher, whether the whole or part of the material is concerned, specifically the rights of translation, reprinting, reuse of illustrations, recitation, broadcasting, reproduction on microfilms or in any other physical way, and transmission or information storage and retrieval, electronic adaptation, computer software, or by similar or dissimilar methodology now known or hereafter developed.
The use of general descriptive names, registered names, trademarks, service marks, etc. in this publication does not imply, even in the absence of a specific statement, that such names are exempt from the relevant protective laws and regulations and therefore free for general use.
The publisher, the authors and the editors are safe to assume that the advice and information in this book are believed to be true and accurate at the date of publication. Neither the publisher nor the authors or the editors give a warranty, express or implied, with respect to the material contained herein or for any errors or omissions that may have been made.

Printed on acid-free paper

Springer-Verlag GmbH Berlin Heidelberg is part of Springer Science+Business Media
(www.springer.com)

Preface

The basics of photonics, which is an optical high-tech technology, use light sources and photodetectors and different kinds of waveguides for its applications. The bandwidth of products and devices on the market is present in every part of our daily lives. Starting from laser pointers to cash systems in the market halls down to telecommunications and lighting systems for cars and lamps, the technology has matured rapidly giving the world long-term stable devices.

Electrical and optical state-of-the-art technologies to package the devices are important to ensure high-volume, low-cost devices with high performance. So the material price of DVD is higher than the production costs of the high-volume device. Smart companies have developed inexpensive packages and automated mounting technologies for photonic products, and they have to develop robust, special packages for high-performance telecommunications and IT devices.

Optical communications technology has undergone in the last two decades, a breathtaking development that was not even in sight of the most daring augurs. In long-distance networks, fiber-optic technology was indeed already used, but the industry had been hesitant to this new technology. Researchers succeeded in just a few years to increase the transmission data rates over a glass fiber by several orders of magnitude of several gigabits/s to several terabits/s.

At the same time, developers have converted the results within short times into commercially usable products, which were then immediately installed by the major telecom companies in their networks and thus have increased the transmission capacity considerably.

The main reason for this development was two independent events:

The deregulation of the communications market by allowing competing smaller telecommunications companies and the explosive growth of the Internet.

The major telecom providers saw themselves exposed in recent years to growing competitive pressure from the start-up of many small dynamic companies. All competitors invested heavily in expanding their own networks in order to reduce the transmission cost. The suppliers were now required to quickly offer new, more effective network components. This was also only possible with massive investment in highly qualified engineers and novel instruments, which in turn significantly revived the labor market. The exponential growth in the number of Internet users through the increased use of multimedia applications since the mid-1990s has had a strong increase in transmission rates in the data range for the result, which is still unbroken, and will probably continue to increase. Data services have now overtaken voice services in capacity and contribute even more to further the hunger for transmission capacity.

Thus, the Internet has become the medium for the development of data networks. A key concern of this book is to make this new technology transparent to a larger audience both in terms of the development of optical networks as well as on the optoelectronic structure of the fiber-optic components used.

Since the connection of optoelectronic components is a complex issue for the construction technology to the ever-increasing transfer rates, I have tried to provide readers with the construction technical side, with the theoretical foundations, as well as from the practical point of view in detail. For in-depth studies, there are numerous literature references.

A central space takes the description of the interface fiber to the chip, which is a problem of extreme importance for the understanding of optical coupling and their adjustment. After an excursion into the traditional electronic package technology with their different bonding techniques, the optical connector and the production of optical fiber lenses are described. The measurement of the near and far fields is an important part for optimal fiber–chip coupling. The different spot-size measurement techniques and their advantages and disadvantages are described in detail.

In the following section, we will consider what adjustment techniques to today's users are available. The trend is from the active techniques for self-adjusting passive adjustment, which promises a miniaturized and much more cost-effective mounting technology. The conclusion is a section with many examples of different module structures and the description of reliability tests as part of the ISO standard.

This book will serve both students of electrical engineering and related engineering courses as a teaching aid as well as professionals who need an application-oriented reference book on optical setup and connection technology for their daily work.

The manuscript for this book was created between 2012 and 2014 and the people involved have tasted a few gray hairs. I would like to thank Matthias Haupt, Günter Bühler, and Krzysztof Nieweglowski especially since they have supplied Chaps. 10 and 12 to this book. I would also like to thank my assistant, Marika Ozegowski in a special way, because she has done it again and again to bring the formats in the right order and gave the outline the finishing touches.

Mr. Truschkewitz I thank for the processing of the images. Also, for all the staff of my group of Photonic Communications Lab here not mentioned by name, I would like to thank you for your input. Without their help, this book would never have been realized.

Wernigerode
June 2014

Ulrich H.P. Fischer-Hirchert

Contents

Abbreviations

Formulas in Latin

a	Thickness of the wave-guiding layer, film thickness, core thickness
b	Width of the wave-guiding layer
$\mathbf{B}$	Magnetic flux density
c_0	Vacuum speed of light $c_0 = 299{,}792 \times 10^6\ \mathrm{ms}^{-1}$
$c.c.$	Complex conjugate extension
$\mathbf{D}$	Electric flux density
e	Elementary charge $e = 1602 \times 10^{-19}$ As
$\mathbf{E}$, E	Electric field strength
E_0	Amplitude of the electric field intensity
E_{c0}, E_{s0}	Amplitude of the electric field strength in the cladding and the substrate
E^{i}	Incident electric field intensity at wave-optical viewing
E^{r}	Reflected electric field intensity at wave-optical viewing
E_N	Normalized electric field strength
E^{t}	Transmitted electric field intensity at wave-optical viewing
E_{TE}	Electric field strength of the TE wave
E_{TM}	Electric field strength of the TM wave
E_x, E_y	Transverse component of the electric field strength
E_{x0}, E_{y0}	Amplitude of the transverse component of the electric field strength
E_z	Longitudinal component of the electric field strength
f	Frequency of a wave
h	Planck's constant $h = 626 \times 10^{-34}\ \mathrm{Ws}^2$
$\mathbf{H}$	Magnetic field strength
H_0	Amplitude of the magnetic field strength
H_z	Longitudinal component of the magnetic field strength
I	Intensity
I_p	Photocurrent of the photodiode
I_x, I_y	Intensity in the transverse direction
$\mathbf{j}$	Current density

k, k	Wave vector, phase constant of a wave in a vacuum
k_r	Radial component of the phase constant
l_0	Optical path
m	Mode index in the x direction
n	Mode index in the y direction
n, n_1, n_2	Refractive index, refractive index of a medium
n_c	Refractive index of the cladding of a waveguide
n_{eff}, $n_{eff}^{(m)}$	Effective refractive index of the mode
n_f	Refractive index of the guiding layer of a waveguide
n_s	Refractive index of the substrate of a waveguide
P	Optical power
r	Radial coordinate, radial distance
r_{TE}	Amplitude reflection coefficient of the TE wave
r_{TM}	Amplitude reflection coefficient of the TM wave
R	Radius of curvature of the phase fronts of the Gaussian beam
R_p	Responsivity of photodiode
R	Power reflection factor, reflectivity
S, S	Poynting vector
S_z	Poynting vector in the longitudinal direction
t_{TE}	Amplitude transmittance of the TE wave
t_{TM}	Amplitude transmittance of the TM wave
T	Power transmission factor, transmission
T	Coupling efficiency (logarithmic)
T_{mm}	Coupling efficiency of mode mismatch
T_s	Coupling efficiency of transverse offset (in the x- or y-direction)
T_z	Coupling efficiency on longitudinal displacement in the Z-direction
T_θ	Coupling efficiency of angular displacement
t	Time
v	Normalized frequency
v_{ph}	Phase velocity of the propagating mode
w	Field radius and beam radius of the Gaussian beam
W	Gaussian far-field width
w_0	Mode-field radius, spot size of the fundamental mode
$w_{0x,y}$	Mode-field radius in the x- and y-direction
w_{01s}	Mode-field radius in the x- or y-direction of the coupling Butt-fiber
w_{02s}	Mode-field radius in the x- or y-direction of the coupling WL-tapers
x	Cartesian coordinate
y	Cartesian coordinate
z	Cartesian coordinate
z, z	Direction of wave propagation
z_0	Confocal parameter of the Gaussian beam
Z_0	Free space wave impedance

Formulas in Greek

α_c	Transverse attenuation constant in the cladding
α_s	Transverse attenuation constant in the substrate
$\beta, \beta^{(\mathrm{m})}$	Propagation constant in the z-direction (direction of propagation)
β_f	Propagation constant in the x-direction
Δ	Normalized refractive index difference, normalized refractive index jump
Δx	Vertical offset in the x-direction
Δy	Horizontal offset in the y-direction
Δs	General transverse offset
Δz	Longitudinal displacement in the z-direction
$\Delta\theta$	Angular displacement
γ	Propagation angle relative to the axis of the fiber waveguide
ε_0	Permittivity $\varepsilon_0 = 885 \times 10^{12}$ As/Vm
ε_r	Relative dielectric constant
η	Coupling efficiency (not logarithmic)
η_{mm}	Coupling efficiency at mode mismatch
η_s	Coupling efficiency of transverse offset (in the x- or y-direction)
η_z	Coupling efficiency on the longitudinal displacement in the z-direction
η_θ	Coupling efficiency of angular displacement
λ	Wavelength in vacuum
λ_{Q}	Band edge wavelength
μ	Mode index in the radial direction
μ_0	Absolute permeability $\mu_0 = 1257 \times 10^{-6}$ Vs/Am
μ_r	Relative permeability
μ_s	Spot-size ratio in the x- or y-direction
v	Mode index in the azimuthal direction
ρ	Electric charge density
ϕ, ϕ_1, ϕ_2	Incidence and the angle at beam optical consideration
ϕ_{B}	Brewster angle
ϕ_{G}	Critical angle of total reflection
$\phi^{\mathrm{i}}, \phi^{\mathrm{i(m)}}$	Incidence angle at wave-optical consideration
θ	Divergence angle of the Gaussian beam
ϕ^{r}	Reflection angle at wave-optical consideration
ϕ^{t}	Transmission angle at wave-optical consideration
ϕ_x, ϕ_y	Phase angle of the transverse component of the field
$\phi(z)$	Additional phase of the Gaussian beam
Φ_c	Phase shift of wave of total reflection at the cladding
Φ_{TE}	Phase shift of TE wave in total reflection
Φ_s	Phase shift of wave of total reflection at the substrate
Φ_{TM}	Phase shift of the T wave in total reflection
ψ	Azimuth angle of inclination of the plane of incidence
ω	Angular frequency of the wave

Shortcuts

Al_2O_3	Ceramics
AlF_3	Aluminum fluoride glass
ANSYS	FEM program
APD	Avalanche photodiode
AR	Anti-reflective coating
AVT	Assembly and interconnection technology
AWG	Arrayed waveguide grating
BaF_2	Barium fluoride glass
BFY	Butterfly package
BIAS	Control and operating current
BL	Bitrate-length product
BLM	Bump-limiting metallization
BNC	RF connector type up to 2 GHz
BPM	Beam propagation method
Butt-fiber	Straight-cut fiber end facet
C4	Controlled collapse chip connection
CAD	Computer-aided design
CaF_2	Calcium fluoride glass
Cladding	Covering layer, cladding of a waveguide
CVD	Chemical vapor deposition
DBR-laser	Distributed Bragg reflection laser
Demux	Demultiplexer
DFB-laser	Distributed feedback laser
DIL	Dual inline package
DOF	Focal length
DUT	Device under test
DWDM	Dense wavelength division multiplex
EDFA	Erbium-doped fiber amplifier
EELED	Edge emitting LED
EH	Hybrid wave, almost transverse magnetic wave
EIM	Effective-index method
EO-PCB	Electro/optical printed circuit board
EXZ	Excitons
FC	Flip-chip
FC/PC	Fiber connector with physical contact
FDTD	Finite-difference time-domain method
FE	Iron
FEM	Finite element method
FIT	Failure in time
FP	Fabry–Perot

FWHM	Full width half maximum
GaAs	Gallium arsenide
GGL	Gain-guided laser
GRIN	Graded index
HE	Hybrid wave, fast transversal electrical wave
HHI	Heinrich Hertz Institute, Berlin
HPC	High-performance computing
IBSE	Ion beam shadow etching
IGL	Index-guided laser
InGaAsP	Indium gallium arsenide phosphide
InP	Indium phosphide
ITU	International Telecommunication Union
IVD	Inside vapor deposition
LD	Laser diode
LED	Light emitting diode
LIGA	Lithography, electroplating, molding
$LiNbO_3$	Lithium niobate oxide
LP	Linear polarized
LWL	Optical waveguide
MCVD	Modified chemical vapor deposition
MFD	Mode-field diameter
MM	Multi-mode
MOMBE	Metal-organic molecular beam epitaxy
MOVPE	Organometallic vapor phase epitaxy
MP3	MPEG3 data compression code
MSM	Multi-segment or finger diode
Mux	Multiplexer
NA	Numerical aperture
Nd	Neodymium
NTC	Negative temperature coefficient
OADM	Optical add/drop multiplexer
OEIC	Optoelectronic integrated circuit
OIC	Optical Integrated Circuit
OST	Optical spot-size transformer, mode field transformer, Taper
OTDM	Optical time division multiplex
OTDR	Optical time domain reflectometer
OVD	Outside vapor deposition
PAVT	Photonic assembly and interconnection technology
PC	Polycarbonate
PCB	Printed circuit board
PCVD	Plasma-activated vapor deposition
PD	Photodetector
PDFA	Praseodymium-doped fiber amplifier

PDL	Polarization-dependant loss
PI	Power/current
PIN	Layer sequence in fast photodiodes with neutral zone (positive-insulator-negative doped zone)
PMMA	Polymethylmethacrylate (acrylic glass)
POF	Polymer optical fiber
POF-LWL	Polymer Optical Fiber
PON	Passive optical network
PS	Polystyrene
PTFE	Teflon
PVD	Physical vapor deposition
$Q_{1,06}$	InGaAsP ($In_{0.90}Ga_{0.10}As_{0.24}P_{0.76}$) mit λ_Q = 1,06 μm
$Q_{1,3}$	InGaAsP ($In_{0.75}Ga_{0.25}As_{0.57}P_{0.43}$) mit λ_Q = 1,3 μm
RF	Radio frequency
RIBE	Reactive ion beam etching
RIE	Reactive ion etching
SiNx	Silicon nitride
SiO_2	Silicon oxide (silica quartz glass)
SLED	Surface-emitting LED
SM	Single-mode
SMA	RF connector up to 18 GHz
SMF	Single-mode fiber
SMT	Semiconductor manufacturing technology
SOA	Semiconductor optical amplifier
Spot-Size	Focal diameter, double the amount of mode-field radius
TE	Transversal electrical
TEM	Transversal electromagnetic
TIR	Total internal reflector
TM	Transverse magnetic
TMM	Ceramic thermoset polymer composites
TO	Thermo-optic
TPA	Two-photon absorption
TSM	Top surface metallization
TTG-laser	Tunable twin guide laser
VAD	Vapor axial deposition
VCSEL	Vertical surface emitting laser
WDM	Wavelength division multiplex
WL	Waveguide
WL-Taper	Waveguide taper

WWW	World Wide Web
YAG	Yttrium–aluminum–garnet
YF_3	Yttrium fluoride glass
ZrF_4	Zirconium fluoride glass

Chapter 1
Introduction into Photonic Packaging

Abstract In the last 20 years, fiber optic systems show a huge development in terms of expanded data rates of 40–160 Gb/s in the telecom core network. Additionally, the data rates in the consumer PC segment such as internal front side bus or external bus structures such as USB 3.0 or Thunderbolt (Intel_Corporation 2011) are also expanding extremely. In automotive applications, the bus speed becomes also in the region of several hundred megabits (MOST 3.0). The advantage of optical fibers depends on its almost infinite transmission bandwidth, but still has strong disadvantages in the field of handling of the fibers and in the fiber–chip coupling, respectively. In this work, the fiber-chip coupling technology will be analyzed, while new coupling techniques and cost optimization basics are presented. Thus, at the beginning, an overview of the technology drivers of optical communications systems will be discussed. Further, many examples of photonic packaging and interconnection technology are presented in depth. All actual adjustment and fixation techniques are focused. A deeper analysis of the optical interconnection technology with basic theory of waveguide coupling and loss mechanisms concludes the introductory part of work. Then, an overview on the modular technology of major photonic components in the field of high-rate fiber optic networks is presented, followed by the analysis of the fiber–chip coupling in these different applications. In the conclusions, a summary and an outlook on the further development of the technology of photonic packaging and interconnection technology will be given.

1.1 Optical Transmission Systems

Optical communications systems and furthermore the optical packaging technology have achieved a very high significance in a relatively short time (approx. 20 years) for general worldwide communications. For commercial applications, these systems are introduced in the mid-eighties of the twentieth century. Today, all international

© Springer-Verlag Berlin Heidelberg 2015

U.H.P. Fischer-Hirchert, *Photonic Packaging Sourcebook*,
DOI 10.1007/978-3-642-25376-8_1

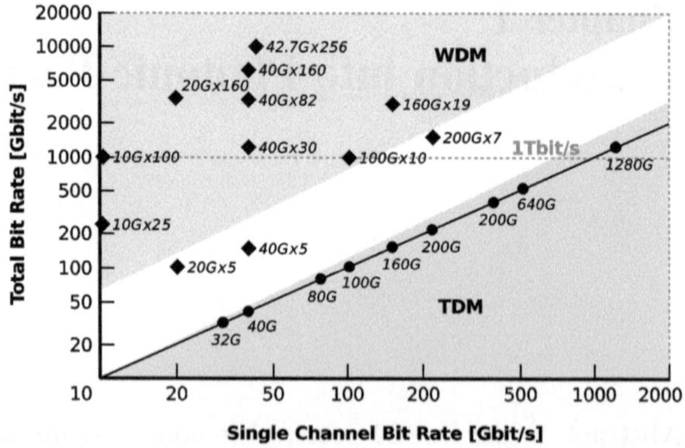

Fig. 1.1 Evolution of transmission data rates

and national long-distance connections are based on fiber optic systems with bandwidths from 2.5 to 100 Gb/s.

Future high-bit-rate transmission optical communication systems will apply wavelength-division multiplexing in combination with high data-rated time-multiplexed single channels to use the almost unlimited bandwidth of the optical fiber. Current estimates (Network 2002; O'Mahony et al. 2006; Gnauck et al. 1996), as collected in Fig. 1.1, show that it will be possible to transfer several tens of terabits via a glass fiber. In particular, the WDM technology makes it possible, to expand the over all data rates up to more than 10 Tb/s with 280 channels which have a single data rate of 40 Gb/s. The electrical time-division multiplexing technique (ETDM) is actually characterized by difficulties in the technological development of RF transmitters and receivers. Therefore, ETDM did not progress as fast as the WDM technology.

As an extension of the electrical time-division multiplexing technology, which could be realized up to now only to 100 Gb/s, the optical time-division multiplex technique (OTDM) is rapidly establishing (Ramaswami 2004). It has been shown by the Heinrich Hertz Institute in Berlin (Ludwig et al. 2007, 2008) that an optical modulation at 160 Gb/s DPSK and 400 km can already transmitted virtually error-free with the use of only three optical amplifier stages. To utilize the full potential of OTDM, it is necessary to develop optical switching elements, which can realize all-optical switching of data packets or even single bits. This function will be used in network nodes, which have the task to distribute incoming optical data packets to the output interconnect ports. However, it is expected that the optical network nodes are expected be brought to technical maturity after 2015 (Japan 2006).

The rapid development of optical transmission technology made it possible that all actual wide area networks are based on optical fibers. These optical systems are forming the backbone of all network operators. In optical point-to-point transmission, links are, however, still part of today's opto-electronic networks. Transmission occurs via the optical fiber, while the exchange between networks and network nodes takes place electrically. Although many companies are working on the realization of all-optical networks, but currently a combination of optical and electronic shows major advantages (Späth 1999).

A fundamental requirement in the use of semiconductor-based optoelectronic elements is to ensure an efficient and reliable coupling to the transmission medium, the so-called fiber–chip coupling. Typically, the coupling from the IO-chip to the fiber shows high insertion losses. The losses are due to the different field distributions in the fiber and the waveguide of the device. Therefore, at the beginning of the international development of photonic technology modules many scenarios have been carried out to improve the coupling efficiency (Saruwatari and Nawata 1979; Saruwatari and Sugie 1981; Mickelsen 1997; Marcuse 1977).

The incorporation of optoelectronic devices (OEIC's) in the transmitter and receiver modules into the designated module housing with the appropriate interfaces, which is called photonic packaging (photonic packaging and interconnection technology—PAVT) has to play a key role. Without optical adaptation, it is hardly possible to couple more than 10 % of the light into the fiber. This insufficient efficiency can be approved only by the adaptation of the optical mode field of the glass fiber to that of the chips. One often realized opportunity can be the use of microlenses or fibers with lens-shaped end surfaces (fiber taper). But in the case that more fibers are fixed in arrays, it is extremely difficult to be coupled to a device with several optical ports. The adjustment of such a couple of fibers is not possible to be individually focused. The alignment tolerances are reduced to a few tenths of a micrometer. The result are hugely increasing costs for the coupling and housing of OEIC's.

For this reason, there is great interest in optical coupling elements, which enhance both, the coupling efficiency, as also the alignment tolerances. This demand can be realized by the implementation of so-called integrated mode-field transformers, and on the glass fiber side by individual lenses or lens systems. However, it appears that the coupling procedure using waveguide mode-field transformer is clearly preferable instead of the complex coupling procedure with the use of simple lenses (Fischer et al. 2002).

Using waveguide mode-field transformer, the structure of the waveguide on the OEIC is realized by a ramp-shaped cross-sectional change. This structure improves the coupling between fiber and OEIC up to 90 %. The field distribution is expanded, and increased alignment tolerances are the result. This offers the possibility to more cost-reducing passive adjustments, such as Flip-chip-bonding or fixing in V-grooves to fix the glass fiber in optimal position in front of the OEIC (Fischer 2002a).

In contrast, the lens solutions always forces closer adjusting tolerances for high coupling efficiencies. The narrowing of the mode-field constrained by the lens transformation forces smaller coupling sub-micron tolerance for lateral adjustment. Additionally, the working distance of the fiber to the chip is also reduced to a few microns. Thus, the mechanical direct fiber-chip contact possibility of both components increases dramatically and the possibility to damage the OEIC end face. However, these discrete fiber–chip coupling techniques via lens systems are most used internationally and steadily improved technologically. In addition, all of these techniques need very labor-intensive processes and should therefore be replaced in the next years by self-aligned processes.

These important aspects of the simplification and improvement of the fiber–chip coupling with respect to the possible improvement and simplification of the coupling process are the focus of this work.

Furthermore, it is necessary to transmit the high data rates of the source to OEIC and back to the data sink. This also means that within the established data modules, the bandwidth of the transmitted data must remain intact. Here, solutions for the high-frequency (RF) packaging based on own work will be pointed out. Finally, the development of packaging technologies of the next generation will be presented.

1.2 System Applications of Optical Communications

1.2.1 Optical Telecommunication Systems

The individual components of an optical communication network are first presented in a summary, and then, the individual elements are discussed in detail in connection with the problems of photonic structures and connection technology. In Fig. 1.2, the essential basic elements of an optical connection are shown in a simplified form as follows:

- Data source
- Modulator/Coder
- Electro-optical converter (e.g., laser diode)
- Multiplexer
- Fiber
- Amplifier
- Coupler
- Switching Matrix
- Regenerator
- Demultiplexer
- Opto-electrical transducer (e.g., photodiode)
- Demodulator/decoder
- Data sink

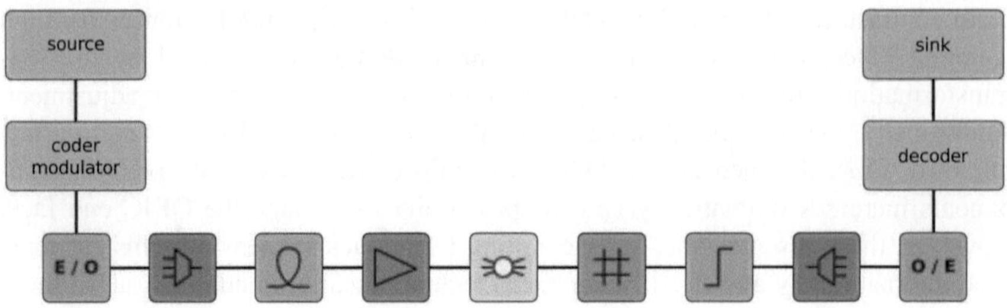

Fig. 1.2 Elements of an optical transmission path

The first and last two elements are contained as basic elements in each communication system and thus no special elements of an optical network. All of these route elements partly in their structure are very different from each other. Thus, transmitter and receiver OEICs, although usually associated with only one fiber, must be provided due to the different waveguide structure on Indium phosphide (InP) or GaAs with lenses or other mode-field transformers (Honecker 2000). Without these additional parts, an efficiency of fiber–chip coupling of only about 10 % can be achieved. In addition, the transmit module, a temperature of OEICs, is often necessary because the output wavelength of the transmitter would otherwise drift too much and would thus disturb a nearby lying transmission channel sensitive. Both transmitter and receiver OEICs use high-frequency connections to the modulation/demodulation of the data.

Optical amplifiers generally work with doped rare earth ions in the fiber. These fiber devices can be spliced very simple to fibers and hardly need any complex setups. Only the integration of the laser driver in the construction of the high heat dissipation is of particular interest (Bjarklev 1993).

Optical multiplexer (d) and demultiplexer (j) are mostly used in the wavelength division multiplex systems. Here, the following essential two variants are used:

- Arrayed waveguide gratings (AWGs) (Kuhlow et al. 1999) and
- Interference components in the form of dielectric filters (Scobey and Spock 1996)

The opto-mechanical design of the dielectric filter is purely passive. In transmission, the filtered wavelength is coupled into the output fiber, while the rest of the transmission comb is imaged on the input side in reflection on the second fiber. This structure is extremely simple to implement cost-effective and long-term stability. However, the losses added from all components are very strong when a large number of wavelengths must be separated. Typically, these components are only realized up to about 16 wavelengths.

AWGs offer of up to 128 separable wavelengths (Przyrembel and Kuhlow 1999) to implement an integrated optical way to implement the wavelength-division multiplexing function in a component.

But, the problem of coupling the optical fiber array with high -fiber-count is still a major challenge to photonic packaging. Both, the deflection of the fiber arrays and of the AWGs, complicate the optical coupling. In addition, the summation of the position errors of the fibers in the V-grooves increases the attenuation of the optical coupling. Furthermore, the filter response of the AWG is strongly temperature-dependent (Ishii et al. 1998), which makes a temperature control of the chips necessary. However, Siemens (Albrecht 1994) has a structure proposed which minimizes the temperature dependence to a negligible amount by using a length adjustment part supporting the fiber array in front of the lens element of the so-called waveguide slabs. However, these patented mechanics by Siemens have been realized only in few modules so far.

Optical switch in the form of an integrated optical switching matrix obtained an outstanding importance in the field of application of the switching nodes (Fig. 1.3).

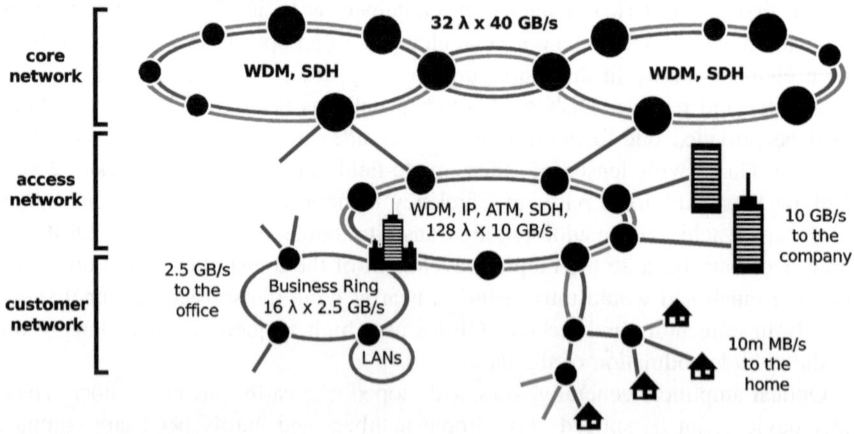

Fig. 1.3 Basic concept for optical networks

Here, we distinguish between slow thermal switches (ms range) and fast components, which, as already mentioned in the introduction, will route in the near-future data bursts (Düser et al. 2002) or even individual data bits in the range of ps time scale (Jahn et al. 1996; Jahn 1996; Rosin et al. 1998). The thermal switch matrices use analogous to the AWG—structures a multi-fiber coupling via fiber arrays and must be thermally stabilized in most cases. Thus, the structures and coupling techniques are very similar to those of the AWGs having the same adjusting problems as the multi-fiber coupling. In contrast, the fast switches are mostly single switches with only a few fibers for optical coupling. However, these components operate on the basis of InP with very asymmetric and not adapted fields to the glass fiber optic cables. The optical fields must therefore be adapted with suitable lens systems (Fischer 2002b) in order to ensure a high coupling efficiency. The packaging process becomes more complex and more expensive enormously.

An optical or photonic network consists essentially of at least three mutually nested grid levels, which are referred to as follows in Fig. 1.3 combined with a basic concept for optical networks:

- Core network,
- Access network, and
- Customer network

Important components within the core area are the so-called switching node (X-connects). These are complex combinations of components for switching, amplifying, and splitting/merging of single-ended channels. A typical X-connect without wavelength conversion is shown in Fig. 1.4. From the left-most input side, the collected optical transmission signals will come in. Here, the signals with the aid of the demultiplexers (DEMUX) are first divided into their individual channels. Then, the separated signals will be conducted with the help of optical switching elements to the corresponding outputs. Within the multiplexer (MUX), all signals are

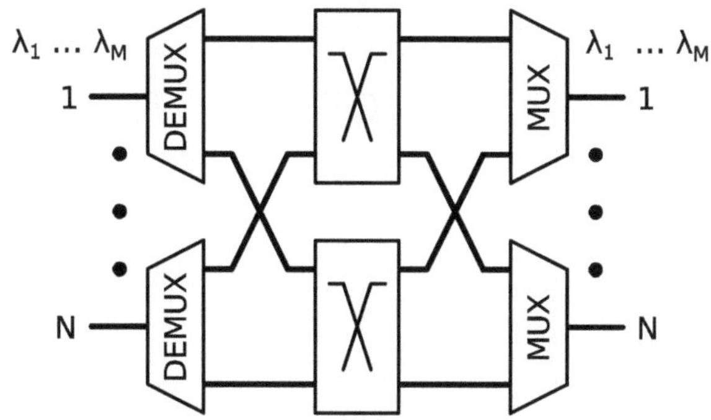

Fig. 1.4 Optical X-connect without wavelength conversion

combined and transmitted according to a gain stage. A central monitoring unit supervised at the same time all incoming and outgoing signals to the correct wavelength, intensity, and routed direction.

In the core network portion connecting the cities, states, and continents together, one expects very high data rates, but few slow switching processes between the individual network nodes. To transfer the safety, nets are designed so that the signals in the event of a cable break can always reach their destination on an alternative route. To do this preferably, a so-called ring topology of the network, as illustrated in Fig. 1.3 will be realized. In central areas of the network, components with the highest performance values are used. Here, the price of the components plays only a very minor role, while the performance is clearly in the foreground. The use of innovative photonic OEICs, e.g., for all-optical clock regeneration or OTDM MUX (Brox et al. 2004; Tekin et al. 2000) when using data rates above 40 Gb/s calls on the photonic packaging technique for innovative as costly solutions out (Ehlers et al. 2002; Agrawal et al. 1997; Rosin et al. 1998).

A special variant of the photonic networks is the all-optical network. Here, not only the bits to be transmitted are transparently send by the sender to the recipient, but the optical bits around the running track are purely routed and amplified, without electronic transformations. The bit transparency is not possible with the components immediately available. In most cases, completely new integrated optical components must be designed, because so far there is no equivalent in the electrical network, such as the optical clock recovery. For the next few years, this type of network is a challenge for the international research and development society.

In the local and metropolitan area networks, one expects lower data rates than in the core network, but with the request to send the branched off from the core network signals to many end nodes as inexpensive as possible. Thus in this network area other less expensive components are used than in the wide area parts. These facts will afford polymer components and either simple transmitters or receivers with data rates of several Gb/s.

Another price jump is necessary for the components to be used in the Subscriber area. Here, it is on products that must be produced for a mass market and so may

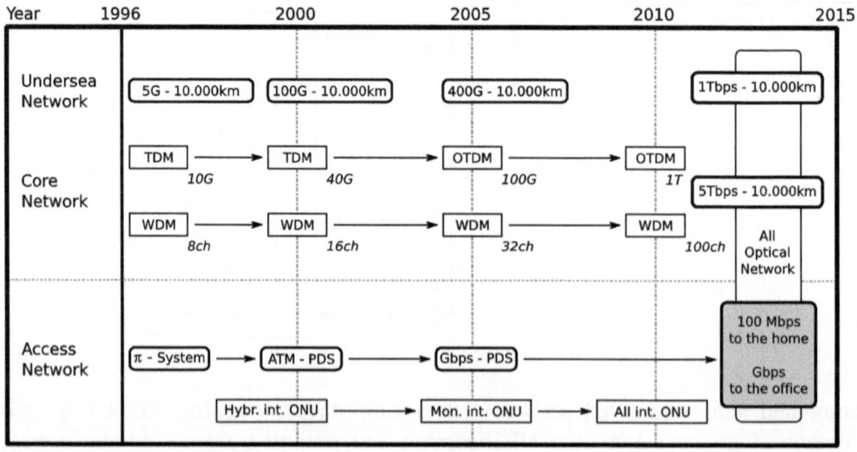

Fig. 1.5 Japanese road map for the development of optical transmission systems

only cost a fraction of the price of the upper network levels. An optical transmitter and receiver must then be in any phone or TV, as final consumption merchandise only an item for a few cents. This is a very big challenge for industrial production. So far, the costs for simple transmitter and receiver are up to now still by two orders of magnitude too high. One expects a strong support from the microsystem technology in conjunction with the photonic packaging. Some technical possible attemps are described in Chap. 9, especially the passive adjustment.

An estimate of the further development of telecommunications networks is based on the increase in the bit rate for a transmitted channel rate of 10 Gb/s in 1996 to 40 Gb/s in 2000 and 160 Gb/s in 2005 which was predicted from the Japanese telecommunications authority OITDA (Fig. 1.5).

However, a delay of approximately five years has been found especially in the introduction of the time-division multiplexing technology, so 160 Gb/s is expected from today's point of view only in 2015 for real used transmission systems. In conjunction with the wavelength-division multiplexing (WDM) technique, a combination of several individual transmission channels to a transmission bundle transmission rates of several hundred terabits by the year 2020 will be possible.

1.2.2 Optical Datacom Systems

In addition to the historically grown telecom networks with their core network-optical structure for data transmission, the market for the datacom systems has increasingly developed with optical transmission media in recent years. Coming from the consumer market, the FDDI transmission system was in competition with the 10 Mb/s Ethernet [IEEE802.3, (IEEE 2011)].

The Fiber-Distributed Data Interface [FDDI (ANSI 1994)] is a late 1980s developed 100 Mb/s standardized network architecture for local area networks [ANSI X3T9.5. (ANSI 1990, 1992, 1994, 1998)]. As a medium, fiber optic cable with a double counter-rotating ring executed with so-called token access. As transmission media multi-mode glass fibers (MMF) and single-mode glass fibers (SMF) were used with a distance between adjacent stations of 2 km and a ring length of up to 100 km. FDDI was the designated successor for the old 10 Mb/s Ethernet (IEEE 1987) in the 1990s. However, it could not prevail for reasons of cost comparing with Fast Ethernet and even Gigabit Ethernet.

In 2003, a new standard for Ethernet transmission has been published by the IEEE 802.3 group, which brought the advantages of optical data transmission up to 10 Gb/s (802.3ae (IEEE 2005) into the Ethernet hierarchy. This made it possible to transmit 10-Gb signals according to the Ethernet protocol using MMF fibers up to 300 m in length and up to 40-km-long SMF (IEEE 2005). In Fig. 1.6, an appropriate scenario for the interplay of the different Ethernet LANs in the metro area is outlined with a link to a WDM telecom core network. Now it is possible to use Ethernet technology not only in the LAN area of 10 Mb/s but also on 100/1000 Mb–10 Gb/s without using transponders. In particular in campus networks, this technique will now enforce steadily and replace the old stand-alone solutions.

A major advantage of the new Ethernet technology is the fact that you could rely to manage high data rates based on the experience of the optical telecommunications industry. The transmitting and receiving elements for 10-Gb Ethernet are based on the STM-64 components for telecommunications networks. Even the data rates differ only slightly: STM64 with 9.9532 Gb/s (ITU G.803 2003) to 10.3 Gb/s (IEEE 2005).

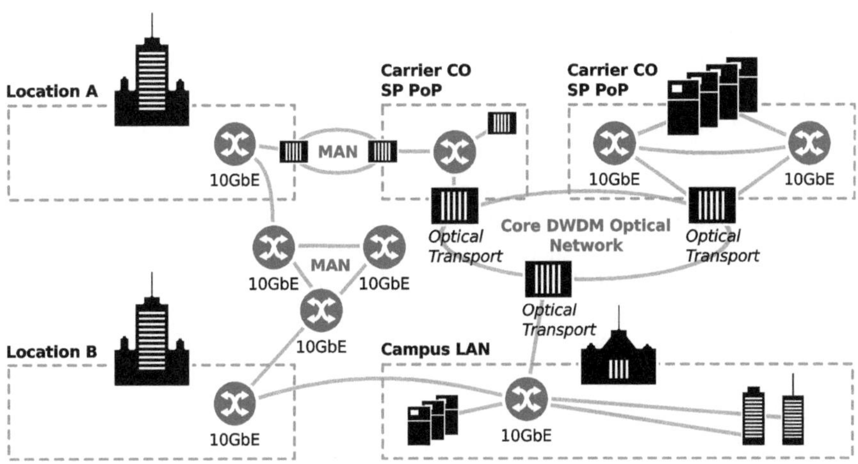

Fig. 1.6 10-Gb Ethernet applications in LAN and metropolitan area

Midscale with increasing suitability to mass market: To apply for the photonic structure technology of Ethernet modules, the same laws as for telecommunications modules up to 10 Gb/s is applicable. Thus, the Ethernet technology will open up an additional large user market for optical modules of the metro network in the coming years. This will also lead to price jerk in the module area that will also encourage the telecom sector.

It had existed since the early 1990s in the mainframe area a great need for fast parallel data connections over short distances. Here, a niche market has developed in which 30 Gb/s can be transferred by data modules with up to 2.5 G/s and 12 parallel optical data lines (Plickert and Melchior 1999). With anticipated future processor configurations at data rates, exceeding 500-Mb/s copper lines are unsuitable because of their high shielding expenses relating bridgeable distances and costs.

The transmission in these parallel systems was realized with lasers at 850 nm wavelength and by an array of vertically arranged source emitting lasers (VCSEL). The modules are connected with fiber ribbon cables that can be up to 75 m long. To use these systems in complex mainframe arrays, rack-to-rack transmission in the form of an optical bus was arranged. Since 1997, data transmission between mainframes and their memory areas was realized with 1 Gb/s Fiber Channel systems (FCIA 1997), which allows to connect multiple hosts to multiple storage subsystems. Many storage area networks are based today on the implementation of the Fibre Channel system (Fibre Channel Industry Association 2014). The achieved data transfer rates are now up to 16 Gb/s, mostly used in full-duplex mode for data transfer rates of up to 1.6 GB/s. As transmission media, copper cables (up to 30 m) and fiber optic cables are used to connect the storage systems of up to 10 km distance.

For data transmission within mainframes, one expects in a few years a data rate of over 100 Gb/s by the processor to the cache memory and between multi-processor architectures that are highly integrated in a stack. The development of new backplane architectures—GIGASCALE—is driven in the computer field (Zolper 2008; Borkar 2004) on one side by the progress of system integration on the board and on the other hand by the extreme increase in the demand for bandwidth in the inter-chip area and also by the increase of clock frequency on the processor chip.

Physical and technological limiting factors are to find in the conventional electrical signal transmission on the one hand, the sensitivity to external electromagnetic radiation at high clock rates in the bus system, and further by the required high number of data lines to and from the processor chip (pin count). The high pin count limits the bandwidth due to parasitic high-frequency problems that show up in the additional increase of reflection and cross talk between the data channels at the required clock frequencies in the GHz range. The high pin count on one hand limits the bandwidth and on the other hand produces high frequency parasitic problems that show up in the additional increase in reflection and crosstalk between the data channels at the required clock frequencies in the GHz range.

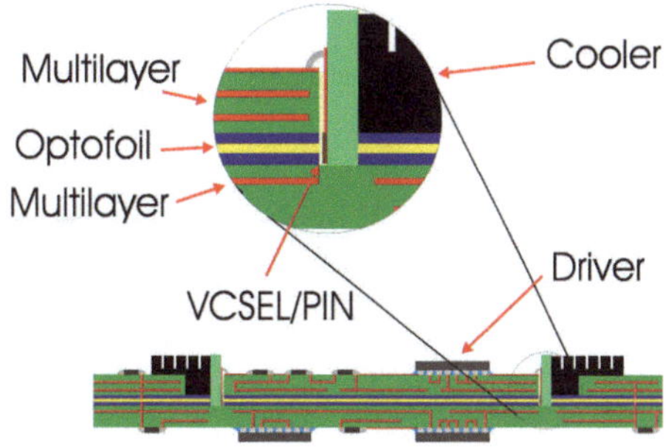

Fig. 1.7 EOCB optical module with fiber optic coupling (Scheel 2003)

Already in 1990, the first generation of a discrete optical fiber connection in the form of a Motherboard rack system (Moisel 2000) was presented. By 1995, the second generation of these systems is implemented in the form of optical flex foil compounds of DaimlerChrysler in a test bed (Kibler et al. 2001). Since the year 2000, the third working generation was developed which integrates optical waveguides into printed circuit board (Scheel 1999; Duzinksi et al. 2004). Here, especially new techniques of embedding the optical waveguide in the motherboard are required. This is called the fifth PCB generation of electro-optical printed circuit board EOCB (Schmieder 2002), (Nieweglowski and Wolter 2004; Krabe and Scheel 1999; Krabe et al. 2000). Here, the data lines are implemented within a new layer within the multilayer printed circuit board, starting from the processors as multi-mode optical waveguides, which are typically 50 μm × 50 μm in cross section (Krabe and Scheel 1999; Krabe et al. 2000).

In Fig. 1.7, such a possible structure is schematically presented (Scheel 2003). The construction concept with integrated waveguides—EOCB—should be for manufacturing reasons compatible with conventional manufacturing technologies of multilayer printed circuit boards. It provides thereby the lamination in the standard printed circuit board in order to achieve a SMD compatible assembly. Passive optical coupling and decoupling of the optical data signals in the waveguide should allow the use of commercial process materials. As waveguide, only multi-mode film waveguide with 50 μm up to 300 μm (Rieske 2003; Rieske et al. 2003; Nieweglowski and Wolter 2004) will come into consideration; otherwise, the alignment requirements for the optical coupling of OEICs to the waveguides would be significantly too high. Since the transmission distances are in the range of only a few tens of centimeters, data rates of several Gb/s can be easily achieved with the multi-mode technology.

1.2.3 Optical Systems in Cars and In-house Areas

Cars are equipped since more than 50 years with radio to entertain the passengers with music. The development of radio to multimedia systems began 50 years ago and has now reached up to MP3/DVD players and a host of advanced information

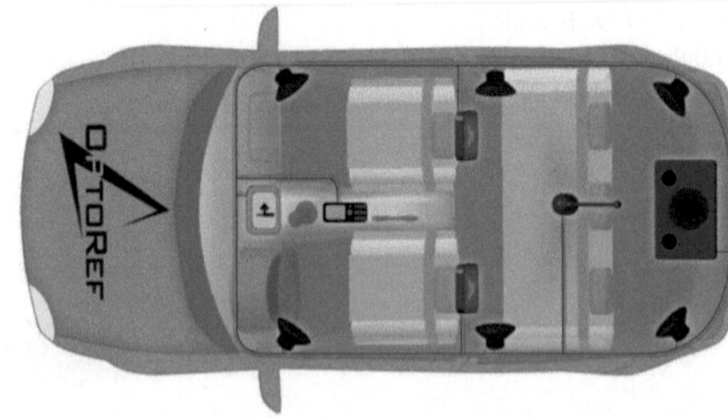

Fig. 1.8 Data routes in the automotive

systems, which are connected with each other and with the driver and must interact between all (Fig. 1.8). Nowadays, many motor cars are equipped with GPS navigation systems that can work in connection with a safety system to locate a stolen car. The safety during driving requires the user to focus on controlling the car instead of on the dependences of the individual components. During telephoning via the car phone, the stereo system should reduce in the volume when a call is made. Voice control and hands-free speaker need a microphone to digitize the voice and large display systems are required for navigation information and DVD playback. For this reason, the multimedia systems can work with the drivers, the presentation of audio and visual information in a wide variety of formats. The information of the driver and at the entertaining of the passengers is to be realized in parallel.

In this case, the data must be processed in a secure way. Additionally, the network should use international standards and program interfaces to ensure complete interoperability.

Other functions should also be added in an easy way, if the network provides the infrastructure to transfer information secure between the sub-devices. This is implemented in the international standardized multimedia optical bus "MOST" (MOSTCooperation). "MOST" was developed by European car manufacturing companies who wished to replace electrical bus systems CAN, because the low bandwidth of 21.2 Mb/s which is too low for real-time multimedia systems and has too much problems with hum loops and alternators whistle due to galvanic connections.

Optical waveguides as a transport medium can overcome these problems. A multimedia fiber network, optimized for automotive applications, can increase the data rate by orders of magnitude compared to the old electrical bus systems like CAN. At the same time, the components of the optical bus can be manufactured at low cost and can deliver a low-cost interface for microphones and speakers and switches. At the same time, more intelligent devices expand the features and functions, provided by other system components on the network.

This can make detailed control units in order to avoid unnecessary inattentions, while other sub-systems try to send important information data for vehicle guidance. The properties of MOST admit it that the bus can be used inside or outside the car to transmit multimedia information along with real-time data and control functions.

To get around such sets as space, weight and repair problems, DaimlerChrysler consistently implemented the Plastic Optical Fiber (POF) in the MOST system. Polymeric fibers are very light and the weight savings in comparison with copper cables are between 80 and 90 %.

In addition, the ring topology of the local multimedia network reduces thc cabling effort, because not each device is individually connected via a separate cable to the control unit in the cockpit. The digitized information uses only one optical line. Such a POF cable is sufficient using two optical fibers for the back and forth direction of the entire data and signal transfer among all connected devices. With the use of optical fibers, the electromagnetic interference problem is reduced strongly. Optical fiber cables are completely insensitive to strong electromagnetic fields. Ground loops that cause "alternator whistle" on audio gear cannot arise. The POF transmission systems are neither EMC nor influence them susceptible to electromagnetic interference to other components in the network.

However, it is a potential bottleneck in the data density and in the long-term reliability of the POF cable. Currently, the bus with a gross data rate of 22.579 Mb/s is operated, which is derived from a multiple of the CD sample rate of 44.1 kHz/s. This rate, however, is already reached for use as DVD bus on its border, so that a simultaneous transmission of several video sources is not possible. Therefore, work is already in further progress to extend the specifications of the MOST for data rates of 150 Mb/s (Grzemba 2011). In 2014, Mercedes introduced the first car with MOST 150 technology (Poferl et al. 2010).

A link in the MOST bus comprises a point-to-point link, as illustrated in Fig. 1.9. The maximum individual lengths of the POF fiber are limited to 10 m in length, due to the high dispersion and attenuation of the POF in the wavelength range of 650 nm. The transmitters (Fig. 1.10) consist of low-cost LED chips that can transmit via 10 m POF up to 100 MHz bandwidth. In the receiver area, simple Si diodes are used (Infineon_Corporation 2010). The coupling between POF and transmission elements (Grzemba 2011) is realized very simply by coupling the fiber end face in contact with the OEIC. Although the efficiency is low, the cost of the compounds is also very low due to the simple workability of the POF by simple knife-cut method. The module costs are in the automotive industry at the lowest

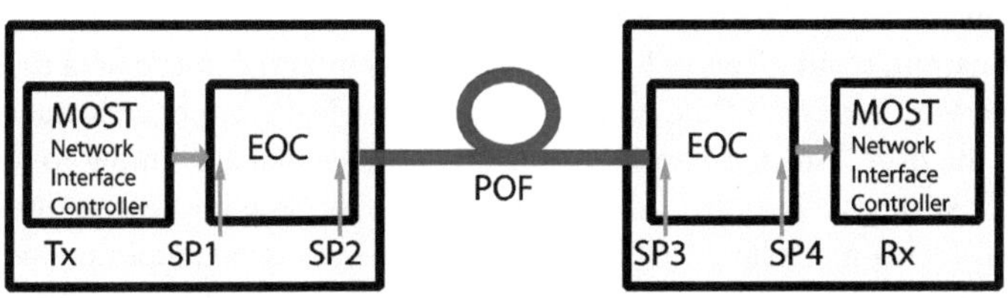

Fig. 1.9 MOST Point-to-point optical link

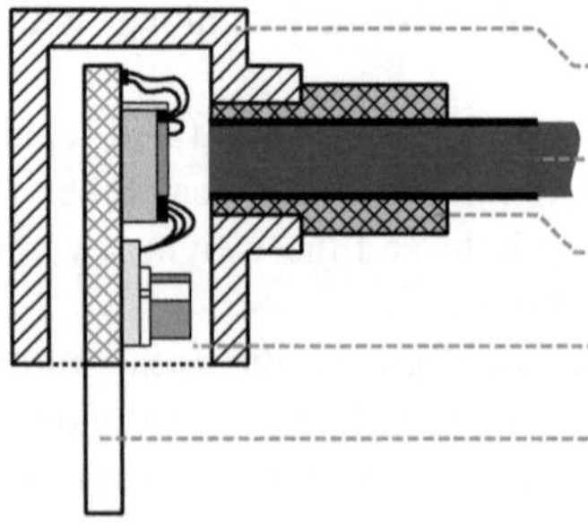

Fig. 1.10 MOST transmitter and receiver modules

part of the cost of all-scale photonic modules of optical transmission technology. Prices of a few euros per module are common here and necessary.

At the cable television network, over 22 million households are connected in Germany today. In addition, in many large residential satellites reception in the form of cable television (CATV) signals is distributed. The distribution is usually based on copper cables, as they are inexpensive and also the laying is easy. However, copper cables have drawbacks. Firstly, the TV reception may be disturbed by cross talk, and on the other hand, there are big problems when the copper installation itself emits radio interference. One advantage of the optical transmission is the compatibility with these EMC influences. In the metro and long-range region, the cable television is already optically transmitted. In this case, highly linear laser and single-mode fiber are used. These components are very expensive, and their handling is difficult. This is a great misfit for the use of these components for inhouse CATV distribution. An optical property distribution (Goff 2003) must be able to compete with a copper cable installation, however, in terms of cost and usability (Weber et al. 2003).

The polymer optical fiber (POF) is a cheap usable for short-haul optical transmission medium either in cars or in inhouse applications. However, the pass band of the POF is to the visible region between 400 and 750 nm very limited by the attenuation behavior of the polymer chains (Ziemann 2008). A further advantage is the large core radius of the POF. They allow easy installation and good optical connection with low loss of the optical components. The POF fiber with its excellent installation properties allows the construction of low-cost short-haul optical networks (Weinert 1998) in the field of home distribution of TV signals (Weber et al. 2003). This could be an alternative to copper technology using the wavelength-division multiplexing technology in the visible range, the polymer optical fiber in the future. However, this development has only just begun (Schrader 2004).

The transmitter and receiver for POF transmission in the house area must be able to compete with the cost of the corresponding electrical components and be easy to handle by the user. This necessarily results in modules that are similar to those in the automobile strongly, and their advantages include the low cost of construction and the simple electro-mechanical assembly. The electrical signal amplifier and transmit/receive OEICs for signal transmission should be as compact as possible located in the plug housing. The optical coupling of the POF is simply cut with a knife. Here, the POF fiber can be connected in direct surface contact with the OEIC (butt coupling). Coupling optics that increases the coupling efficiency would be too expensive to implement or only in rudimentary form by plastic lens in front of the OEIC.

1.3 Photonic Packaging and Interconnection Technology

The photonic Packaging and Interconnection technology (PAVT) is the interface between the optical communications engineering and classical electrical connection technology and microsystem technology, as shown in Fig. 1.11. The techniques such as bonding, adhesive bonding, soldering, thick-film/thin-film technology, and the multilayer printed circuit board technology also be used to build up of an optoelectronic device (OEIC), such as some techniques of microsystems technology such as etching of silicon substrates or masks and coating techniques. The application of the OEICs in the communication system determines the system structure of the package that needs to be very cost-effective in extreme cases, e.g., to be used in future optical Ethernet connections. This can be very expensive and time-consuming, if only some specific functions are used in a wide-area range, as has already been explained in the previous chapter to the optical transmission systems.

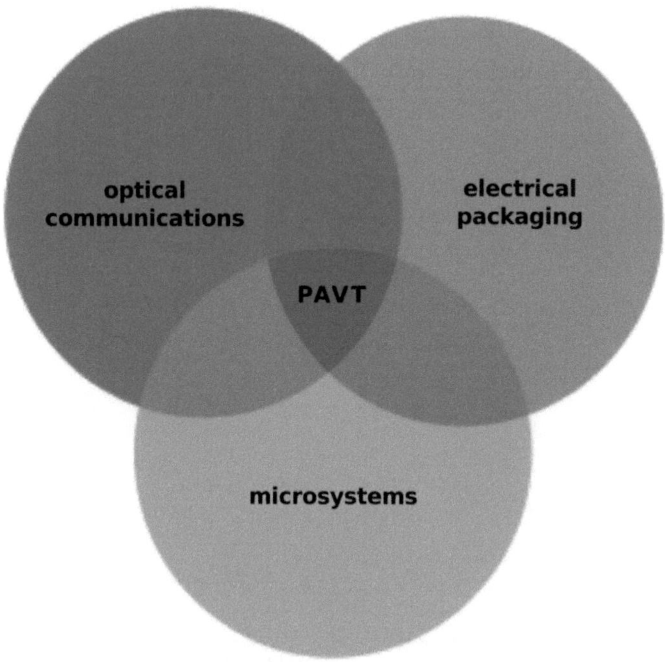

Fig. 1.11 The photonic assembly and interconnection technology as an interface for optical communications technology, microsystems technology, and electrical packaging technology

For clarity and definition of the roles of the photonic assembly and interconnection technology (PAVT), a proper definition is to be introduced at this point:

> The photonic assembly and interconnection technology is used to connect optical and electrical transmission media and electrical supply lines to photonic devices in a stable, environmentally adapted housing.

To develop an OEIC Packaging (Mickelsen 1997), as already mentioned, a combination of very different techniques are necessary:

- High-frequency technology,
- Classical optics/wave optics,
- Precision engineering design and CAD design,
- Cooling/heat management,
- Communications engineering,
- Solid-state physics,
- Etching of silicon substrates,
- Mask techniques,
- Thick-film/thin-film technology, and
- Gluing, soldering, welding, bonding technology

In the PAVT, one can distinguish between different packaging levels. In any case, the process of building starts during separation of the chips from the wafer and examining its properties. The classified ICs will then receive a first-build platform (Lau 1995; Topper et al. 2000a; Topper et al. 2014; Becker et al. 2004; Wolf et al. 2008) in the starting level of the so-called first-level package in single-chip modules or multi-chip modules. In the hierarchy of levels of packaging, (Fig. (1.12) the first-level

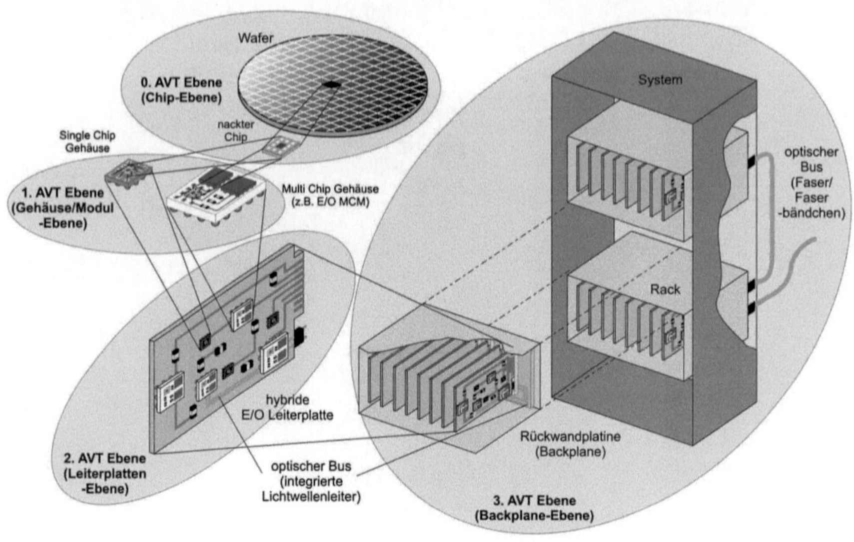

Fig. 1.12 Three levels of packaging (Nieweglowski 2011)

packaging, bare chips are applied to a sub-carrier and electrically connected. In summary:

- 0th-level is chip-level packaging, for example, CSP;
- 1st-level is either single- or multi-chip modules;
- 2nd-level is from the 1st level to the printed circuit board (PCB); and
- 3rd-level is from the card to the motherboard.

Therefore, it is just there is of great importance to use a cost-effective, time-economy bonding methods that should be usable for optical contacting. The flip chip technology will be discussed in more detail below; this promises to be a great potential. However, contrary to electric packaging in the photonic PAVT, the optical chip is usually applied in the first level to a heat sink and then installed in conjunction with the heat sink in a specific photonic housing.

In Fig. 1.13, the main elements of a "first-level package" of the photonic PAVT are shown schematically. To a photonic chip, with lateral dimensions of 300 μm × 500 μm and a height of about 100 μm, to be able to manipulated, it is necessary to implement it in a macroscopically large housing (construction/design). First, the chip on a so-called carrier (chip carrier) is applied. This stabilizes the chip mechanically and realizes good heat dissipation capability for thermal management.

Usually, the leads on a circuit board (etching technique, thick-film/thin-film technology) can be within the housing to be brought close to the chip. At the housing, electrical sockets are mounted, which are used on the outside for the supply and derivation of the electrical signals, and regulating currents for the operation of the chip. These jacks can then easily connect cable or attach the housing to a board. The electrical leads are connectors for the modulated signals (RF) and also for control and operating currents (bias).

The second-level packaging is then carried out with the onset of first-level modules on the chipboard, which should be executed as EOCB in the optical connection, such as optical section in datacom systems has been presented.

The third level is achieved in the form of installation of the individual boards or in the case of the photonic PAVT housing into a larger package on a motherboard. An overview of all module technologies of photonic PAVT has been summarized

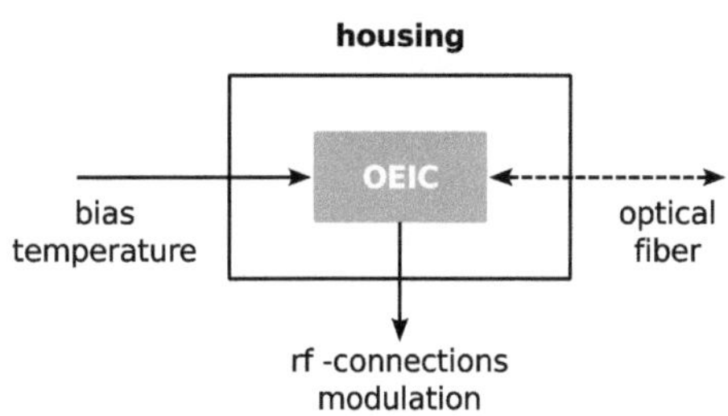

Fig. 1.13 Principle of optical/electrical packaging

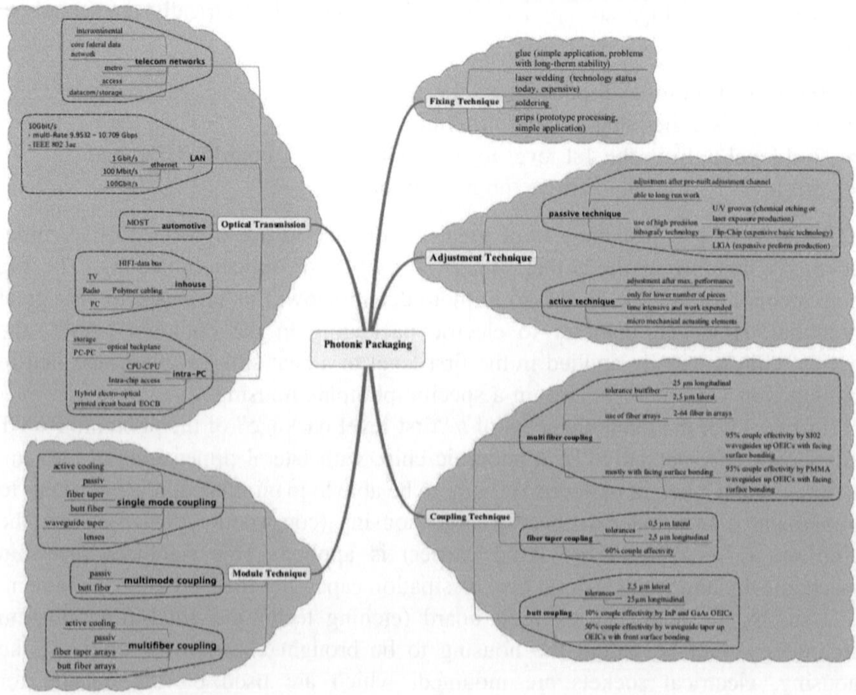

Fig. 1.14 Techniques and systems of photonic PAVT

in Fig. 1.14 with the important optical transmission technologies. While the transmission systems have been considered in the first few sections, in the following section different fixation and adjustment techniques will be discussed in detail.

References

Agrawal, N., Jahn, E., Pieper, W.: Integrated Optics for High Bitrate Systems. Tokyo, Japan (1997)

Albrecht, H.: Photonic integrated device research at Siemens. LEOS Summer Topical Meeting (1994)

ANSI: ANSI X3.139, Media Access Control (MAC) Spezifikation, Token-Passing, Frame-Format, Ringaufbau (1992)

ANSI: ANSI X39.5, Station Management (SMT) Spezifikation, Verbindungs- und Ringaufbau, Fehlererkennung und Beseitigung, Stationsmanagement (1998)

ANSI: ANSI X3T9.5, Physical (PHY) Spezifikation, Codierung der Daten mit Taktinformation (1990)

ANSI: Fiber Distributed Data Interface (FDDI), ANSI X3T9.5, Physical Media Dependent (PMD) Spezifikation, Zugriff auf Medium (LWL, Kupfer). In: X3T9.5. ANSI (1994)

Becker, K.F., Erik, J., Ostmann, A., Braun, T., Neumann, A., Aschenbrenner, R., Reichl, H.: Stackable system-on-packages with integrated components. IEEE Trans Adv. Packag. **27**(2), 268–277 (2004)

Bjarklev, A.: Optical Fiber Amplifiers: Design and System Applications. Artech House, London (1993)

Borkar, S.: Gigascale system design—challenges and opportunities. In: IEEE Proceedings of the International Conference on Computer Design: VLSI in Computers and Processors, 2004 (ICCD 2004). 11–13 Oct 2004, p. 2

Brox, O., Bauer, S., Biletzke, M., Ding, H., Kreissl, J., Wünsche, H.-J., Sartorius, B.: Self-pulsating DFB for 40 GHz clock-recovery: impact of Intensity fluctuations on jitter. Paper Presented at the Proceedings of 29th Optical Fiber Communication Conference (OFC 2004), Los Angeles CA, USA

Coskina, P., Ehrmann, O., Reichl, H.: Wafer-level chip size package (WL-CSP). IEEE Trans. Adv. Packag. **23**(2), 233–238 (2000)

Düser, M., De Miguel, I., Bayvel, P.: Timescale analysis for wavelength-routed optical burst-switched (WR-OBS) networks. Paper presented at the OFC 2002, Anaheim, USA (2002)

Duzinksi, M., Rieske, R., Wolter, K.-J., Patela, S.: PCB integradet waveguides-launching light into highly multi mode structures. In: 4th International spring seminar on electronics 27th technology pp. 171–176 (2004)

Ehlers, H., Schlak, M., Fischer, U.H.P.: Packaging of integrated mach-zehnder interferometers for optical communication systems. J. Opt. Commun **23**, 799 (2002)

FCIA: FCIA Roadmap. http://www.fibrechannel.org/roadmaps (1997)

Fibre Channel Industry Association. Fibre Channel—T11 Specifications (2014). Retrieved from http://www.fibrechannel.org/fibre-channel-features.html

Fischer, U.H.P., Honecker, J., Umbach, A., Trommer, D., Eckhardt, T.: Optical receiver modules with up to 45 GHz modulation bandwidth for photonic systems. In: Conference on Networks and Optical Communications (NOC 2002), Darmstadt, Germany 2002

Fischer, U.: Optoelectronic Packaging. VDE Verlag, Berlin (2002a)

Fischer, U.H.P.: Optoelectronic Packaging. Wiley, Hoboken (2002b)

Gnauck, A.H., Chraplyvy, A.R., Tkach, R.W., Zyskind, J.L., Sulhoff, J.W., Lucero, A.J., Sun, Y., Jopson, R.M., Forghieri, F., Deroshier, R.M., Wolf, C., McCormick, A.R.: One terabit/s transmission experiment. In: OFC'96, IEEE, San Jose, CA, (1996)

Goff, D.R.: Fiber Optic Video Transmission: The Complete Guide. CRC Press, Oxford (2003)

Grzemba: MOST. The Automotive Multimedia Network. From MOST25 to MOST150. Franzis, Poing (2011)

Honecker, J.: Untersuchung der Koppeleffizienz und der Reflexionen zwischen Monomode-Fasern und integrierten Modenfeldtransformern auf InP für LWL-Empfänger im Bereich 40-160 Gbit/s. University of Applied Sciences, HTW, Berlin (2000)

IEEE: IEEE 802.3i Clause 10 (early IEEE 802.3a1988) 10BASE-T 10 Mbit/s (1.25 MB/s) over twisted pair, 802.3a,b,c, and e-1988 IEEE Standards for Local Area Networks: Supplements to Carrier Sense Multiple Access With Collision Detection (CSMA/CD) Access Method and Physical Layer Specifications. IEEE Standards Association (1987). doi:10.1109/IEEESTD. 1987.78883

IEEE: IEEE 802.3ae (auch 802.3ae-2002). In: 10GBASE-XX. (2005) http://www.ieee802.org/3/ae/index.html

IEEE: IEEEb 802.3.bg 40 Gb/s Ethernet (2011) http://www.ieee802.org/3/bg/index.html

Infineon_Corporation. http://www.infineon.com (2010

Intel_Corporation: Thunderbolt™ Technology. https://thunderbolttechnology.net/tech/how-it-works (2011). Accessed 1 Feb 2012

Ishii, M., Hibino, Y., Hanawa, F., Nakagome, H., Kato, K.: Packaging and environmental stability of thermally controlled arrayed-waveguide grating multiplexer module with thermoelectric device. J. Lightwave Technol. **16**(2), 258 (1998)

ITU G.803.: ITU-T G.803, Architecture of transport networks based on the synchronous digital hierarchy (2003). USA. Retrieved from http://www.itu.int/ITU-T/recommendations/rec.aspx?rec=4955

Jahn, E., Agrawal, N., Ehrke, H.-J., Ludwig, R., Pieper, W., Weber, H.G.: Monolithically integrated asymmetric Mach-Zehnder interferometer as a 20 Gbit/s all-optical add/drop multiplexer for OTDM applications. Electron. Lett. **32**, 3 (1996)

Jahn, E.: Monolithically integrated semiconductor laser amplifier based interferometers for optical signal processing. Ph. D. Thesis, Technical University of Berlin (1996)

Japan, O.: Evolution of transmission data rates. Optoelectronics Industry and Technology Development Association (2006)

Kibler, T., Rode, M., Moisel, J., Zeeb, E.: Folienvernetzungstechniken mit Lichtwellenleitern für Automobilanwendungen. In: ITG-Workshop Photonische Integration und Aufbautechnik, Heinrich-Hertz-Institut Berlin (2001)

Krabe, D., Ebling, F., Arndt-Staufenbiel, N., Lang, G., Schee, W.: New technology for electrical/optical systems on module and board level: the EOCB approach. In: Proceedings of 50th ECTC, Las Vegas, NV, pp. 970–974 (2000)

Krabe, D., Scheel, W.: Optical interconnects by hot embossing for module and PCB technology—the EOCB approach. In: 49th ECTC Electronis Components and Technology Conference, San Diego, CA, USA, pp. 1164–1166 (1999)

Kuhlow, B., Przyrembel, G., Pawlowski, E., Ferstl, M., Fürst, M.: AWG-based device for a WDM overlay PON in the 1.5 μm band. IEEE Photonic Technol. Lett. **11**(2), 3 (1999)

Lau, J.H.: Flip Chip Technologies. McGraw-Hill, New York (1995)

Ludwig, R., Huettl, B., Hu, H., Schmidt-Langhorst, C., Schubert, C.: 107 Gb/s RZ-DPSK transmission over 320 km dispersion-managed Fiber with balanced detection ETDM integrated receiver. In: Optical Fiber Communication/National Fiber Optic Engineers Conference, 2008. (OFC/NFOEC 2008), 24–28 Feb 2008, pp. 1–3

Ludwig, R., Weisser, S., Schmidt-Langhorst, C., Raddatz, L., Schubert, C.: 160 Gb/s RZ-DPSK OTDM-transmission over 480 km using 160 km repeater spans and advanced forward-error-correction. In: Optical Fiber Communication and the National Fiber Optic Engineers Conference, 2007. (OFC/NFOEC 2007). Conference on, 25–29 March 2007, pp. 1–3

Marcuse, D.: Loss analysis of single-mode fiber splices. Bell Syst. Tech. J. **56**, 15 (1977)

Mickelsen, A. R.: Optoelectronic packaging. John Wiley and Sons, New York (1997)

Moisel, J.: Optical backplanes with integrated polymer waveguides. Opt. Eng. **39**(3), 673–679 (2000)

MOSTCooperation: MOST—Media Oriented Systems Transport—bus. http://www.mostnet.de (2001). Accessed 04 April 2012

Network, O.T.: IST roadmap for optical communications. In. Network project OPTIMIST (Optical Technologies in Motion for the IST Programme), (2002)

Nieweglowski, K.: Beiträge zur Aufbau- und Verbindungstechnik für optische Kurzstreckenverbindungen. Verlag Dr. Markus A. Detert, Templin (2011)

Nieweglowski, K., Wolter, K.J.: Optical Interconnections on PCBs—recent developments. In: International Students and Young Scientists Workshop "Photonics and Microsystems", Wroclaw/Szklarska Poreba Poland, pp. 29–33 (2004)

O'Mahony, M.J., Politi, C., Klonidis, D., Nejabati, R., Simeonidou, D.: Future optical networks. J. Lightwave Technol. **24**(12), 4684–4696 (2006)

Plickert, V., Melchior, L.: Fertigungsgerechte Koppeltechnik für Module zur parallen-optischen Datenübertragung mit Vertikalemittern. In: ITG-Workshop Photnische Integration und Aufbautechnik, HHI, Berlin (1999)

Poferl, S., Becht, M., De Pauw, P.: 150 Mbit/s MOST, the Next Generation automotive infotainment system. In: 12th International Conference on Transparent Optical Networks (ICTON), (2010) doi:10.1109/ICTON.2010.5549160

Przyrembel, G., Kuhlow, B.: AWG-based devices for a WDM overlay PON. In: WDM Components, vol. 29. OSA Trends in Optics and Photonics, pp. 87–94. Optical Society of America (OSA) (1999)

Ramaswami, R., Sivarajan, K.: Optical Networks. Academic Press, San Diego (2004)

Rieske, R., Nieweglowski, K., Wolter, K.-J.: Cost effective Method for spatially resolved characterisation of board level optical waveguides. In: Proceedings of ISSE, Stara Lesna, Slovakia 2003, pp. 222–226

Rieske, R.: PCB goes optical—technologies, trends, Inspection. In: International Students and Young Scientists Workshop "Photonics nand Microsystems", Wroclaw, Szklarska Poreba, Poland 2003

Rosin, T., Bornholdt, C., Hoffmann, D., Burghardt, R.: Opto-electronic packaging for broadband high speed (40 Gb/s) optical demultiplexer chip. Paper Presented at the LEOS 98

Saruwatari, M., Nawata, K.: Semiconductor laser to single-mode fiber coupler. Appl. Opt. **18**, 9 (1979)

Saruwatari, M., Sugie, T.: Efficient laser diode to single-mode fiber coupling using a combination of two lenses in confocal condition. IEEE J Quant. Electron. **6**, 7 (1981)

Scheel, W.: EOCB, die 5. Leiterplattengeneration. In: ITG-Workshop Photnische integration und Aufbautechnik, Hochschule Harz, Wernigerode (2003)

Scheel, W.: Baugruppentechnologie der Elektronik. Verlag Technik, Eugen G. Leutze Verlag Saulgau, Berlin (1999)

Schmieder, K.: Aspekte der Aufbau- und Verbindungstechnik, Elektro-Optischer Verdrahtungsträger. Detert (2002)

Schrader, M.: Aufbau einer optischen Übertragungsstrecke mit Polymer-Faser für die Hausverteilung breitbandiger analoger TV-Signale. Diploma work, Hochschule Harz, FHG IIS Nürnberg (2004)

Scobey, M.A., Spock, D.E.: Passive WDM Components using microplasma optical interference filters. Paper presented at the OFC'96, San Jose (1996)

Späth, J.: Mehr Licht: Photonische Netze, die Zukunft der Kommunikation. c´t mag. für com. technik **1**, 6 (1999)

Tekin, T., Schlak, M., Brinker, W., Maul, B., Molt, R.: Monolithically integrated MZI comprising band gap shifted SOAs: a new switching scheme for generic all-optical signal processing. Paper presented at the ECOC 2000, Munich, Germany

Topper, M., Auersperg, J., Glaw, V., Kaskoun, K., Prack, E., Keser, B., Coskina, P., Jager, D., Fetter, D., Ehrmann, O., Samulewicz, K., Meinherz, C., Fehlberg, S., Karduck, C., Reichl, H.: Fab Integrated Packaging (FIP): a new concept for high reliability wafer-level chip size packaging. In: Proceedings of 50th Conference Electronic Components and Technology (2000a), pp. 74–80

Topper, M., Fehlberg, S., Scherpinki, K., Karduck, C., Glaw, V., Heinricht, Fibre channel industry association. Fibre Channel—T11 Specifications (2014). Retrieved from http://www.fibrechannel.org/fibre-channel-features.html

Weber, N., Tschekalinskij, W., Junger, S., Plötz, A.: Transmission of CATV-signals with high carrier frequencies using step index POF. In: 12th International POF Conference 2003, Seattle 2003, pp. 240–242

Weinert, A.: Kunststofflichtwellenleiter: Grundlagen, Komponenten Installation. Publicis-MCD-Verlag, Erlangen (1998)

Wolf, M.J., Ramm, P., Klumpp, A., Reichl, H.: Technologies for 3D wafer level heterogeneous integration. In: Symposium on Design, Test, Integration and Packaging of MEMS/MOEMS, 2008. 9–11 April 2008, pp. 123–126

Ziemann, O.: POF handbook : optical short range transmission systems, 2 edn, Springer, Berlin (2008)

Zolper, J.C.: The GSRC's Role in meeting tomorrow's design challenges. Des. Test Comput IEEE **25**(4), 366–367 (2008)

Chapter 2
Optical Waveguides

Abstract This section describes the characteristics of optical waveguides in various material systems. The main focus will be on the properties of glass fiber whose different types and production methods will be described. The material systems of optical waveguides, which are described in this chapter, are displayed in the following: (1) in fibers, (2) in SiO_2, (3) in Polymethylmethacrylate (PMMA), (4) in GaAs, and (5) in InP.

2.1 The Most Important Optical Laws

Light can be described both, as an electromagnetic wave [wave theory (Webb 2005)], as well as particles [corpuscular (Demtröder 2010)]. Heinrich Hertz was the first who experimentally discovered the photoelectric effect but Einstein was the one who explained this effect later. The dualism of light is reflected in the combination of two fundamental laws of physics, and the energy of a photon W can be described by its frequency or its mass m_p:

$$W = hf \quad \text{(Plancks law)} \tag{2.1}$$

$$W = m_p c_0^2 \quad \text{(Einstein)} \tag{2.2}$$

c_0 speed of light in vacuum and
h Plancks constant.

By combining both equations, one can determine the DeBroglie-wavelength (Demtröder 2010) which allocates every particle of mass m and the velocity v with an adequate wavelength:

$$\lambda_0 = h/(mv) \quad \text{(DeBroglie)} \tag{2.3}$$

© Springer-Verlag Berlin Heidelberg 2015

U.H.P. Fischer-Hirchert, *Photonic Packaging Sourcebook*,
DOI 10.1007/978-3-642-25376-8_2

2.1.1 Homogeneous Plane Wave

A wave can be described by a sinus function. First, we shall confine to the stationary time-dependent treatment:

$$g(t) = a\cos(2\pi ft + \phi) \tag{2.4}$$

a	amplitude
$(2\pi ft + \phi)$	phase

or in complex numbers:

$$g(t) = a\cos(t + \phi) = a\,\mathrm{Re}\left\{e^{j(t+\phi)}\right\} = \mathrm{Re}\left\{Ae^{jt}\right\} \tag{2.5}$$

with complex amplitude; $A = ae^{j\phi}$

More important optical basics:

$$\omega = 2\pi f \quad \text{frequency} \tag{2.6}$$

$$t_p = 1/f \quad \text{time of oscillation} \tag{2.7}$$

$\lambda_1 = c_1 t_p = c_1/f$	light speed in wave length 1
$c_1 = c_0/n_1$	light speed in material 1
$n_1 = c_0/c_1$	Refractive index in material 1
$k_1 = k_0 n_1 = 2\pi n_1/\lambda_0$	wave number in material 1

Example: Frequency of the optical wave at 1.55 μm

$$f = c_0/\lambda = \left(3\times 10^8\,\mathrm{m/s}\right)/\left(1.55\times 10^{-6}\,\mathrm{m}\right) = 1.935\times 10^{14}/\mathrm{s} = 193.5\times 10^{12}\,\mathrm{Hz}$$

An overview of the exact name of m, s, and oscillation frequency is listed in the following Table 2.1.

Furthermore, the propagation direction e of the wave has to be determined when an additional spatial dependence of the wave is used as well. The coordinate system with the radius vectors $r(x, y, z)$ is chosen in such a way, that e is coincident with the z-direction as shown in Fig. 2.1.

$$\mathbf{r}e = \text{const} = \mathbf{r}\cos(\mathbf{r}, e) \tag{2.8}$$

This generates a plane that is perpendicular to the z-direction. The phase of the wave vector of a plane wave is constant within a site-fixed or time-fixed plane!

Table 2.1 Time and frequency units

Frequency units 1 Hz = [1/s]		
kHz	Kilohertz	10^3 Hz
MHz	Megahertz	10^6 Hz
GHz	Gigahertz	10^9 Hz
THz	Terahertz	10^{12} Hz
PHz	Petahertz	10^{15} Hz
Time units in seconds [s]		
ms	Millisecond	10^{-3} s
µs	Microsecond	10^{-6} s
ns	Nanosecond	10^{-9} s
ps	Picosecond	10^{-12} s
fs	Femtosecond	10^{-15} s
as	Attosecond	10^{-18} s

$$c_1 t \pm (re) = \text{const}$$

Complex:

$$\mathbf{E}(t) = \mathbf{A}e^{jt}$$

or with the spatial dependence:

$$\mathbf{E}(t) = \mathbf{A}\exp\{j(wt - rk)\} \tag{2.9}$$

This is the general description of the electric field vector **E**. The magnetic field **H** can be deduced analogously.

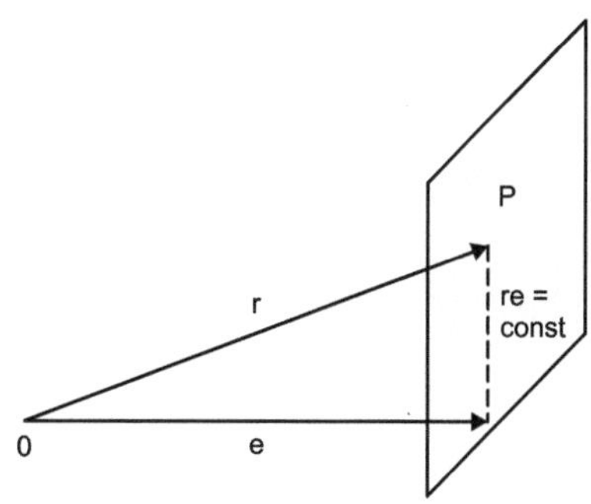

Fig. 2.1 Propagation of an electromagnetic wave

2.1.2 Phase and Group Velocity

The phase velocity (Demtröder 2010) of light can exceed the group velocity, especially in dispersive media. The behavior of group and phase velocities in quartz (Brooker 2003) is shown in Fig. 2.2. The group velocity is corresponding to the derivation of the refractive index of the medium in which the light propagates. The group refractive index describes the propagation of pulses in the medium, which may contain many frequency components according to Fourier (Bloomfield 2000). The group refractive index is usually higher than the phase refractive index, as shown in Fig. 2.2.

$$c_{ph} = w/k_1 = c \quad \text{phase velocity} \tag{2.10}$$

$$v_{gr} = \mathrm{d}w/\mathrm{d}k_1 = -\lambda^2 \mathrm{d}\omega/\mathrm{d}\lambda + c \quad \text{group velocity} \tag{2.11}$$

2.1.3 Reflection

The reflection is a process in which a beam of light which hits a reflective surface is reflected at the same angle, as at the incidence (see Fig. 2.3):

Angle of incidence = angle of reflection; $\alpha_1 = \alpha_2$

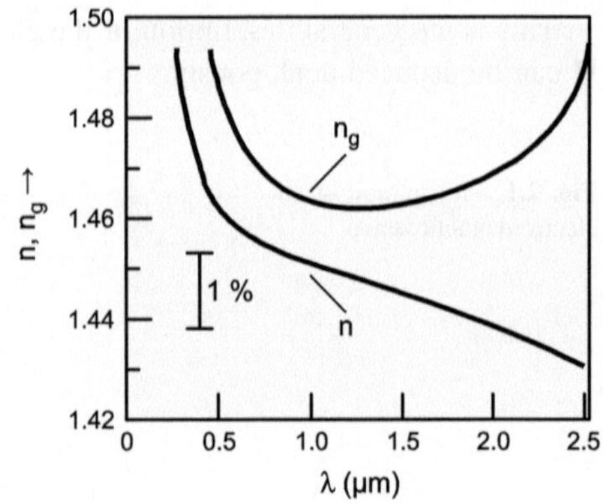

Fig. 2.2 Phase and group velocity of light in quartz

Fig. 2.3 Reflection of light

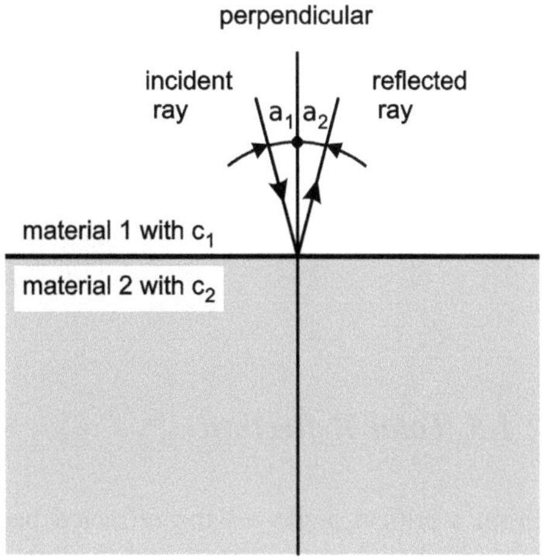

2.1.4 Refraction

Refraction (Fig. 2.4) eventuates when the light beam from a medium with a lower refractive index reaches a medium with higher refractive index.

There, it is broken off from the incidence perpendicular, due to the fact that the exit angle is larger than the incidence angle. The refractive indices are in medium $1 n_1$ and in medium $2 n_2$.

c_1 velocity of light medium 1
c_2 velocity of light medium 2

Fig. 2.4 Optical refraction

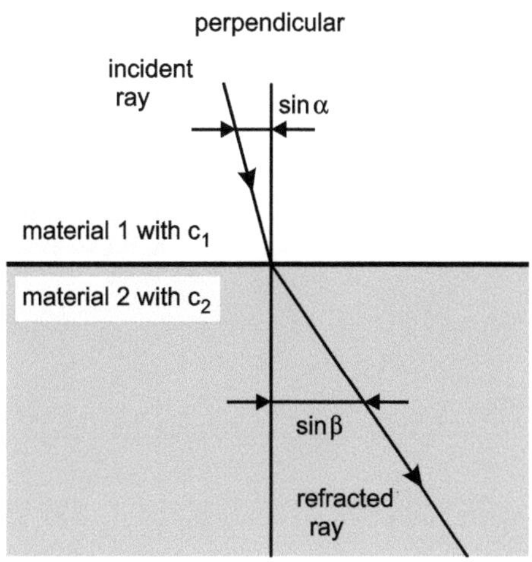

Example $n = 1.5$ of glass fibers

$$c_1 = c_0/n = \left(3 \times 10^8\,\text{m/s}\right)/1.5 = 2 \times 3 \times 10^8\,\text{m/s} = 200/\mu\text{s or } 5\,\text{ns/m}$$

This leads to Snell's (Brooker 2003) law of refraction:

$$\frac{\sin\alpha}{\sin\beta} = \frac{c_1}{c_2} = \frac{n_2}{n_1} \tag{2.12}$$

2.1.5 Total Reflection

From a critical angle α_0, the refracted beam 2 (Fig. 2.5) proceeds parallel to the interface between medium 1 and medium 2. During the transition from an optical denser medium to an optical thinner medium, the following condition applies to the total reflection angle (Brooker 2003):

$$\sin\alpha_0 = n_2/n_1$$

Example Refractive index of water = 1.333 against air $n = 1$

$$\sin\alpha_0 = 1/1.333 = 0.75 \quad \text{and} \quad \alpha_0 = 49°$$

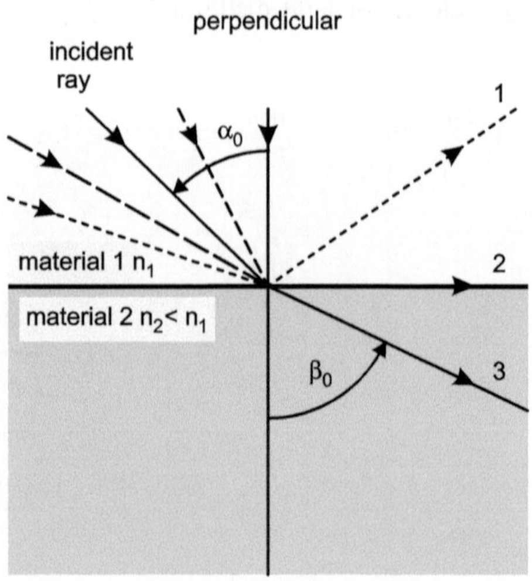

Fig. 2.5 Total optical reflection

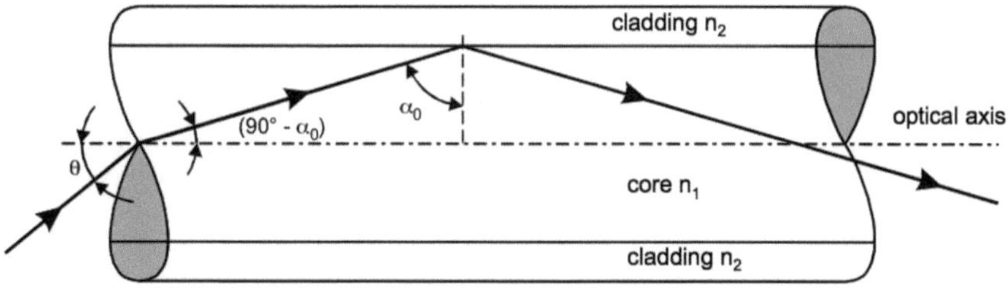

Fig. 2.6 Guided light in the fiber

2.1.6 Numerical Aperture

In the fiber, the light is guided through different refractive indices of the shell and the core, which suffice the following equation (see Fig. 2.6).

$$\sin \alpha_0 = n_2/n_1$$

To inject a light beam into the fiber, which can be guided in the core, the following equations are useful:

$$\sin \Theta / \sin(90° - \alpha_0) = n_1/n_0;$$

with $n_0 = 1$ and sin(90∅−x) = cos(x) one has

$$\sin \Theta = n_1 \cos \alpha_0 = n_1 \mathrm{SQRT}\left(1 - \sin^2 \alpha_0\right)$$

from $\sin \alpha_0 = n_2/n_1$ follows the numerical aperture (NA) (Demtröder 2010) which can be seen in Fig. 2.6:

$$\sin \Theta = \mathrm{SQRT}\left(n_1^2 - n_2^2\right) = \mathrm{NA} \tag{2.13}$$

2.2 Optical Fiber Profiles

When we consider the refractive index n of an optical fiber as a function of radius r, there is talk of the refractive index profile (Brooker 2003) of the optical fiber. Thereby, it is described the radial change of the refractive index of the optical fiber axis in the core glass to the outside in the direction of the coat glass (Fig. 2.7):

$$n = n(r)$$

$$n^2(r) = n_1^2[1 - 2\Delta(r)^g] \quad r < \text{a in core} \tag{2.14}$$

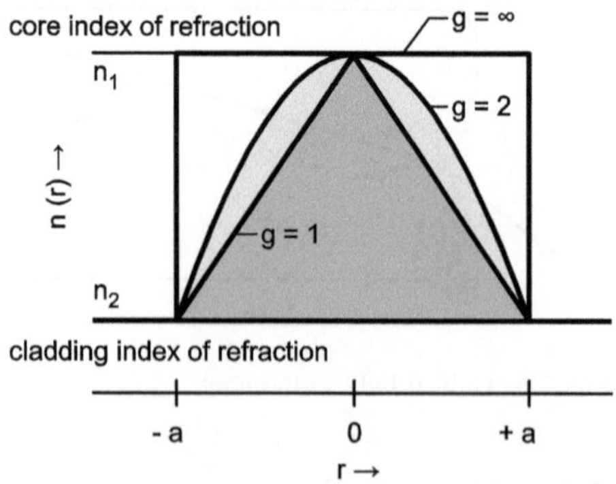

Fig. 2.7 Refractive indexprofile of waveguides

$$n^2(r) = n_2^2 = \text{const} \quad r > \text{a in cladding} \tag{2.15}$$

n_1 Refractive index at the fiber axis in microns ($r = 0$)
Δ normalized difference of the refractive index
r distance from fiber-optic cable axis in μm
a nuclear radius in μm
g potency, also called profile exponent
n_2 Refractive index of the cladding

The normalized difference in refractive index can be expressed as follows:

$$\Delta = \frac{(\text{NA})^2}{2n_1^2} = \frac{n_1^2 - n_2^2}{2n_1^2} \tag{2.16}$$

Special cases:
$g = 1$ Triangle profile
$g = 2$ Parabola profile
$g \to \infty$ Step profile (marginal case)

Only in the latter case—the step profile—is the refractive index $n(r) = n_1$ = constant in the core glass. In all other profiles, the refractive index $n(r)$ increases in the core glass gradually (gradually increasing) of the value of n_2 of the cladding glass to the value n_1 of the fiber-optic axis (Fig. 1.7).

In case that you want to calculate the number of modes in a waveguide, you have to calculate at first the **V-parameter** (Marcuse 1974) or calculate the normalized frequency of the light in the waveguide:

$$V = \left(\frac{2\pi a}{\lambda}\right)\sqrt{(n_1^2 - n_2^2)} = k \cdot a \cdot \text{NA} \tag{2.17}$$

a core radius,
λ wavelength,
NA numerical aperture,
k wavenumber, and
Δ normalized difference of the refractive index.

Example for the calculation of the V-parameter

$2a = 50\ \mu m$
$\lambda = 1550\ nm$

$$V = \left(\frac{2\pi 50\ \mu m}{1.55\ \mu m}\right)\sqrt{(1.48^2 - 1.46^2)} = \pi(50\ \mu m/1.55\ \mu m)0.242 = 24.52 \quad (2.18)$$

The number of modes is generally calculated as follows:

$$\text{Step-index fibers:} \quad N = V^2/2 \quad (2.19)$$

$$\text{Graded-index fibers:} \quad N = V^2/4 \quad (2.20)$$

Example of calculating the number of modes:
$g = 2$
Graded-indexfibers $N = 82^2/2 * 2/4 = 1681$ modes
Step-index fibers $N = 82^2/2 = 3362$ modes

2.2.1 Step Profile

Figure 2.8 displays a typical multi-mode step waveguide. Typical dimensions for a multi-mode fiber with a stepped profile are the following internationally standardized (ITU 1994) dimensions:

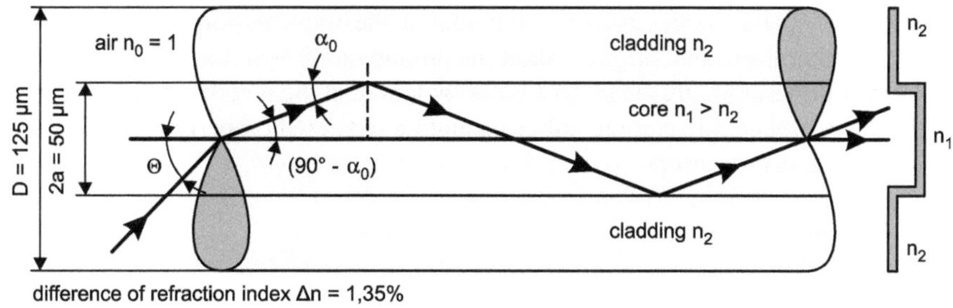

Fig. 2.8 Step-index waveguide

Core diameter	$2a$	50 mm
Cladding diameter	D	125 mm
Core number	n_1	1.48
Refractive index of the cladding	n_2	1.46
Difference of the refractive index	Δn	1.35 %

The critical angle α_0 characterizes the total reflection, i.e., the smallest angle to the perpendicular incidence at which a light beam is guided in the core glass and not broken in the cladding glass, which is shown in this example:

$$\sin \alpha_0 = \frac{n_2}{n_1} = \frac{1.46}{1.48} = 0.9865$$

$$\alpha_0 = 80.6°$$

$$\sin \alpha_0 = \sqrt{n_1^2 - n_2^2} = \sqrt{1.48_1^2 - 1.46_1^2} = 0.242$$

$$\theta = 14°$$

Since the sinus of the acceptance angle is defined as NA, the following is determined as:

$$\text{NA} = \sin \theta = 0.242$$

Example:
An optical fiber with step profile of 1 km length is traversed in about 5 μs of light. The *time difference* (Ramaswami and Sivarajan 2004) Δt between the mode with the fastest and slowest distance in the conductor can be approximately estimated by the difference of the refractive index between core and cladding as shown below:

$$\Delta t = 5\,\text{ms} \times \Delta n = 5 \times 10^{-6}\,\text{s} \times 0.01 = 50\,\text{ns}$$

If we assume a data rate of 1 MHz, the pulse will have a maximum allowable width of one μs. The distortion of the duration of the individual modes is called mode dispersion. In this case, this would be 1/20 of the data rate. This causes that the different spatial modes arrive at the end of the transmission line at different times and therefore, the original short incoming pulse will be extended at the outcome. Furthermore, this leads to a pulse widening at the receiver and this results in that the consecutively coming pulses cannot be separated properly which leads to an increasing rate of error.

2.2.2 Monomode Glass Fibers

The one-mode fiber-optic cable is also known internationally as single-mode fiber (SMF) (Ramaswami and Sivarajan 2004). Typical dimensions for a SMF are shown in Table 2.2.

The next Fig. 2.9 illustrates the path of a beam. To describe the light propagation of these small waveguide dimensions exactly, it is essential to use the wave theory of light.

In such a perspective the light propagates in the waveguide with a radial intensity distribution, which can be described by a very good approximation with a Gaussian function. This picture displays an intensity profile for a bimodal wave propagation (Ramaswami and Sivarajan 2004) which is shown in Fig. 2.10. It should be noted that the intensity distribution reaches up to the cladding.

Table 2.2 Single-mode fiber parameters

Mode field diameter	$2d$	8.5 μm
Outer diameter	D	125 μm
Core index	n_1	1.46
Index difference	Dn	0.003 = 3 ‰

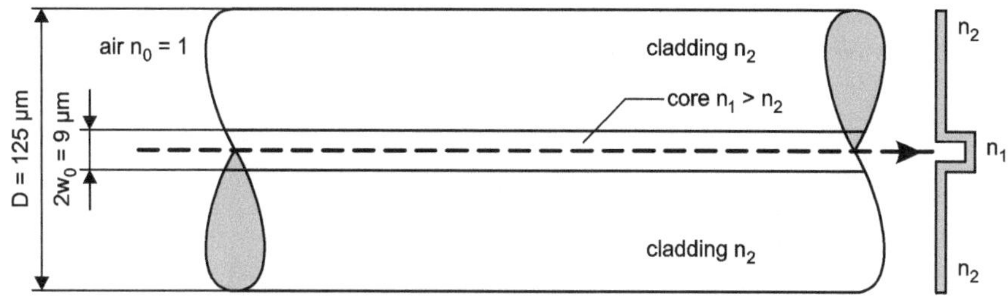

Fig. 2.9 Single-mode fiber

Fig. 2.10 Radial distribution of the fundamental mode LP01 at 1300 and 1550 nm

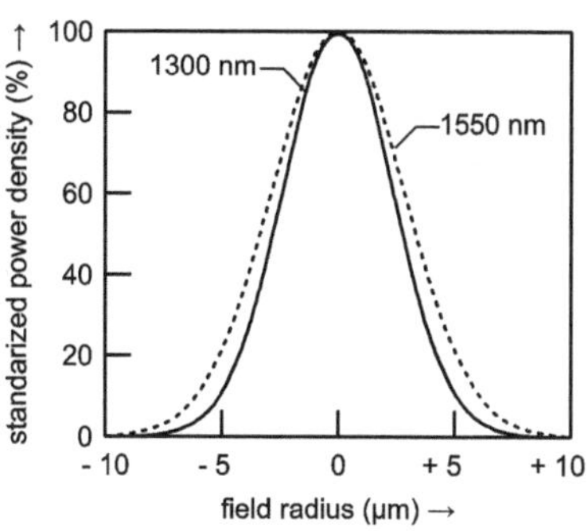

For a more detailed description of the Gaussian wave propagation in waveguides, see Chap. 3. A typical single-mode fiber-optic cable has the following NA:

$$\mathrm{NA} = n_1\sqrt{2 \times 0.003} = 0.113$$

Therefore, the acceptance angle Θ is calculated as follows:

$$\sin\theta = \mathrm{NA} = 0.113$$
$$\theta = 6.5^\circ$$

It should be noted that not only the core diameter, but also the NA, and therefore the angle of acceptance compared to the multi-mode-stepped profile fiber is considerably smaller whereby the coupling of light in the single-mode fiber-optic cable (see Chap. 4) is relatively difficult. The marginal wavelength, at which the SMF becomes multimodal, is generated in the following way (Mahlke and Gössing 1994; Ramaswami and Sivarajan 2004):

$$\lambda_c = \pi\frac{2a}{V_c}\mathrm{NA} = \pi\frac{8.5\ \mu\mathrm{m}}{2.405}0.113 = 1.255\ \mu\mathrm{m} \tag{2.21}$$

Example:
A single-mode optical fiber with step profile and a core diameter $2a = 8.5\ \mu$m and a marginal-wavelength $\lambda_c = 1255$ nm ($Vc = 2.405$) has a field diameter $2 \cdot w_0$ at the wavelength $\lambda = 1300$ and 1550 nm (see Table 2.3):

$$2 \cdot \omega_0 = \frac{2.6 \cdot \lambda}{V_c \cdot \lambda_c}2a \tag{2.22}$$

$$\lambda = 1300\,\mathrm{nm}:\quad 2\omega_0 = \frac{2.6 \cdot 1300\,\mathrm{nm}}{V_c \cdot 1255\,\mathrm{nm}}8.5\ \mu\mathrm{m} = 9.5\ \mu\mathrm{m}$$

$$\lambda = 1300\,\mathrm{nm}:\quad 2\omega_0 = \frac{2.6 \cdot 1550\,\mathrm{nm}}{V_c \cdot 1255\,\mathrm{nm}}8.5\ \mu\mathrm{m} = 11.3\ \mu\mathrm{m}$$

Table 2.3 Mode field diameter at different wavelengths ($2a = 8.5\ \mu$m)

Wavelength in nm	Mode field diameter in µm
1550	11.3
1300	9.5
1060	6.5
850	5.2
630	3.9

2.2.3 Gradient profile

Fibers with gradient profile have no step in the process of the refractive index but further parabola-shaped profile (Fig. 2.11). The advantage of this fact is that all the light beams in the waveguide need the same time over the transmission length because the product of the refractive index and optical path length can be kept constant. It does not matter whether the light beam is injected more flat or at a higher aperture. That means a significant increase of bandwidth for the signal in comparison to the step-index multi-mode fiber. The pulse that has to be transmitted does not run temporal wide apart because of the coordination of the propagation speed of the individual local modes as in the step-index fiber. However, the technical meaning of the gradient fibers is moved in the background because of the invention of the single-mode fiber with an improved transmission quality. Typical dimensions for a fiber-optic cable with a gradient profile (see Table 2.4) are :

For a typical fiber-optic cable with gradient profile, the NA is given as follows (Fig. 2.12):

$$\mathrm{NA} = \sin\theta = \sqrt{n_1^2 - n_2^2} = \sqrt{1.46^2 - 1.606^2} = 0.206$$

$$\theta = 11.9^\circ$$

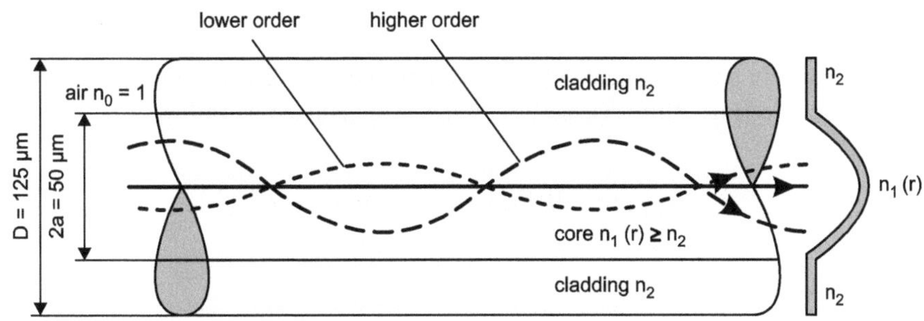

Fig. 2.11 Waveguide with parabolic refractive index profile

Table 2.4 Parameters of multi-mode glass fibers with gradient index profile

Core diameter	$2a$	50/62.5 µm
Cladding diameter	D	125 µm
Max. refractive index of core	n_1	1.46
Difference of refractive index between core and cladding	Δ	0.010

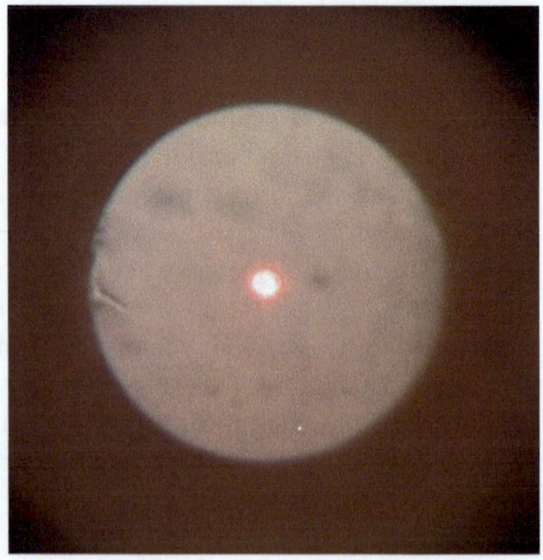

Fig. 2.12 Front facet of a single-mode glass fiber

2.2.4 Phase-space Diagrams: (sin²Θ *and* r²)

A phase-space diagram describes the graphical representation of the square of the NA as a function of radial expansion of the waveguide (Fig. 2.13). It is a convenient tool for determining the coupling efficiencies (Marcuse 1977, 1978) between gradient fibers. The area below the marginal curve for the maximum angle of acceptance Θ_{max} is proportional to the fed-in light power in the core.

It is obvious that with the same NA and the same core radius a, this power is twice as large in a fiber-optic cable with step profile (Fig. 2.14) as in a fiber with a gradient profile (Fig. 2.15). These losses can be calculated as follows:

$$L_{MM} = -10\log\left[\frac{a_1}{a_2}\right]^2 \quad \text{with } a_1 > a_2 \tag{2.23}$$

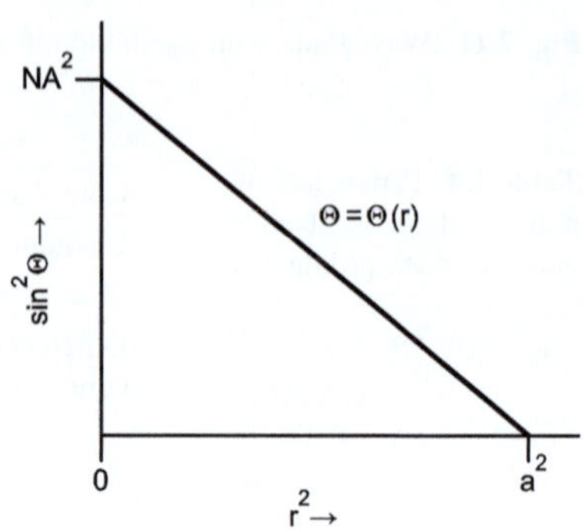

Fig. 2.13 Phase diagram of a gradient indexglass fiber

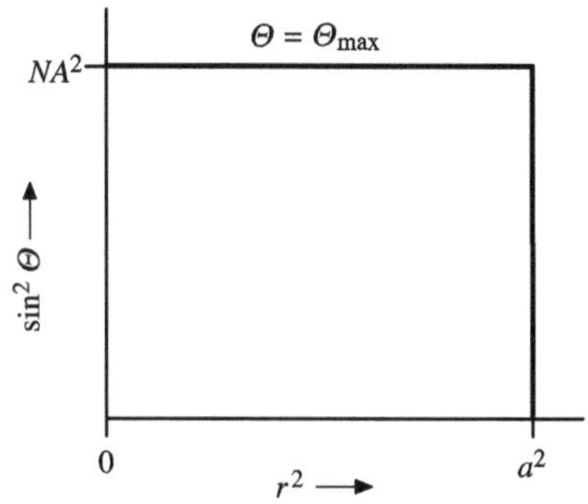

Fig. 2.14 Phase diagram of a waveguide with step-index profile

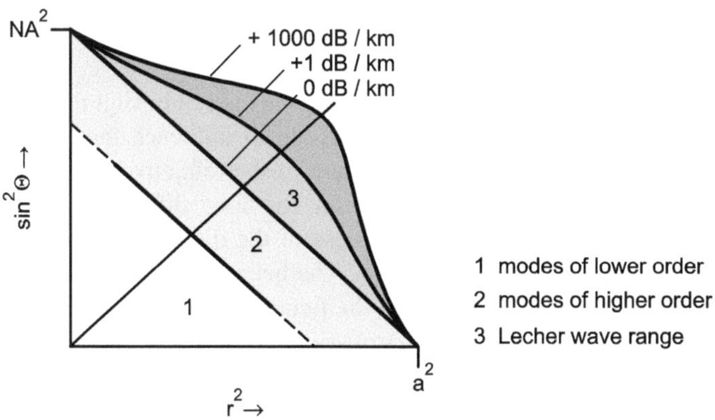

Fig. 2.15 Different mode propagation losses shown by a phase diagram

The behavior of the propagable modes in the multi-mode fiber-optic cable is plotted in Fig. 2.15. Modes with a lower order spread more in the core center, while the higher order modes spread out more near by the cladding. Leaky wave modes (Geckeler 1990) are also indicated, but have an extremely high attenuation of 1000 dB/km.

2.3 Dispersion

One has to consider two types of dispersion in glass fibers:

(a) Material dispersion
(b) Waveguide dispersion

Both types of dispersion together are called chromatic dispersion. In the wavelength range larger than 1300 nm, both fused silica dispersions have an

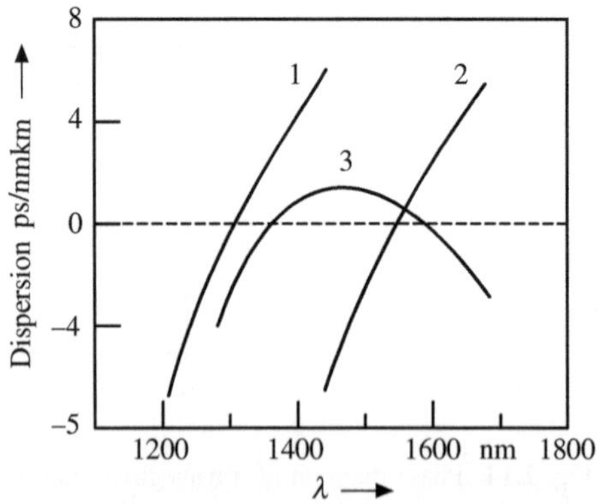

Fig. 2.16 Chromatic dispersion as a function of wavelength in glass fiber

1 w/o dispersion shift 2 with dispersion shift
3 with dispersion flattening

opposite sign. The material dispersion can be changed only slightly by other glass endowments. On the other hand, it is possible to influence the wavelength dispersion essential through a different structure of the refractive index profile.

Figure 2.16 displays dispersion behavior for three differently endowed glass fibers. Moreover, curve 1 shows the process of the dispersion of a standard fiber according to Mahlke and Gössing (1994). The behavior of a suspended dispersion fiber is shown in curve 2. Therefore, in the frequency range of 1550 nm, there is a very low dispersion to increase the transmission rate at this point. Curve 3 illustrates the process of another variant of optical fibers, the dispersion smoothed fiber. Thereby, one can reach a preferably low dispersion in a wide spectral range that can be advantageous in a wavelength multiplex system because many transmission channels see the same low dispersion.

2.4 Attenuation

The attenuation of the fiber is made up of several absorption parts, which can be seen in Fig. 2.17. A large proportion of the absorption of short wavelengths is based on the Raleigh-dispersion (Brooker 2003). The reason for this is the scattering of light by dipole molecules of the glass material. Moreover, the dispersion increases with the increasing frequency of light ($\sim \omega^4$). In addition, the glass absorbs strongly in the UV region because the offshoots of the absorption edges still need to be considered, but are negligible in the range of 1.55 μm. Roughness of the surface zone of the waveguide between core and cladding increase the attenuation slightly by a few thousandths of a dB per km. At higher wavelengths, the infrared absorption of the glass material limits the transmission range at 1.6 μm wavelength. Additionally, absorption peaks at 1.4 μm wavelength are recognized which are caused by OH-absorption of the glass.

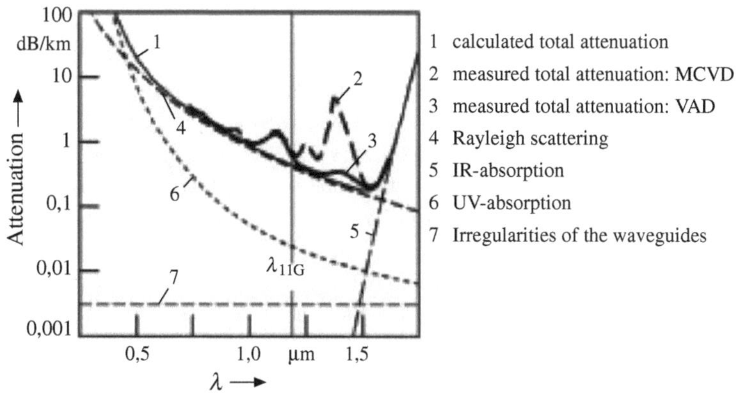

Fig. 2.17 Attenuation curves of glass fiber

In more recent fibers, it has been successfully implemented that these OH-radicals of the core region of the waveguide are squeezed out by a special tempering regulation that cause the disappearance of the absorption. These fibers are called "water free".

If one wants to extend the transmission range to longer wavelengths, it will be necessary to use other glass systems. Halogen glasses such as BaF_2-CaF_2-YF_3-AlF_3 and BaF_2-ZrF_4 GdF_3 or chalcogen glass (GeS_3) shows infrared absorption at wavelengths starting from 3 μm. Figure 2.18 shows the attenuation characteristic of these glasses (Grau and Freude 1991). The refractive index is presented with n_D regarding to the associated kind of glass which is measured at a wavelength of 589 nm. It is obvious that in contrast to SiO_2, it is possible to reduce the expected attenuation by two dimensions. However, at the moment, the high manufacturing costs of these glasses prevent the further spread outside the framework of the experiments in laboratories. Moreover, the long-term stability leaves a lot to be desired. Over time, these compounds gather water, which diffused into the fibers and in this way they pollute and change the characterizations massive.

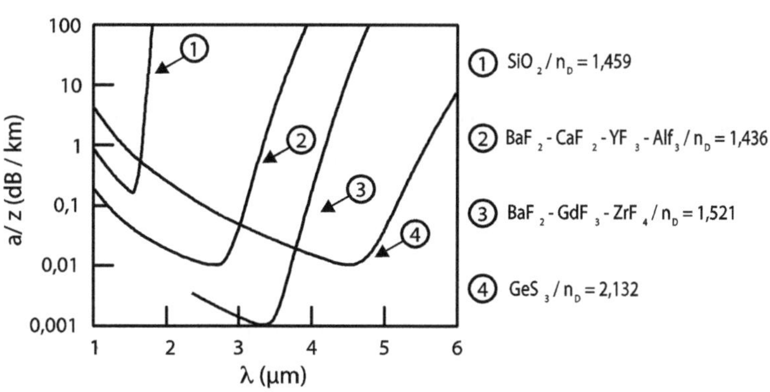

Fig. 2.18 Attenuation curves of halogen and chalcogen glasses

2.5 Polymeric Fibers

Indoors, in the automotive industry and in the range up to 100 m cable length, the following fibers are used: PMMA, polystyrene (PS) and polycarbonate (PC). The advantages in comparison with glass fibers are the low cost, the high flexural strength, and the easier treatment of the material. Apart from these values, the standard VDE 0888 part 101 describes more values such as mechanical, optical transmission, and technology values.

Minimum attenuation values (Fig. 2.19) of 80/km are achieved at a wavelength of 570 nm. At higher wavelengths, the attenuation values increase to more than 1000 dB/km, so that these fibers can be used only at short wavelengths and short distances due to the extremely high attenuation values.

Nowadays, recent variants with endowed PMMA reach attenuation values below 30 dB/km (D-PMMA). The material Cytop and mixtures of PMMA and SiO_2 already show attenuations below 5 dB/km, but the manufacturing is very expensive. The structure of the fiber is also very different compared with the glass fiber types; the core is very large but the cladding is very thin with a high refractive index jump of 0.5 %. The following characteristics describe this fiber type (Table 2.5).

In polymeric fiber systems, the NA is fixed to 0.5. This represents an aperture angle of 30°. Comparing POF to glass fibers, one can see the difference of the core

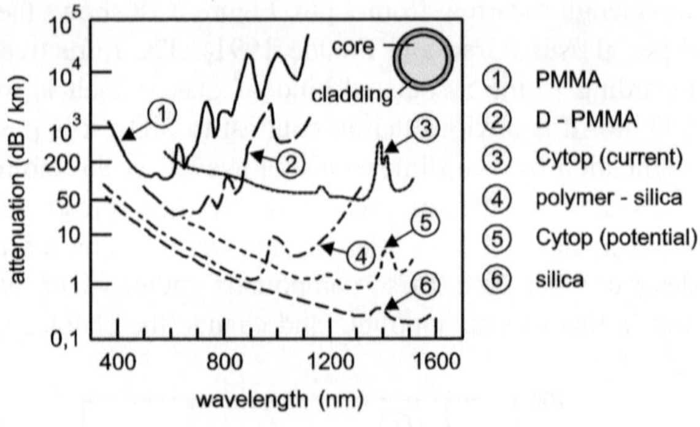

Fig. 2.19 Attenuation of different polymeric fibers

Table 2.5 Parameters of polymeric fibers for short-range optical transmission systems

Core diameter	480, 720, 980 μm
Refractive index	$n_{core} = 1.492$, $n_{clad} = 1.412$
Cladding diameter	500, 750, 1000 μm
NA	0.5
Transport and storage temperature	−40 bis 85 °C
Installation temperature	5 bis 40 °C
Operating	−40 bis 85 °C
Minimum attenuation	80 dB/km (PMMA/570 nm)

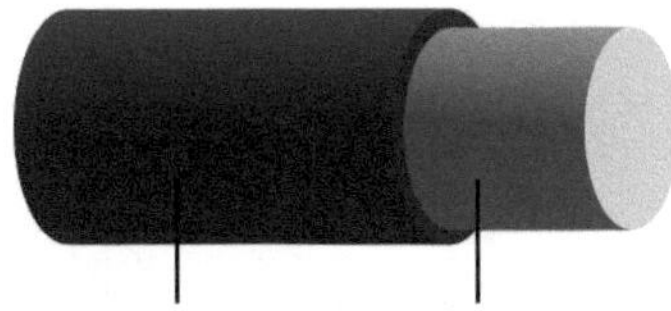

Fig. 2.20 Polymeric step-index fiber

and cladding refractive indices is 5 %. The NA is correlated by the normalized parameter V. The V-parameter is a correlation to the number of optical modes in the optical waveguide. The number of the modes, which are confined in a fiber, can be determined by the relationship between the wavelength of the light passing through the fiber, the core diameter, and the material of the optical waveguide. This relationship is known as the normalized frequency parameter, or V number shown in Eq. (2.17).

Standard single-mode fibers typically have a V number that is about 2.405. Here, the light will propagate in only one single mode. On the other hand, in multi-mode step-index POF fibers (Fig. 2.20), the V number is 2.799, at of yellow/550 nm, core radius of 490 μm, and NA of 0.5. This is more than 10^3 times larger than for single-mode fibers. In this case, the light will propagate in many paths or modes via the fiber. The count of optical modes for step-index POFs will be extracted to $N = V^2/2 = 3.917$ Mio modes. Using higher wavelengths, the number of modes will reduce to 2.804 Mio modes at red/650 nm. By this reasons, POFs are only introduced in short-range transmission systems using distances between 1 m and 100 m.

2.6 Optical Waveguides in InP, GaAs, PMMA, and SiO_2

In addition to the radial symmetric waveguides, which are discussed in the previous sections, optical waveguides are mainly used in semiconductor technology. The reason behind this is that the coatings and endowments are always applied in layers in the semiconductor material, and the radial symmetric structures can only be realized with great effort.

The propagation in a waveguide layer (Fig. 2.21) is realized by reflection on the two marginal surfaces, similar to a radial waveguide. The side of the boundary of the waveguide gives the extension in the width. Different forms of layer waveguides are given as follows:

- rib waveguides
- buried waveguides
- inverse rib waveguide.

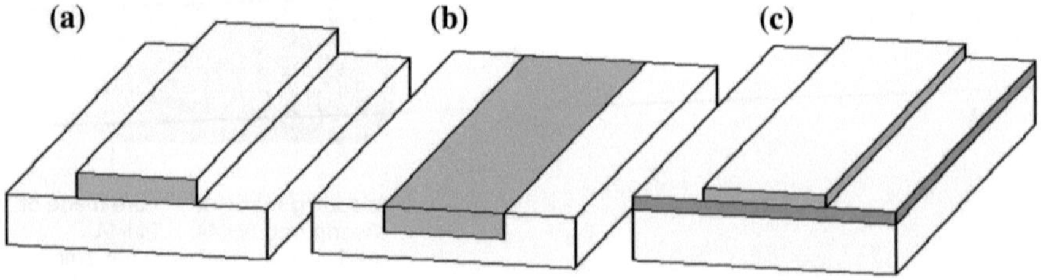

Fig. 2.21 **a** Rib waveguide, **b** buried waveguide and **c** inverse rib waveguide

Active elements are produced in the quaternary material system consisting of III–V semiconductors: indium, gallium to aluminum, arsenic, and phosphorus. Passive elements are made of glass, SiO_2, and lithium niobate ($LiNbO_3$). The latter is widely used in the manufacturing of optical modulators in the gigahertz range.

A visual representation of the modus guide in a layer waveguide is shown in Fig. 2.22. One can see that the guide can be explained analogous to the glass fibers. The calculation of the modus propagation in the layer waveguides is nowadays done with the aid of complex programs, the so-called beam-propagation programs (He and Shi 2010; Paltani and Medhekar 2010) (BPM, see Chap. 12). These programs calculate the propagation of the electromagnetic wave with very small steps in the waveguide and display the intensity distribution of the wave over the entire waveguide.

A typical example for the calculation of a Mäandercoupler is illustrated in Fig. 2.23. The field distribution of a rip waveguide is calculated in several local modes. Here, too, the share of the wave is guided in the cladding, which is used in the optical coupler. Thereby, two waveguides are brought together so close, in which the light wave can be coupled from one waveguide to another. The Mäandercoupler uses several serpentine approximations of two waveguides to achieve the goal that the light of the right waveguide is completely transmitted to the left waveguide.

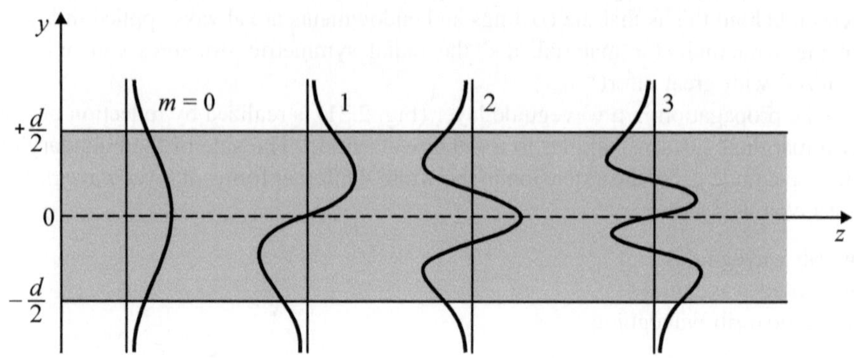

Fig. 2.22 Guided modes in a rib waveguide

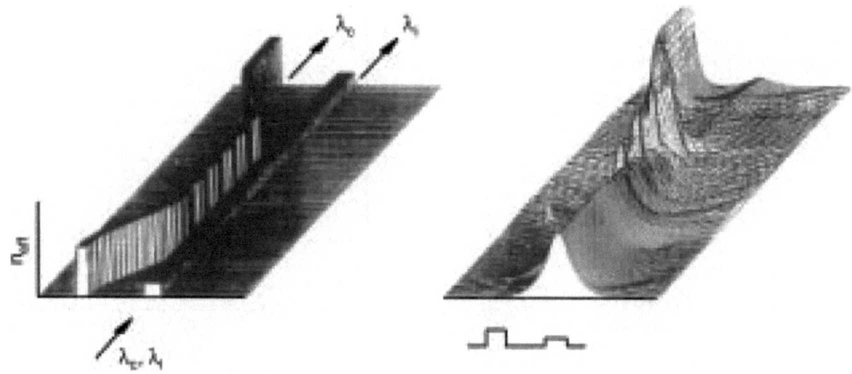

Fig. 2.23 Simulation of an optical waveguide in a 3 dB splitter, *left* effective index profile, *right* BPM simulation

2.6.1 Geometry of Integrated Waveguides

The waveguide of a layer structure in the InP system has the typical dimensions of 0.2 μm × 2 μm and generates at the waveguide end a highly asymmetrical elliptical field (Fig. 2.24). Therefore, it is necessary to realize a field adjustment to the fiber through a lens or a lens system. An example of the calculated field of a rib waveguide in InP is shown in Fig. 2.25.

In contrast to the rib waveguide, the buried waveguide (Fig. 2.25) field distribution in a rib waveguide with thickness of 0.11 μm, width 3 μm.

Figure 2.26 guides the wave through the middle of the waveguide, which makes the distribution of intensity of the field symmetrical. The field distribution of the rib waveguide is asymmetric concerning the y height. The main intensity of the light is not guided in the rib, but in the material underneath the rib, the bulk material. Curvatures of the waveguide without large radiation losses are impossible for the rip waveguides in contrast to the buried waveguides. Therefore, one prefers buried waveguides for monolithic integrated circuits because smaller curvatures allow a higher packing density in the circuit and thereby the required area and costs for a component decrease.

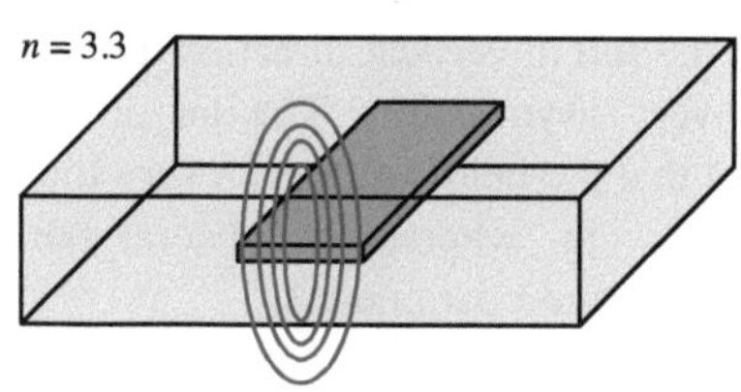

Fig. 2.24 Optical field distribution of a buried waveguide

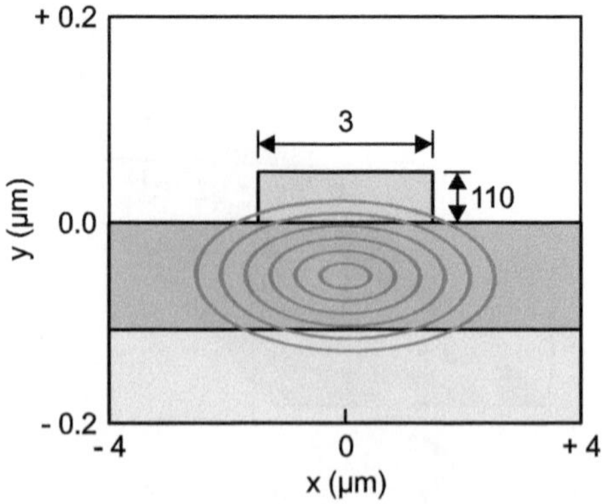

Fig. 2.25 Field distribution in a rib waveguide with thickness of 0.11 µm, width 3 µm

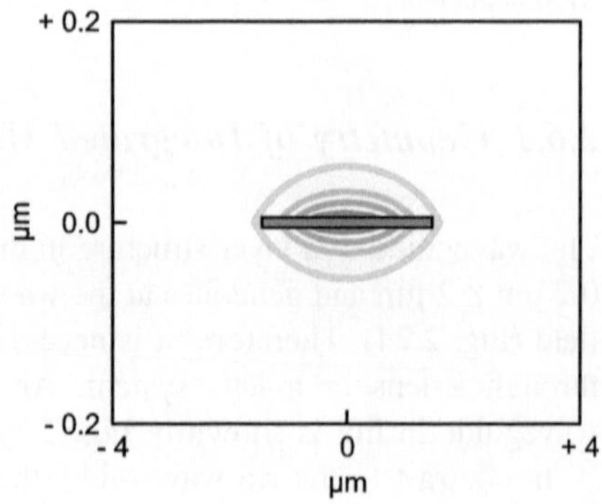

Fig. 2.26 Field distribution in a buried waveguide with width of 3 µm and thickness of 110 nm

2.6.2 Semiconductor Laser

The layer structure of a semiconductor laser is illustrated schematically in Fig. 2.27. The figure shows a buried waveguide in which lies simultaneously the active zone. In this zone, the charge carriers are generated, which emit light when they are recombined (Fischer 2002). The slit edges of the chip form the laser mirrors at the same time which reflect approximately 30 % air in InP components. In the present example, the boundary of the active region is caused by the size of the electrical connection field. At the same time, the active zone has a thickness of 0.2. The optical field of such a laser on the surface of the gap plane is shown in Fig. 2.28.

With an expansion of 4.7 µm in vertical direction and 2.5 µm in lateral direction, the field is distributed very asymmetric. The larger dimension in the vertical direction is due to the fact that the field still breaks very low in the bulk material and is transmitted in the following, whereas the lateral boundary is retained well because of the mechanical-defined parameters.

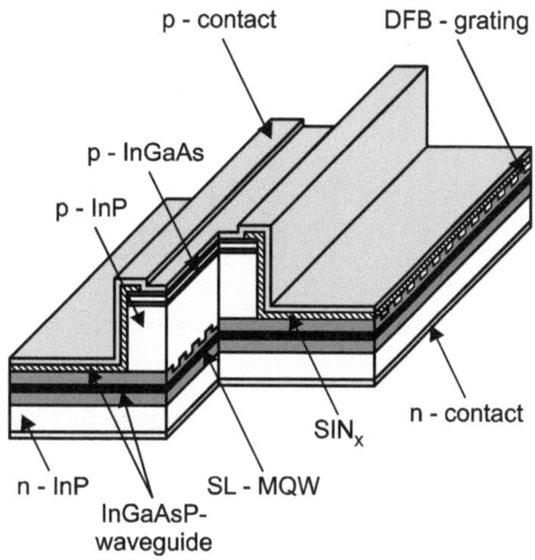

Fig. 2.27 Semiconductor multi-quantum well (MQW) laser with distributed feedback grating (DFB)

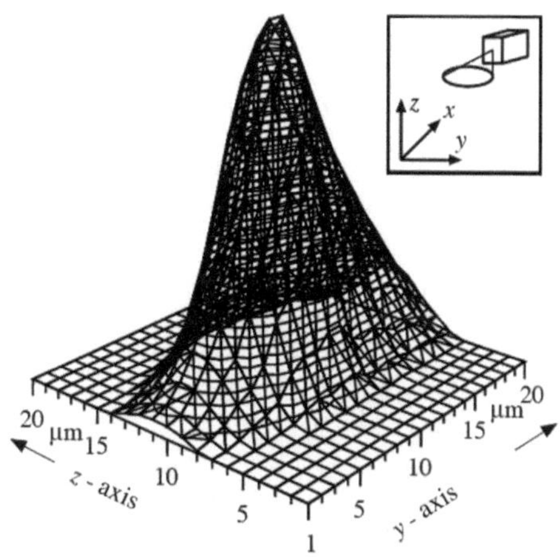

Fig. 2.28 Near-field distribution of a DFB laser diode with mode field widths of 4.7 μm horizontal and 2.5 μm vertical

2.6.3 PMMA-integrated Waveguides

This type of waveguide is used for passive structures. Figure 2.29 shows the layer structure. Out of it, one can manufacture optical splitters, switches, etc. The production of this kind of layer waveguides is achieved by hurling in the centrifuge.

The liquid material is applied to the base material, in this case silicon, and then hurled with about 1000 revolutions per minute. Now, it is formed a thin polymer film whose thickness is a direct function of the rotational speed and time. Therefore, the thickness of 1–30 μm can be applied well-controlled layer by layer. Many different optical devices like AWGs, thermal switches or multi coupler structures have been shown by several groups in Germany (Keil et al. 1996, 2001; Yao et al. 2002) and Korea (Kim et al. 2007a, b; Lee 2012).

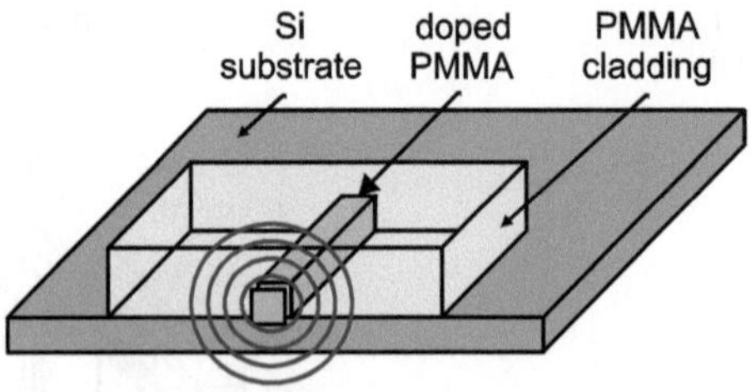

Fig. 2.29 Sandwich layer setup and optical field distribution of an integrated PMMA waveguide

The lateral structure is achieved by photolithography. Therefore, a chrome mask is placed on the substrate, in which the waveguide structure is kept free. In the following, the photophobic material is exposed with UV light, which leads to an increasing of the refractive index at the exposed places. A subsequent thermal treatment results in an evaporation of the remaining photosensitive endowment material. In a further step, it is possible to hurl up a cladding layer with a lower refractive index than the waveguide material of the waveguide layer. All working steps are shown in Fig. 2.30.

The characterizations of the resulting waveguides in PMMA are listed below:

- Waveguide 8 μm × 8 μm,
- Symmetrical field,
- Good field adaptation to optical fiber with a
- Refractive index difference of $n_{\text{Clad}} = 1.51$–$n_{\text{core}} = 1.53$.

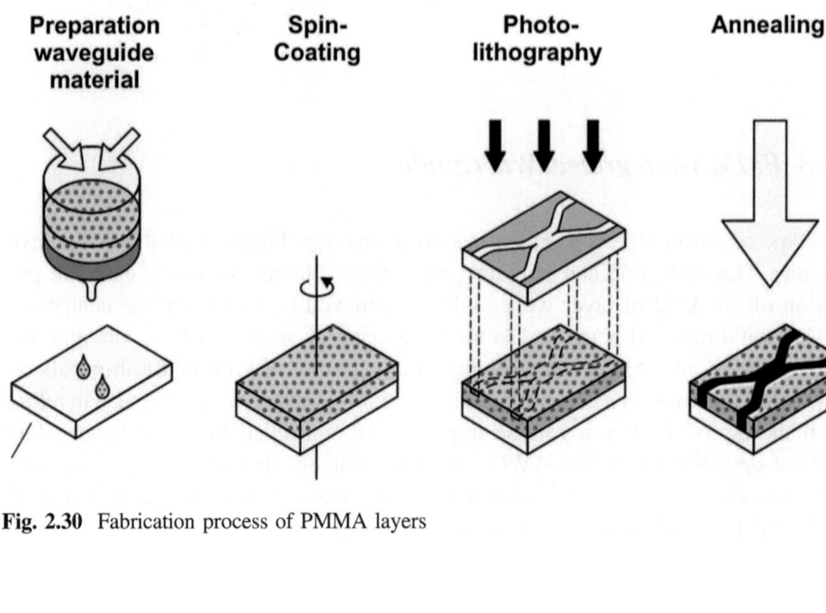

Fig. 2.30 Fabrication process of PMMA layers

2.7 SiO_2-Optical Waveguides

SiO_2 waveguides (Kilian et al. 2000) are also used at layer waveguides for passive structures. The waveguide dimensions are 6 μm × 6 μm in width and height, with a refractive index jump of 4 % at a core refractive index of n_{core} = 1.447. The symmetry of the waveguide leads to a very symmetrical field with a good field adaptation to glass fiber.

For the production of these waveguides, the base material is again the silicon substrate. Upon this, we evaporate a layer of silica (SiO_2) by flame (Kawachi et al. 1983; Kominato et al. 1990; Bebbington et al. 1993; Kilian et al. 2000; Garcia-Blanco et al. 2004). Therefore, a mixture of the required materials is heated in an oxyhydrogen container. In the following, Si and O_2 are bonded to SiO_2 by chemical reaction and with other endowment materials, which are required for refractive index settings, they settle themselves down. After that, a mask is applied to the wafer that is connected to the waveguide strip (negative mask). Through these openings, it is possible to skim the material through reactive ion-etching. After removing the mask, the rest of the waveguide plank stands alone. In a further step, another layer of glass (the cladding layer with about 30–50 μm thickness) is evaporated by flame hydrolysis on this surface. The result corresponds to the sequence of events which are in Figs. 2.31 and 2.32.

Thus, prepared SiO_2 waveguides have an expansion of 6 μm × 6 μm, a symmetric field and also a good adaptation to the field optical fiber with a refractive index of n_{Clad} = 1.447 with 0.7 % refractive index jump. Average losses in the fiber-chip coupling are less than one dB.

The attenuation at 1.55 μm is analogous to the optical fibers very low. However, the process is currently not satisfying to lead because water is still left in the material after the production process which influences the attenuation negative. Typical attenuation values with different endowment materials are given as follows:

$$SiO_2 - TiO_2 (Dn = 0.25\ \%): \quad 0.27\ dB/cm$$
$$SiO_2 - GeO_2\ (Dn = 0.75\ \%): \quad 0.04\ dB/cm$$

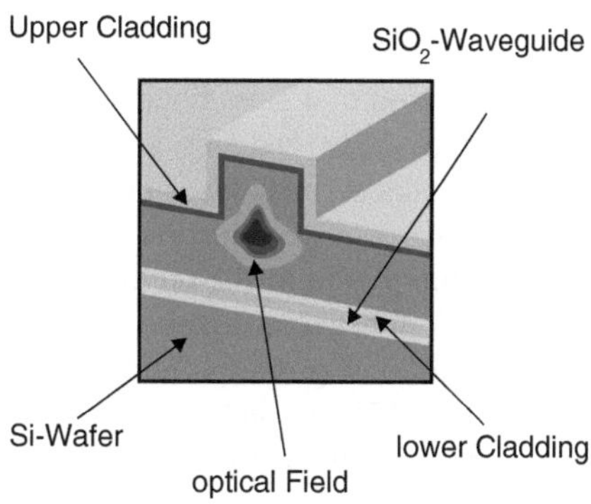

Fig. 2.31 Cross section of a silica waveguide with near-field distribution

Fig. 2.32 Fabrication process of SiO_2 layers using flame hydrolysis

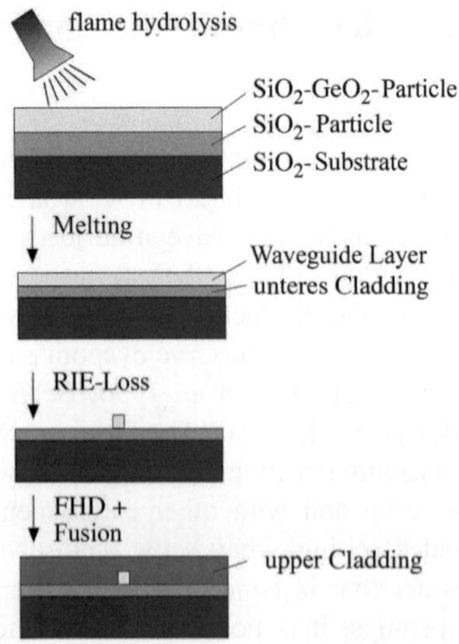

In summary, the properties of the silica waveguides in planar construction are listed below:

- Very low propagation losses (<0.05 dB/cm)
- Low temperature dependence of the refractive index ($dn/dT = 1.1 + 10^{-5}$) and therefore low thermal drift at ambient temperature.
- Thermally and mechanically stable
- Polarization-independent components can be realized
- Low-loss optical fiber coupling with non-critical alignment tolerances (±2 μm at 1 dB additional loss)
- It is possible to do an Er-doping of SiO_2 for the production of integrated optical amplifiers

2.8 Production of Optical Fibers

The production of a glass fiber consists—independent of the used method—always of two stages. The first step is the preparation of a preform which is shaped with easy mechanical methods (milling and spinning) and which is coated in an endowment machine with the refractive index of core and cladding. When we consider the profile of the preform, one can see a true to scale enlargement of the mechanical dimensions and of the refractive profile of the future glass waveguide.

In the following, the preform is stretched through a fiber stretch machine to a thickness of 125 μm and a plastic layer is immediately applied (primary coating). The necessary endowment of SiO_2 monomode fibers is done with additional

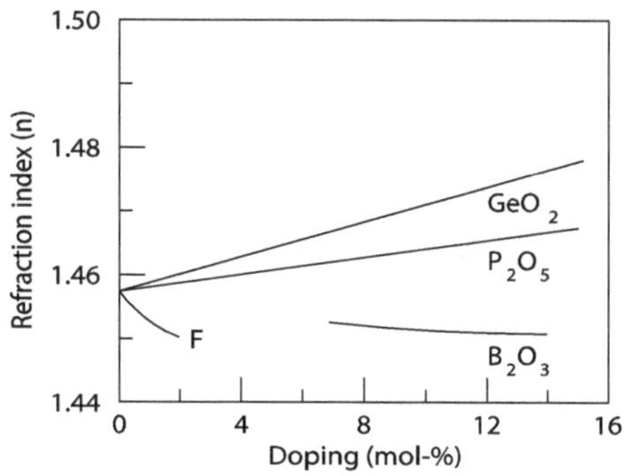

Fig. 2.33 Variations of refractive indices using germanium, phosphorus, or boron oxide

material to modify the refractive index of the core and cladding. Figure 2.33 illustrates the change of refractive index as a function of the endowment concentration of germanium, fluorine, and phosphorus oxide.

2.9 Gas Phase Methods

The manufacturing methods of extremely low-vapor glass fibers normally used today are achieved with the aid of separating processes from the gas phase that were developed in 1970 (Schultz 2010) by Corning scientists Dr. Robert Maurer, Dr. Peter Schultz, and Dr. Donald Keck, Corning Inc., USA. A distinction is made between three different methods of separation. If the separation is on the outside of a preform, the process is called the OVD method (Schultz 2010; Murata 1996) (outside vapor deposition). A process that carries out the separation on the facing side of a preform is called the VAD method (vapor axial deposition).

If the materials are deposited on the inside of a rotating glass rod, this is referred to as the IVD method (inside vapor deposition). In the latter IVD method, a distinction is made between the process where the required melt energy is supplied by oxyhydrogen burners, the MCVD method (modified chemical vapor deposition), and the method where the energy is supplied by a microwave source, the PCVD method (plasma-activated chemical vapor deposition).

In all methods, the deposition results from the decomposition of high-purity compounds in an oxyhydrogen flame. The gases required for this are listed in (Fig. 2.34) and belong to the group of chlorides, which, in turn, belong to the group of metal halides. Oxygen is fed into the burner together with the required doping agents and converted there into the corresponding oxides.

Oxygen is allowed to flow through vessel (bubblers), which contain liquid compounds at room temperature. Bubbles are formed and the oxygen, as the carrier gas, takes up the chlorides (Fig. 2.35). Due to the high level of heat energy supplied in the burner, the compounds are then converted into the corresponding oxides, which separate as a fine coating on the preform. If the burner is then also moved

Fig. 2.34 Gas application for glass fiber preform fabrication

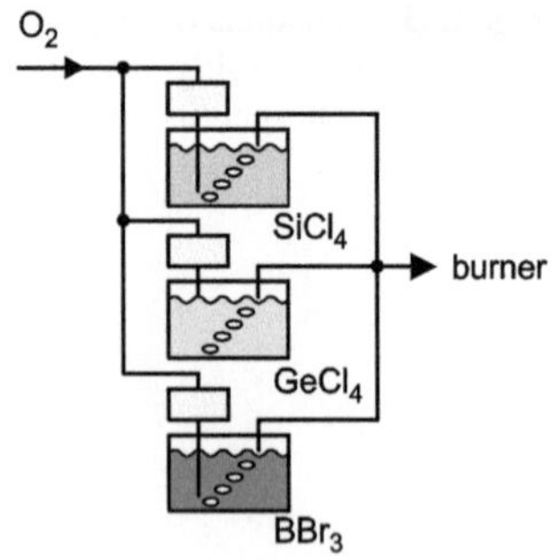

Fig. 2.35 Outside vapor deposition (OVD) technique

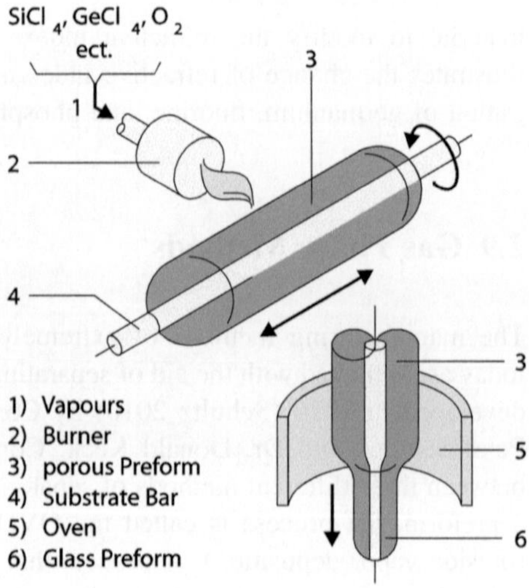

back and forth, a porous glass coating is formed in layers where each layer can be doped differently by varying the concentration of gases supplied. The outside vapor deposition method (OVD) is described below:

$$SiCl_4 + O_2 \rightarrow SiO_2 + 2Cl_2$$

And for germanium as the doping agent:

$$GeCl_4 + O_2 \rightarrow GeO_2 + 2Cl_2$$

The layers are deposited on the outside of the preform, hence the name "Outside." In the VAD method, a rotating glass rod is heated on the facing side with the burner and the gases separate there (Fig. 2.36).

The resulting form is slowly drawn upward where the distance from the burner to the form must always be kept constant. Several burners can also be used at the same time in order to coat the core and sheath simultaneously. Subsequent to the

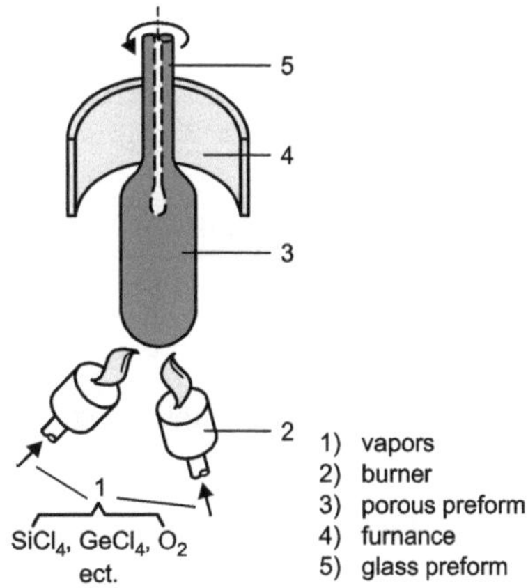

Fig. 2.36 Vapor outside deposition (VAD)

separation, the shrinking of the preform is then carried out via an annular heating arrangement. Chlorine gas is rinsed around the form for "drying," i.e., removal of water residues.

In MCVD (Fig. 2.37) and in the PCVD processes (Fig. 2.38), the gas flow is directed through a hollow preform so that the deposition occurs inside the form. The resulting temperatures are around 1600 °C which in the case of MCVD is generated by an oxyhydrogen burner and in PCVD by a microwave heater. Heating by microwave is particularly homogeneous and the layer sequence especially easy to control. In both methods, after applying the desired core layers, the glass is heated (2000 °C) until the form melts, thus shrinking it. No special drying processes are required, as no oxyhydrogen gas has been used inside the preform in either of the inner coating methods.

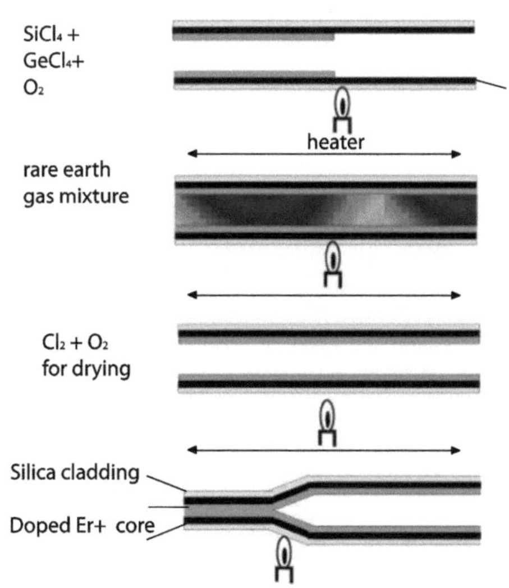

Fig. 2.37 MCVD technique (modified chemical vapor deposition)

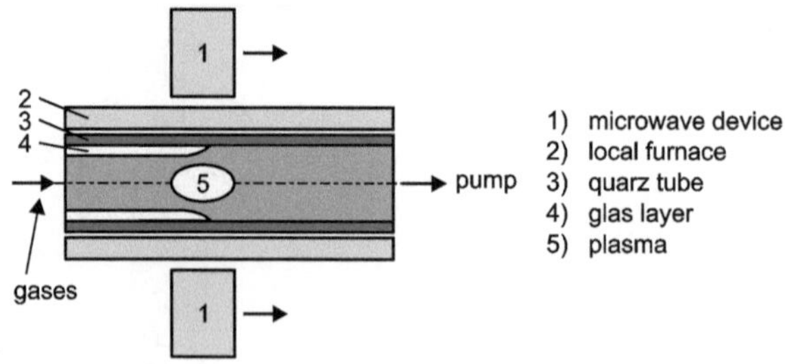

Fig. 2.38 PCVD process (plasma chemical vapor deposition)

2.9.1 Drawing of Glass Fibers

For extending the fiber, the finished base body (see Fig. 2.39) is hanged in a pull tower (Gehrke et al. 2009; Mahlke and Gössing 1994). The retainer is adjustable by *x*-, *y*-, *z*-positioning units. The height of such a tower is approximately 15 m. The end of the preform is heated to 200 °C and the next step is to grab the end and pull the fiber. The diameter of the fiber must be very consistent at 125 μm (±2 μm). To ensure this, the fiber diameter is measured online and therefore the pulling velocity and the heating temperature controlled accurately. During the drawing process, the geometric relationships of the preform remain stable so that a reduced image of the fiber preform is formed within the fiber with a corresponding refractive index profile.

Fig. 2.39 Fiber drawing tower

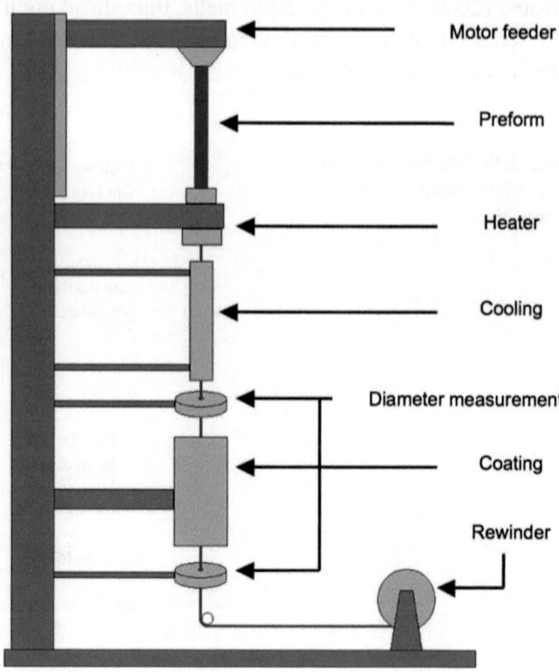

Aspect ratios of 400:1 are possible. Straight after the measurement of the fiber cross section, a first protective layer of PMMA is applied and immediately cured by ultraviolet light. This primary cladding layer is supposed to absorb the harmful environmental factors immediately after the manufacturing. Thus, the fiber diameter is summed up to a thickness of 250 μm. Then, the fiber is measured with a roller system, to test the tensile strength, and in the following, the fiber is wound up onto a cable drum.

2.9.2 Types of Fiber-Optic Cables

The finished fiber-optic cable needs further protection layers for the use in the field against bending, breaking, torsion, and tension. Therefore, the optic fiber is coated with a further protection layer, which can be varied depending on the requirement profile. The form and nature of the envelope is achieved through construction of cable technology. These are special structures of the fiber cladding and the installation of train and compression elements in the entire cable, such as steel wires or gel fillings.

Figure 2.40 illustrates several types of layer cables. There is a talk of a stranding at which rope elements are concentric organized around a central element. Cables with single elements such as glass fiber strands, copper strands, or others are called as plies cables. On the other hand, one speaks of bunched cable when the soul of a cable is setup of bunches of different elements. In the long-distance networks, there are mainly used plies cables and in the local network are used bunched cables.

Another variant of cables is ribbon cables. Therefore, a number of individual fiber-optic cables are fixed together to a flat band structure. Depending on the type of application, one distinguishes several different types, such as the ones in Fig. 2.40. These ribbon cables can be stacked in several layers and so you will reach a very high packing density.

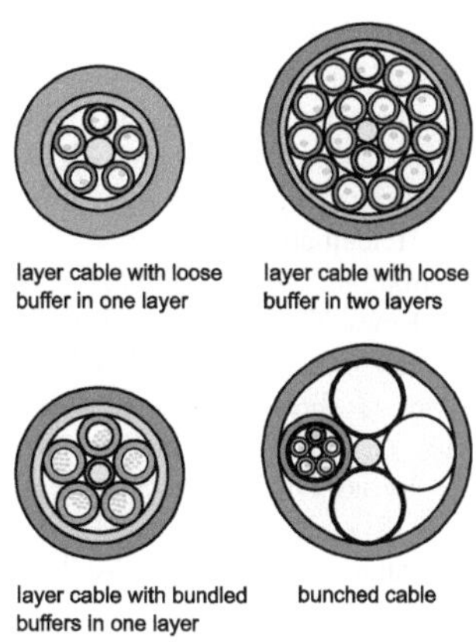

Fig. 2.40 Types of fiber-optic cables

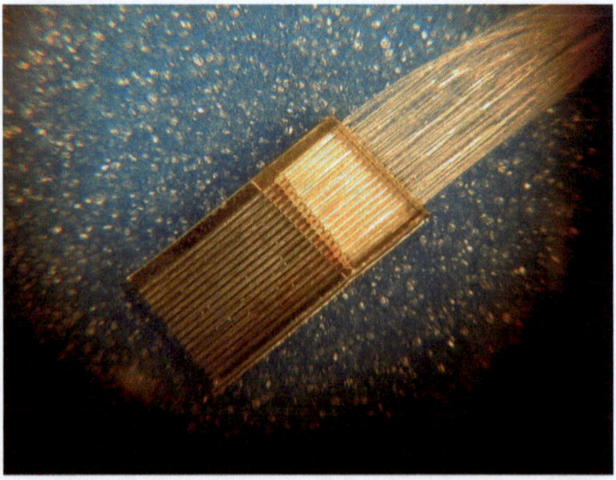

Fig. 2.41 Fiber-optic ribbon cable

The main application of this cable is located indoors at the direct connection to optical modules and components, such as Arrayed Waveguide Grating (AWG) modules. In such modules, the ribbon fiber is embedded and fixed to an array of silicon V-groove. This has the advantage to get a stable matrix of fibers with strong defined distance to the fiber-chip coupling (Fig. 2.41).

References

Bebbington, J.A., Barbarossa, G., Bonar, J.R., Aitchison, J.S.: Rare earth doped silica waveguides on Si fabricated by flame hydrolysis deposition and aerosol doping. Appl. Phys. Lett. **62**(4), 337–339 (1993)

Bloomfield, P.: Fourier Analysis of Time Series: An Introduction. Wiley Series in Probability and Statistics. Wiley, New York (2000)

Brooker, G.A.: Modern Classical Optics (Physics). Oxford University Press, Oxford (2003)

Demtröder, W.: Atoms, Molecules and Photons: An Introduction to Atomic-, Molecular- and Quantum Physics. Graduate Texts in Physics. Springer, Berlin (2010)

Fischer, U.H.P.: Optoelectronic Packaging. VDE, Berlin (2002)

Garcia-Blanco, S., Alfaro-Cid, E., De La Rue, R.M., Aitchison, J.S.: Genetic algorithm-based characterization of the optical properties of flame-hydrolysis deposited silica waveguides. J. Lightwave Technol. **22**(12), 2801–2807 (2004)

Geckeler, S.: Lichtwellenleiter für die optische Nachrichtenübertragung. Springer, Berlin, Heidelberg (1990)

Gehrke, N., Casper, A., Otto, O., Heinemann, R., Tockhorn, M., Fischer, UHP: Studie Zur Realisierung Eines Ziehturms Für Polymeroptische Fasern. In: 10 NachwuchsWissenschaftlerKonferenz, HS Merseburg (FH). HS Merseburg 2009

Grau, G., Freude, W.: Optische Nachrichtentechnik: Eine Einführung. Springer, Berlin (1991)

He, Y., Shi, F.: Beam Propagation Method and Microlens Design for optical Coupling. VDM Verlag Dr. Müller, Saarbrücken (2010)

ITU RG: Transmission Media Characteristics: Definition and Test Methods for the relevant Parameters of Single-Mode fibres. T94. ITU (1994)

Kawachi, M., Yasu, M., Edahiro, T.: Fabrication of SiO_2-TiO_2 glass planar optical waveguides by flame hydrolysis deposition. Electron. Lett. **19**(15), 583–584 (1983)

Keil, N., Yao, H.H., Zawadzki, C.: Polymer waveguide optical switch with < -40 dB polarisation independent crosstalk. Electron. Lett. **32**(7), 655, 657 (1996)

Keil, N., Yao, H.H., Zawadzki, C., Bauer, J., Bauer, M., Dreyer, C., Schneider, J.: Athermal all-polymer arrayed-waveguide grating multiplexer. Electron. Lett. **37**(9), 579–580 (2001)

Kilian, A., Kirchof, J., Przyrembel, G., Wischmann, W.: Birefringence free planar optical waveguide made by flame hydrolysis deposition (FHD) through tailoring of the overcladding. J. Lightwave Technol. **18**(2), 193–198 (2000)

Kim, J.T., Ju, J.J., Suntak, P. Myung-Hyun, L.: O/E integration of polymer waveguide devices by using replication technology. IEEE J. Sel. Top. Quantum Electron. **13**(2), 177, 184 (2007a)

Kim, J.T., Ju, J.J., Suntak, P., Koo, S., Kim, M.-S., Myung-Hyun, L.: UV-embossed polymer optical bench for integration of polymer waveguide devices. Digest of the IEEE/LEOS Summer Topical Meetings. pp.111–112 (2007b)

Kominato, T., Ohmori, Y., Okazaki, H., Yasu, M.: Very low-loss GeO_2 doped silica waveguides fabricated by flame hydrolysis deposition method. Electron. Lett. **26**(5), 327–329 (1990)

Mahlke, A., Gössing, P.: Fiber Optic Cables. Fundamentals, Cable Engineering, Systems Planning. Publicis Corporate Publishing (1994)

Marcuse, D.: Theory of Dielectric Waveguides. Academic Press, New York (1974)

Marcuse, D.: Loss analysis of single-mode fiber splices. Bell Syst. Tech. J. **56**, 15 (1977)

Marcuse, D.: Gaussian approximation of the fundamental modes of graded-index fibers. J. Opt. Soc. AM **68**(1), 103–109 (1978)

Murata, H.: Handbook of Optical Fibers and Cables, 2nd edn. Optical Engineering. Marcel Dekker Inc., New York (1996)

Paltani, P.P., Medhekar, S.: Optical Beam Propagation in Non-Linear Media and All-Optical Devices: Investigation of Light Beam Propagation in Non-Linear (NL) Media For All-Optical Devices (AODs). VDM Verlag Dr. Müller, Saarbrücken (2010)

Ramaswami, R., Sivarajan, K.: Optical Networks. Academic Press, San Diego (2004)

Schultz, P.C.: Title: Making the first low loss optical fibers for communications. In: 36th European Conference and Exhibition on Optical Communication, ECOC. Torino (2010)

Webb, R.H.: Elementary Wave Optics. Dover Books on Physics (2005)

Yao, H.H., Zawadzki, C., Keil, N., Bauer, M., Dreyer, C., Schneider, I.: Polymer waveguide devices. Lasers and Electro-Optics, 2002. CLEO '02. Papers Presented at the Technical Digest. Summaries. **1**, 489–490 (2002)

Chapter 3
Optical Mode-field Adaptation

Abstract The optical coupling between the various optical components requires low coupling loss and low reflection. Very often is the application of ray optics not useful, but the wave observation of light must be used. In most cases, the optical modes of the components (laser, fiber, waveguide) will be described in the form of a Gaussian distribution, and all coupling efficiencies will be calculated with the aid of an overlap integral. The biggest problem occurs by mechanical adjustment of the components and the long-term stability of the coupling.

Waveguides

Light is the part of the electromagnetic spectrum and can be described in terms of an electric and a magnetic field, which propagate together through the space. Maxwell's equations describe how this electromagnetic wave stores its energy in the electric and magnetic fields.

The equations that Maxwell (1982, 2001; Maxwell and Boltzmann 1895, 2008) developed based on the Faraday's investigations describe not only the propagation of the electromagnetic wave but also the coupling of the electric and magnetic fields, and the interaction of the fields with their environment.

For many basic effects in the optical communications, it is sufficient to use ray optics to represent those optical effects. However, to have a deeper understanding of the optical and electro-optical effects in solids, a fundamental analysis of the propagation of the electromagnetic waves in solids is required. Only then, the design of photonic devices such as lasers, waveguides, modulators, or even integrated waveguide lenses, which are found in this book, can be simulated realistically. The basics of wave optics will not be discussed further here. Instead, references to the corresponding literature are made. However, the calculation of the fiber–chip coupling efficiency will be described in detail because it represents a core topic of this work.

Fundamental works on the propagation of optical frequencies in glass-like media were published simultaneously with the production of the first glass fibers. Among them, the work of Kao and Hockham (Mahlke and Gössing 1995), published in 1966, was awarded with a Nobel prize. The propagation of light in optical fibers was analyzed in a variety of books. Marcuse (1972, 1974) and Kogelnik (Kogelnik

© Springer-Verlag Berlin Heidelberg 2015

U.H.P. Fischer-Hirchert, *Photonic Packaging Sourcebook*,
DOI 10.1007/978-3-642-25376-8_3

and Li 1966; Kogelnik 1965, 1975) have summarized the topic at a very early stage, and their works are considered as reference works even today. Green (1993), Snyder and Love (1983), Neumann (1998), Opielka (1995), and Geckeler (1990) further developed waveguide theory based on the different application scenarios of the Maxwell's wave equations. The work from März (1995) particularly deals with the use of waveguiding in integrated optical components and their design. A large portion of analytical calculations for fiber–chip coupling is also presented in his work. Some of those calculations will also be used here. Agrawal (1992) focused his work on the wave theory to explain the influence of the chromatic dispersion on the transfer function.

3.1 Theory of Optical Mode-field Adaptation

Kogelnik published first fundamental works on the propagation of Gaussian beams (Kogelnik 1965) between dielectric waveguides with lenses. Interest in this topic was based on the fact that newly developed laser diodes were able to produce a radiation field that had almost perfect Gaussian distribution. Therefore, for the coupling of laser diode to single-mode fibers (SMF), a Gaussian approximation could be used to explain the coupling of mode profiles. In another fundamental work, Kogelnik (1964) considered the general problem of mode matching at the transition of the optical wave from one to another medium, and by introducing so-called coupling coefficients, he gave an analytical solution for the transfer of optical power. The results indicate the limits of the coupling efficiency that can be easily calculated through the field radii of the waveguides that are mechanically optimally aligned. Based on the results from Kogelnik, Marcuse (Ishii et al. 1998) included the effects of different beam widths, as well as of the lateral, longitudinal, and angular mechanical offset in the description of the coupling efficiency. In the following sections, the key terms of the optical mode coupling and coupling efficiency will be presented. The fundamental theory of wave propagation in dielectric media is given first. Further on, the coupling efficiency will be analyzed only in the case of direct coupling between two waveguides. The analysis of the optical elements, such as lenses, lying on the optical path will not be performed, but corresponding references will be given (Kuhmann 1998; Saruwatari and Sugie 1981; Kawano and Saruwatari 1985).

3.2 Definition of Field Radius

The definition of a characteristic field parameter, which allows approximate description of the transverse amplitude profile of the fundamental mode, turns out to be very problematic. Partly for historical reasons and partly for the extraction of the results of different measurement methods, a mode-field radius w_0 was introduced. Its definitions and errors resulting out of them will be described more in detail in this chapter. This parameter plays a very important role when describing the radiation behavior of the fiber and faces. Since mode-field radius directly affects the coupling behavior between different optical components, significant attention will be paid on its analysis.

The transverse amplitude profile of the field distribution in the optical waveguide can be approximated by a Gaussian profile. Marcuse (1978) clearly showed that the approximation is valid for both the step-index and the graded-index light waveguides. Solving the Helmholtz differential equation (Reider 1997) results in a family of solutions, which are determined differently in the two areas of the core and the cladding glass, shown in Eq. (3.1).

The solution in a case of a single-mode propagation in a SMF are Bessel functions of a first kind and νth order J for the core area, and modified Bessel functions of the second kind and νth order K for the cladding region. The first resembles a cosine or a sine function with radially attenuated amplitude. The second refers to a so-called Hankel function, which decays approximately exponentially:

$$\frac{\partial^2 E}{\partial r^2} + \frac{1}{r}\frac{\partial E}{\partial r} + \left[n^2(r)k^2 - \beta^2 - \frac{\nu^2}{r^2}\right]E = 0 \tag{3.1}$$

With the use of fibers that have a step-index profile and are applied not directly to the so-called cutoff frequency, which determines the transition between single-mode and multi-mode operation, there are typically large deviations in determining the mode-field radius. It is therefore necessary to develop a definition based on the measurement method used to determine this parameter and find mathematical approximation that sufficiently accurately supports this definition. A very important value for the characterization of the optical intensities is defined by the ratio of the field intensity at a certain distance with respect to the maximum of the field. Two different methods are accepted:

(a) the so-called full width at half maximum (FWHM) value
(b) $1/e$ drop-off of the maximum field intensity.

The simplest definition is given by the mode-field radius w_0 or the focus diameter (spot size) $2w_0$ (Neumann 1998), which is the width of the near field at which the **normalized electrical field amplitude of the Gaussian approximation has dropped to $1/e$ = 0.368** (Neumann 1998). At that distance, the **normalized intensity decreases to $1/e^2$ = 0.135** of its maximum value.

The boundary value originates from the mathematical characterization of the Gaussian function. Because of the similarity between the Gaussian function and the field distribution in single-mode optical waveguide, the value is also used for the description of the field of the fundamental mode.

Since this method only approximates the real field profile, there are various definitions of mode-field radius in the literature. As a so-called focus diameter of the Gaussian beam, this value can be directly obtained from the measurement of the near-field intensity. However, there are other measurement methods that often provide different measurement results.

The definition of the mode-field radius is mainly based on the six different measurement methods (Andersen 1984; Marcuse 1974; Kogelnik 1975). For this work, the most important are the methods that are based on the near-field measurement. The coupling loss in the near field is a particularly important parameter since it can be easily measured.

Other methods are focused on the description of the far field, but also use the Gaussian approximation. The less common methods are based on the linking of the waveguide dispersion or the attenuation due to microbending with the mode-field radius. All this methods lead to the integral definitions that are just shortly presented here, but can be found with complete descriptions in Marcuse (1978), Saruwatari and Nawata (1979).

For the glass fiber technique, the ITU-T Recommendation G.650 (1994) is very significant. It defines the mode-field diameter $2w_0$ by means of the far-field intensity. Different measurement methods and the resulting definitions are offered. It is explicitly stated that the definition is strictly bounded to the measurement method. Accordingly, the local definition for the present work is only of academic interest.

Because of the difference of the actual profile and the Gaussian approximation at a large distance from the cutoff (v_c) of the fundamental mode, this definition is very limited in use. It only applies to the waveguides with very weak guidance ($\Delta v < 0.01$), which is the case for the glass fibers. In Fig. 3.1, the drop-off of the Gaussian approximation is compared to the drop-off of the actual amplitude for the case of a step-index single-mode fiber.

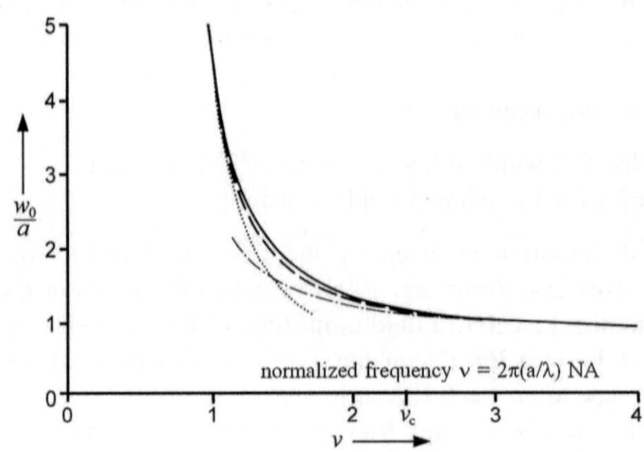

Fig. 3.1 The normalized mode-field radius $w_0/a(v)$ according to the definition of the $1/e$ drop-off of the Gaussian approximation (*dashed line*) and real field amplitude according to the Bessel/Hankel function (*dashed and dotted line*). For comparison the of transverse-offset method (*solid line*) and the approximation according to Geckeler (1990) (*dotted line*) are plotted

Some authors use the same definition but relate the 1/*e* width to the intensity, like in Geckeler (1990), that inevitably results in a different value for w_0, which equals 0.707 w_0 for a Gaussian function. The definition was meanwhile precisely formulated and standardized. Accordingly,

in this work, the 1/*e* value refers exclusively to the field intensity

and from now on will be calculated based on the Gaussian approximation for a single-mode operation.

Another definition was also introduced by Geckeler (1990). This refers to the maximum excitation of the fundamental mode of the fiber in order to determine its mode-field radius. A concentric Gaussian beam with a variable field radius excites the approximate Gaussian fundamental mode of the fiber. A mode-field radius of the fiber is determined as the value w_0 of the beam, at which the coupling of the power reaches its maximum, i.e., the coupling efficiency (transmission) is maximized.

In the first approximation, the phase fronts within the fiber can be considered to be flat. In contrast, the Gaussian beam has highly curved phase fronts as the distance from the fiber end face increases. For maximum power, coupling the phase fronts must match, which is only true for one small range of the Gaussian beam up to the so-called confocal length z_0, which is defined as:

$$z_0 = \frac{\pi \cdot n \cdot w_0^2}{\lambda} \tag{3.2}$$

The condition of the maximum excitation is thus derived via the coupling efficiency. It can be shown that with maximizing the coupling efficiency, the difference in the amplitude distribution between the actual fiber mode and the Gaussian approximation minimizes. Only when the shapes of two curves are identical, the maximum transmission is possible. In order to minimize the error, Anderson (1984) and Neumann (1998) presented an analytical approximation of the definition of w_0. That will be discussed more in detail in the next chapter.

As an alternative to the definition of the maximum excitation, the transverse-offset method was introduced by Streckert and Geckeler (Geckeler 1990; Streckert and Brinkmeyer 1982). For this purpose, two identical single-mode fibers are first ideally end-face-coupled and then shifted across the transverse axis to produce misalignment. The mode-field radius w_0 corresponds to the value of the transverse offset *d* for which the normalized coupled intensity drops off to the 1/*e* of the maximum value. Figure 3.2 shows this drop-off in the case of a transverse offset for different normalized frequencies *n*, on which the definition is based.

Since this method is also used to determine other fiber parameters, a brief explanation is required. The coupling efficiency η, or equivalently, the transmission *T* between two single-mode fibers with the Gaussian field distributions, in case of the transverse misalignment *d* and neglectable reflection, can be described by a simple function (Geckeler 1990)

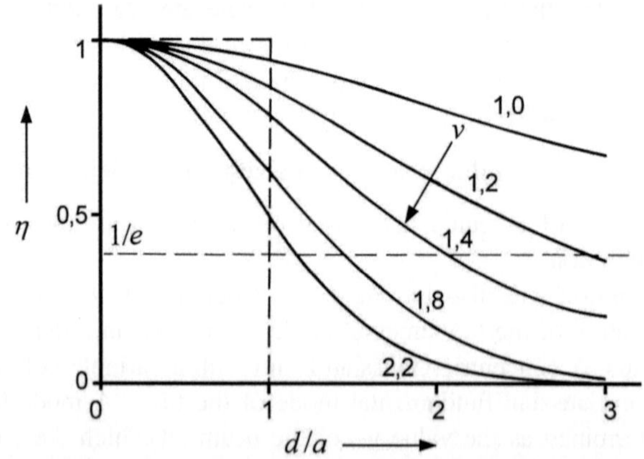

Fig. 3.2 Coupling efficiency $\eta(d)$ between two identical single-mode fibers in the presence of transverse misalignment with the value d, and different normalized frequencies v to derive a definition of the mode-field radius

$$\eta(d) = T(d) = \left(\frac{2w_{01}w_{02}}{w_{01}^2 + w_{02}^2}\right)^2 \exp\left(-\frac{2d^2}{w_{01}^2 + w_{02}^2}\right). \tag{3.3}$$

This function drops to exactly $1/e$ value when offset d equals

$$d = \left(\frac{w_{01}^2 + w_{02}^2}{2}\right)^{\frac{1}{2}}. \tag{3.4}$$

In the case of two identical fibers, the mode-field radii w_{01} and w_{02} are equal, resulting in $d = w_0$.

For the transverse-offset method, the connection can also be made via the coupling efficiency. Therefore, a simple mathematical approximation is also possible.

This definition has several other advantages. On one hand, the value w_0 can be easily obtained by a relatively simple measurement and can be averaged for elliptical mode fields from a minimum and maximum radius (Neumann 1998). On the other hand, the values for w_0 differ only slightly according to the definitions for the maximum excitation and the transverse-offset method. This means that in approximations that have been developed for a w_0 at maximum excitation, the above-defined value of the mode-field radius may be used.

Other definitions go a step further than the transverse-offset method and describe and analyze the factors which fixes the coupling efficiency and therefore the mode-field radius w_0 as an important variable, e.g., for fiber splicing. This brings the integral definition that specifies the value for w_0 and for which Neumann (1998) provides an overview.

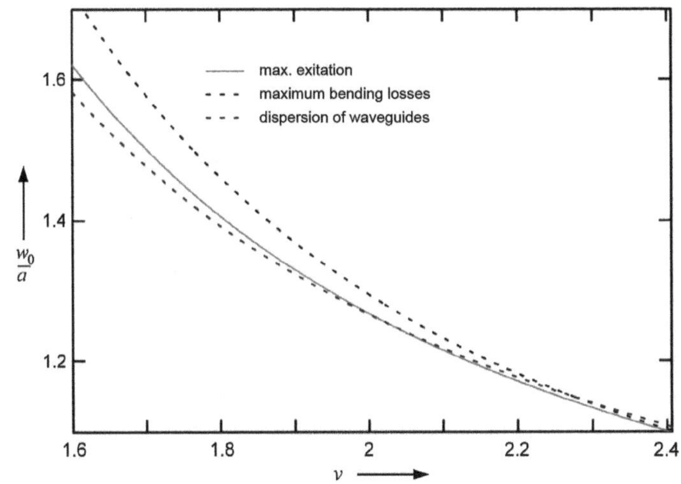

Fig. 3.3 Comparison between different definitions used to derive the mode-field radius w_0

For completeness, two other derivations may be mentioned, but for this work they are not from deeper interest. In the first case, Petermann (1976) gives the definition of the mode-field radius w_0, that is, based on the additional attenuation due to microbending losses. In the second case, according to Petermann (1983), w_0 can be obtained out of the analysis of the waveguide dispersion.

Figure 3.3 shows the difference between the above-mentioned definitions. From the figure, it is clear that three considered possibilities for the definition of mode field provide different values for the mode-field radius w_0. If it is considered that w_0 is a result of an approximation, it becomes clear that the derivations regarding different properties (transmission, microbending losses, dispersion, etc.) cause differences.

For the end of this overview, another method for determination of the mode-field radius is mentioned. The method was proposed by Anderson (1984) and is similar to the definition of the maximal excitation. In this derivation, the minimization criterion for the intensity of the near field must be fulfilled.

Besides from the near field, the derivation of the mode-field radius w_0 is also possible from the far field. The relationship is established by the Hankel transformation and described in detail by Anderson (1984), so that at this point it will not be discussed further.

In the far-field region, a parameter known as FWHM (full width half maximum) width is typically used to describe the amplitude distribution. It indicates the section where the amplitude has dropped to the half of its maximum value (Reider 1997). However, since $1/e$ value is already established when considering the mode-field width, in this work the field amplitude is always derived based on $1/e$ value.

As can be seen, there are different definitions to measure and determine the mode-field radius w_0. As recommended in the ITU-T G.650 Recommendation, the derivation of w_0 should be strictly bounded to the measuring method. Figure 3.3

clearly shows that the definition is determined by the parameter used (transmission, microbending losses, dispersion).

3.3 Approximations to Determine the Mode-field Radius

Besides numerous definitions to determine the w_0, there are also numerous approximations that should approximate this value for a specific region. The biggest problem with these approximations is the fact that the real radial field profile consists of two different functions in the core and cladding (Bludau 1998).

These functions are closely related to the refractive index profile and offer a solution in a closed form only for a step-index waveguide. In addition, a sufficiently good match is done only for a highly localized region of the approximation, which lies below the normalized frequency $v_c = 2.405$. It should be noted here that by strong deviations from the cutoff wavelength, the difference between real and Gaussian curve rises sharply.

In order not to make the approximation unnecessarily complicated, a single parameter is defined as a variable with a limited scope. The normalized frequency or other parameters, which are derived out of it and include all values influencing the field radius, are suitable.

A variety of approximations were examined with respect to their validity range for calculation of the coupling efficiency. Some approximations showed surprisingly good results. The most important of them are shown in Fig. 3.4 and will be shortly explained.

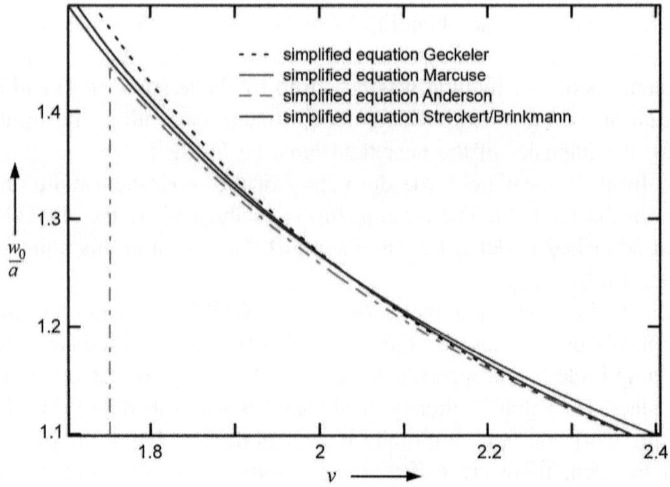

Fig. 3.4 Comparison of the most important approximations for the calculation of the mode-field radius w_0 according to the maximum excitation or the transverse-offset method

The approximations are mostly derived from the measurement results, which obtain the value w_0 based on the maximum excitation of the fundamental mode of the fiber. As mentioned in the previous chapter, these are also applicable to the transverse-offset method. The forms derived from the integral definition from Petermann (1976, 1983) can also be occasionally found.

A simple approximation of the mode-field radius is obtained from the defined field profile. Closer to the border at which the wave becomes multimodal, i.e., in the region of the cutoff $n \approx 2.405$, the $1/e$ value is approximately equal to the value of the mode-field radius w_0. For very small values up to $v < 1.2$, the following approximation according to Marcuse (1972) applies

$$\frac{w_{0a}}{a} \approx \frac{1}{w} \tag{3.5}$$

This approximation is however not suitable for n-values close to the cutoff at $n \approx 1.9$.

It is therefore necessary to modify Eq. (3.5) using suitable simplifications and through recalculation of the coefficients w back to the normalized frequency n, so that the scope of application is expanded. It was shown by Geckeler (1990) that the approximation can be thereby expended, so that w_0 can be well approximated for the values of n close to the cutoff wavelength

$$\frac{w_{0b}}{a} \approx 0.77 \cdot e^{\frac{2}{v^2}} \tag{3.6}$$

It is easy to notice that this but also the following equation offer the advantage that the approximation is independent of the cladding parameter w. Thus, the mode-field radius only depends on n.

Another approximation, which is probably the best known, was formulated by Marcuse (1977):

$$\frac{w_{0\mathrm{d}}}{a} \approx 0.65 + 1.619 \cdot v^{-\frac{3}{2}} + 2.879 \cdot v^{-6} \tag{3.7}$$

It is valid for $n \geq 0.85$ and is often used in other publications, e.g., Andersen (1984) as a comparison. The maximum error of the approximation, which results from the variation of wavelength at a fixed core radius a and normalized refractive index difference Δ, lies below <0.7 %. The error occurs at the cost of the wide range of n, which can be covered by this approximation.

For the smaller range $1.75 \leq n \leq 2.8$, it is possible to simplify the Eq. (3.7) to the form that produces only a small error of <0.2 %

$$\frac{w_{0g}}{a} \approx 0.7993 + 2.6128 \cdot v^{-2.5046} \tag{3.8}$$

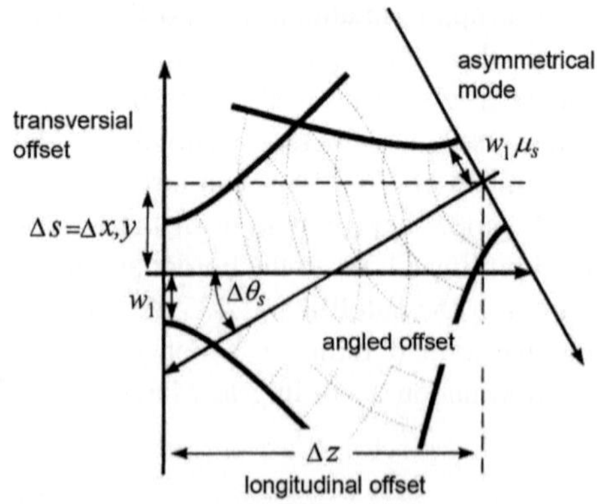

Fig. 3.5 Representation of the loss mechanisms when coupling two Gaussian beams

Anderson (1984) showed the validity of this assumption through a nonlinear regression of the data which were obtained by a numerical analysis of the mode-field radius. **This approximation will be used for subsequent calculations.**

A similar approach based on the Eq. (3.7) was presented by Streckert and Brinkmeyer (1982), taking into account other possible profile shapes. For the step-index profile, the approximation reduces to the following equation

$$\frac{w_{0f}}{a} \approx 0.6043 + 1.755 \cdot v^{-\frac{3}{2}} + 2.78 \cdot v^{-6} \tag{3.9}$$

Thus, a function with values lying between Eqs. (3.8) and (3.7) is obtained. This function provides a very accurate description (see Fig. 3.5).

However, it is evident that the approximation of the field parameter w_0, which includes a variety of fiber properties, always contains a very small approximation error.

It is also clear that a simple calculation of the coupling efficiency, which is based on the Gaussian approximation of the field, can only be approximated. More precise results can be obtained only with the propagation algorithms [BPM (He and Shi 2010)].

3.4 Loss Mechanisms in the Waveguide Coupling

The following sections are devoted to the calculation of the coupling efficiency between OEICs with asymmetrical mode fields and between SMF fibers in the presence of different misalignments. The description of the coupling efficiency will thereby be kept general. Additional coupling structures, in particular a fiber taper, will be introduced in a subsequent chapter.

3.5 Coupling Efficiency in Case of Mode-field Mismatch

When coupling between the optical components is considered, it is often of a great interest that how the field of one component excites the modes of another component. Thereby, a fraction of the total power of the excitation field is coupled to the leading mode of the other component. The relationship between the two variables is called the coupling efficiency or the degree of coupling:

Coupling efficiency = in mode$_{mn}$ coupled power/overall exciting power

The coupling loss and the insertion loss are therefore the inverse of the coupling efficiency.

As already indicated, the coupling efficiency of two optical components can be derived from the coupling between their Gaussian beams. Of interest is the overlapping of the normalized field distributions of both beams (overlap integral), and its solution for different modes. The method was comprehensively described by Kogelnik (1964) and will therefore not be further discussed here.

As the result, Kogelnik and Saruwatari introduce coupling coefficients C, which describe the coupling efficiency for different modes. For the coupling in the fundamental mode applies the following (Kogelnik 1964; Saruwatari and Kawate 1979):

Overlap integral

$$\eta = \left|C_{x0,0}\right|^2 = \left|\int_{-\infty}^{+\infty}\int_{-\infty}^{+\infty} E_{N1}(x,y)\cdot E_{N2}^{*}(x,y)\,\mathrm{d}x\,\mathrm{d}y\right|^2 \tag{3.10}$$

with

$$C_{x0,0} = C_{0,0}\exp\left(-\frac{jk\theta x_0}{2}\right)\exp\left[-\frac{x_0^2}{q}\left(\frac{1}{w_1^2}+\frac{jk}{2R_1}-\frac{jk\theta}{2x_0}\right)\times\left(\frac{1}{w_0^2}+\frac{jk}{2R_0}-\frac{jk\theta}{2x_0}\right)\right]z \tag{3.11}$$

The coupling efficiency η, excluding the Fresnel losses, indicates the transmission T at the coupling point and is typically expressed in logarithmic units.

$$T(\mathrm{dB}) = 10\cdot\log(\eta) = 10\cdot\log\left|C_x\right|^2 \tag{3.12}$$

The work of Kogelnik has been widely interpreted and correspondingly optimized, so that there are different equations to represent the coupling efficiency. Numerous are particularly the approaches that are used to describe fiber splices, such as Marcuse (1977), Geckeler (1990), or Neumann (1998). Due to the radially symmetrical field distribution of the coupled fibers, these approaches can sometimes not be applied to the coupling between the fiber and OEICs. The asymmetry of the field should be recalled here, as for example by both WL tapers.

In this work, the theoretical coupling losses between two **asymmetrical Gaussian fields** are determined based on the explanation given by März (1995). For that purpose, it is necessary to separate the coupling efficiencies for the *x*- and *y*-directions, because of possibly different mode-field radii. The individual coupling efficiencies can be processed separately, and both results can be subsequently merged.

When describing the coupling, four different loss mechanisms should be considered:

(a) *Mismatch of the mode fields* because of the different mode-field radii, field distributions, or because of elliptical mode fields. The mode-field mismatch is designated by μ_s [for definition see Eq. (3.17)]
(b) *Longitudinal offset* between two coupling components by a value Δz, referred to as defocusing,
(c) *Transverse displacement* of the axes of the two coupling elements by a values Δx and Δy, which can be summarized in a single parameter Δs ($\Delta x, y = \Delta s$),
(d) *Angular misalignment* by tilting of the optical axis of a coupling component with respect to the common optical axis by a value $\Delta\theta$.

Figure 3.5 illustrates these loss mechanisms and corresponding error quantities μ_s, Δz, Δs, and $\Delta\theta_s$. The misalignments Δz, Δs, and $\Delta\theta_s$ will be threated individually in order to show the influence of these parameters on the coupling efficiency.

In case of the Gaussian field distribution in two waveguides, it is possible to solve the overlap integral in a closed form (März 1995). In addition, the optical system should be ideal, without any aberrations from mode-field adapters (Lens, etc.) between the fiber and the chip.

$$\eta = \kappa \cdot \exp\left[-\kappa\left\{\frac{x_0^2}{2}\left(1/w_1^2 + 1/w_2^2\right) + \pi^2\theta^2\left[w_1^2(z) + w_2^2\right]/2\lambda^2 - x_0\theta z/w_1^2\right\}\right] \quad (3.13)$$

with

$$\kappa = 4w_1^2w_2^2/\left\lfloor\left(w_1^2 + w_2^2\right)^2 + \lambda^2z^2/\pi^2n_{\text{gap}}^2\right\rfloor \quad (3.14)$$

n_{gap}: Refractive index of the medium in the intermediate space.

This exact solution of the overall coupling efficiency shows clearly the connection between the phase error and the mode-field differences. However, in order to analyze the individual errors more accurately, a separation into three individual efficiencies T_z, T_s, and T and into additional coupling efficiency in case of mode-field mismatch T_{mm} will be made. Each of them will be threated separately, and exact mathematical solution will be given for individual efficiencies. Furthermore, this approach is valid only for small coupling errors. That results in

$$T = T_z + T_s + T_\theta \rightarrow T = T_{mm}(T\_{}_z + T\_{}_s + T\_{}_\theta) \tag{3.15}$$

When applying the approximation Eq. (3.15) on the coupling between the fiber and the waveguide, the separation of the mode-field mismatch from other misalignments is performed. Before the individual losses can be represented, it is necessary to define some important parameters in order to simplify the equations. The desired separate treatment of x- and y-coordinates requires the consideration of the sum of the individual efficiencies

$$T_{(mm,z,s,\theta)} = T_{(mm,z,s,\theta)x} + T_{(mm,z,s,\theta)y} = 10 \cdot \log\left(\sqrt{\eta_{(mm,z,s,\theta)y}} \cdot \sqrt{\eta_{(mm,z,s,\theta)y}}\right) \tag{3.16}$$

It is essential to take this merging into account. Because of different mode-field radii between a single-mode glass fiber and OEICs in x- and y-directions, a focus diameter ratio μ_s is further defined. This parameter specifies the difference of the mode-field radii for each axis, such that

$$\mu_s = \frac{w_{02s}}{w_{01s}} \tag{3.17}$$

Ratio of mode-field radii

The different mode-field radii are designated with w_{01s} for the fiber and with w_{02s} for the OEIC. Most of the integrated optical ICs radiate such elliptical fields. This results in inherent coupling losses because these fields cannot be made symmetrical with conventional lens systems.

The most frequent mismatch in SMF-integrated waveguide coupling arises from the fact that the stimulating mode and the mode to be excited are different. Here, the principle of maximum excitation, which ultimately manifests itself as a reduction of transmission, cannot be followed any more. For the case of mode-field mismatch, in the presence of no other misalignment, the coupling efficiency can be described by the following expression (Wengelink 1996):

$$T_{mm} = 10 \cdot \log \frac{4}{\left(\frac{w_{01x}}{w_{02x}} + \frac{w_{02x}}{w_{01x}}\right)\left(\frac{w_{01y}}{w_{02y}} + \frac{w_{02y}}{w_{01y}}\right)} \tag{3.18}$$

This expression already considers an asymmetry in x- and y-directions. Using the mode-field radii μ_s defined in Eq. (3.17), Eq. (3.18) can be simplified. The coupling efficiency is then given as

$$T_{mm} = 10 \cdot \log\left(\left(\frac{2 \cdot w_{01x}w_{02x}}{w_{01x}^2 + w_{02x}^2}\right) \cdot \left(\frac{2 \cdot w_{01y}w_{02y}}{w_{01y}^2 + w_{02y}^2}\right)\right)$$
$$= 10 \cdot \log\left(\left(\frac{2 \cdot \mu_x}{1 + \mu_x^2}\right) \cdot \left(\frac{2 \cdot \mu_y}{1 + \mu_y^2}\right)\right) \quad (3.19)$$

If the resulting Eq. (3.18) is separated for both coordinate axes, **the coupling efficiency due to the mode-field mismatch** is obtained for each axis (März 1995)

$$T_{mms} = 10 \cdot \log\left(\frac{2 \cdot \mu_s}{1 + \mu_s^2}\right) \quad (3.20)$$

Coupling efficiency

The coupling efficiency in the presence of mode-field mismatch is determined only by the mode-field radii of the coupling elements and thus, only by the geometry of the structure. Neglecting the Fresnel losses, this value indicates the maximum possible transmission.

3.6 Coupling Efficiency in the Presence of Longitudinal Displacement

The second factor that degrades the coupling of the fiber and OEICs is the so-called defocusing caused by longitudinal displacement. According to Fig. 3.5, this displacement introduces the distance Δz between the end faces of the components to be coupled. As the consequence, a free Gaussian beam is formed. The field radius of the beam rises with the increased distance from the end face of the launching component. The cross-sectional area of the field distribution also increases through the divergence of the beam.

It follows immediately that the overlap of the field distribution between the Gaussian beam and the mode to be excited according to Eq. (3.11) is changed and the coupling efficiency is reduced accordingly.

Additionally, a second error is caused by a curvature of the phase fronts of the beam with increasing distance and has a negative effect on the mode to be excited. An effective excitation requires either planar phase fronts, which has a Gaussian beam only in the focus area, or the phase fronts, which are adapted with the help of lenses. If the phase fronts do not match, one obtains an ineffective mode coupling and thus a reduction of the coupling efficiency. This is the so-called **phase front mismatch**.

The influence of both mismatches was derived by Kogelnik (1964) and can be found in the corresponding source. The coupling efficiency in the presence of the longitudinal displacement, after appropriate separation of the axes according to Eq. (3.16), can be represented as:

$$T_z = 5 \cdot \log \frac{\left(4 \cdot w_{01x} w_{01y} w_{02x} w_{02y}\right)^2}{\left(\left(w_{01x}^2 + w_{02x}^2\right)^2 + \left(\frac{\lambda \cdot \Delta z}{\pi \cdot n_z}\right)^2\right) \cdot \left(\left(w_{01y}^2 + w_{02y}^2\right)^2 + \left(\frac{\lambda \cdot \Delta z}{\pi \cdot n_z}\right)^2\right)}$$

$$T_z = 5 \cdot \log \left(\frac{\left(2 \cdot w_{01x} w_{02x}\right)^2}{\left(w_{01x}^2 + w_{02x}^2\right)^2 + \left(w_{01x}^2 \cdot \Delta z / z_0\right)^2} \cdot \frac{\left(2 \cdot w_{01x} w_{02x}\right)^2}{\left(w_{01y}^2 + w_{02y}^2\right)^2 + \left(w_{01y}^2 \cdot \Delta z / z_0\right)^2} \right)$$

which leads to the following combined expression for the loss of the longitudinal displacement

$$T_z = 10 \cdot \log \left(\frac{2 \cdot w_{02x} / w_{01x}}{\sqrt{\left(1 + \left(w_{02x}/w_{01x}\right)^2\right)^2 + \left(\Delta z / z_0\right)^2}} \cdot \frac{2 \cdot w_{02y} / w_{01y}}{\sqrt{\left(1 + \left(w_{02y}/w_{01y}\right)^2\right)^2 + \left(\Delta z / z_0\right)^2}} \right) \tag{3.21}$$

The parameter z_0 is the confocal parameter of the Gaussian beam. Through its use are the beam expansion in the emerging gap with distance Δz and the curvature of the phase fronts of the Gaussian beam taken into account. This results in an equation that includes the effects of the longitudinal displacement Δz, the medium in the resulting gap n_z, and the mode-field mismatch. By using the ratio of the mode-field radii, this equation can be further simplified to:

$$T_z = 10 \cdot \log \left(\frac{2 \cdot \mu_x}{\sqrt{\left(1 + \mu_x^2\right)^2 + \left(\Delta z / z_0\right)^2}} \cdot \frac{2 \cdot \mu_y}{\sqrt{\left(1 + \mu_y^2\right)^2 + \left(\Delta z / z_0\right)^2}} \right) \tag{3.22}$$

If the variables T_{zx} and T_{zy} for both coordinate axes are separated according to the Eq. (3.16), the **coupling efficiency in the presence of defocusing and mode-field mismatch** for each coordinate axis is obtained

$$T_{zs} = 10 \cdot \log \left(\frac{2 \cdot \mu_s}{\sqrt{\left(1 + \mu_s^2\right)^2 + \left(\Delta z / z_0\right)^2}} \right) \tag{3.23}$$

Coupling efficiency

It is often necessary to separate mode-field mismatch in Eq. (3.19) from Eq. (3.20). Because of the logarithmic representation, the quantity of interest is obtained as the difference of the two equations. Thus, the direct comparison with the measured coupling efficiency is possible. The curves start in the origin.

Intensive comparison of the coupling efficiencies is conducted both in Chap. 4 and particularly in Chap. 7 through presented coupling of the straight-cut fiber and the OEICs performed by the mode-field transformer. It was shown that the calculation with the data extracted from the far field is well suited to describe the coupling efficiency with up to 3 dB additional losses due to increasing distance.

3.7 Coupling Efficiency in the Presence of Transverse Offset

According to the Fig. 3.5, the transverse offset between the axes of the two coupling elements by an amount of Δs (Δx or Δy) should be considered as the next error term in the coupling of the optical components. The coupling efficiency decreases because of this offset.

This can be easily intuitively understood since the overlapping of the two field distributions according to the Eq. (3.11) is reduced. If the radial symmetry also does not exist, the difference must be made between both transverse directions. The coupling efficiency in the presence of transverse offset, which will be here presented, includes the issue of asymmetry, but also leads to the known approximate formulas (Geckeler 1990; Neumann 1998) in case of the symmetrical field distribution.

To solve the formulas, Wengelink (1996) formulated a new assumption. This assumption will be used to transpose the simplifications in this work to an easier form. If the coupling efficiency in the presence of the transverse offset and mode-field mismatch will be applied according to Wengelink (1996), then one can obtain the following equation:

$$T_{xy} = T_{mm} + 10 \cdot \log(\mathrm{e}) \cdot \left(\frac{-2 \cdot \Delta x^2}{w_{01x}^2 + w_{02x}^2} + \frac{-2 \cdot \Delta y^2}{w_{01y}^2 + w_{02y}^2} \right) \tag{3.24}$$

After taking into account the ratio of the mode-field radii μ_s according to the Eq. (3.17), the expression can be written in the following form:

$$T_{xy} = T_{mm} + 10 \cdot \log\left(\exp\left(\frac{-2 \cdot \Delta x^2}{w_{01x}^2 \cdot (1 + \mu_x^2)} + \frac{-2 \cdot \Delta y^2}{w_{01x}^2 \cdot \left(1 + \mu_{ys}^2\right)} \right) \right)$$

$$T_{xy} = T_{mm} + 10 \cdot \log\left(\exp\left(\frac{-2 \cdot \Delta x^2}{w_{01x}^2 \cdot (1 + \mu_x^2)} \right) \right) + 10 \cdot \log\left(\exp\left(\frac{-2 \cdot \Delta y^2}{w_{01y}^2 + \left(1 + \mu_{ys}^2\right)} \right) \right) \tag{3.25}$$

In this representation, the coupling efficiencies for both coordinate axes are included. The individual efficiencies can be separated according to the Eq. (3.16), resulting in the following equation:

$$T_s = 10 \cdot \log\left(\frac{2 \cdot \mu_s}{1 + \mu_s^2}\right) + 10 \cdot \log\left(\exp\left(\frac{-2 \cdot \Delta s^2}{w_{01}^2 \cdot (1 + \mu_{sx}^2)}\right)\right) \tag{3.26}$$

The front term is the mode-field mismatch for each coordinate axis according to the Eq. (3.26) and is included below. That is how the **coupling efficiency in the presence of transverse offset and mode-field mismatch** is obtained:

$$T_s = 10 \cdot \log\left(\frac{2 \cdot \mu_s}{1 + \mu_s^2} \cdot \exp\left(-\frac{2 \cdot \Delta s^2}{w_{01}^2 \cdot (1 + \mu_s^2)}\right)\right) \tag{3.27}$$

Based on the Eq. (3.22), the coupling efficiency in presence of transverse offset can be derived. This is how the determination of the mode-field radius according to the transverse-offset method is explained (Andersen 1984). For the coupling of the same fibers with radially symmetrical field distributions, $w_{02} = w_{02x} = w_{02y}$ applies. Thus, the equation reduces to the form given by Eq. (3.3), which is also frequently found in the literature (Geckeler 1990; Marcuse 1977).

The Eq. (3.27) indicates that for the approximation of the efficiency in the presence of transverse offset, only the mode-field radii of the coupling elements and the offset play a role. As a prerequisite, $\Delta z = \Delta\theta = 0$ must be fulfilled.

This requirement cannot be practically maintained. The results in Chap. 4 show that a longitudinal distance between the waveguides of less than 1 μm can be neglected. For an angular misalignment of less than 0.1°, the same statement applies. As a result, the following diagrams provide only the coupling efficiency in the presence of transverse offset for the case of the air gap.

3.8 Coupling Efficiency in the Presence of Angular Misalignment

The final misalignment that affects the coupling efficiency is the angular misalignment. According to the Fig. 3.5, it is caused by a tilt of the optical axis by a value $\Delta\theta$.

This misalignment will also be considered separately from the others. Thereby, it should be borne in mind that theoretically it is not possible to achieve $\Delta z = \Delta s = 0$. A very small longitudinal displacement cannot be excluded. However, for $\Delta z \to 0$, it can be neglected.

According to Kogelnik (1964), the angular misalignment, as well as the translational, causes the mismatch of the phase fronts. The influence is taken into account by means of the introduction of an additional phase of the guided mode in order to describe the overlap of the fields. As a result, modified coupling coefficients, specified by Saruwatari and Nawata (1979), are obtained. This will be calculated in a modified reduction, to part the calculation into distinct axis parts.

The coupling efficiency for the angular misalignment can be threated in the same way as for the transverse offset. For that purpose, an equivalent parameter, which can replace the transverse offset Δs, can be introduced:

$$\Delta s = k_0 \cdot n_z \cdot \sin(\Delta\theta) \cdot \frac{w_{01} \cdot w_{02}}{2} \tag{3.28}$$

This interpretation follows directly from the derivation of the coupling efficiency and has no physical background. However, it simplifies the representation for the case of the same mode-field radii (Geckeler 1990). The size $\Delta\theta$ is the corresponding angular misalignment. The inserted refractive index n_z refers to the propagation medium, typically air. The coupling efficiency for the angular misalignment is then directly obtained from the efficiency in case of transverse offset.

This procedure is carried out for the desired separation of the axes with the help of the overall efficiency. Afterward, applying the elliptical field assumption will perform the separation of the axes. The coupling efficiency for angular misalignment is then for this specific case:

$$T_\theta = T_{mm} + 10 \cdot \log\left(\exp\left(-\frac{w_{01}^2 \cdot w_{02}^2}{2 \cdot \left(w_{01}^2 + w_{02}^2\right)} \cdot k_0^2 \cdot n_z^2 \cdot \sin^2(\Delta\theta)\right)\right) \tag{3.29}$$

To the corresponding extension to the coordinate axes, the following equations are provided with the general indexing s. It should be noted that the separation of the mode-field mismatch is assumed to be known and is omitted below. A preliminary indexing with apostrophe is also necessary to avoid confusion with the below following Eq. (3.30). The following applies

$$T_{\theta s'} = 10 \cdot \log\left(\exp\left(-\frac{w_{01}^2 \cdot w_{02s}^2}{2 \cdot \left(w_{01}^2 + w_{02s}^2\right)} \cdot k_0^2 \cdot n_z^2 \cdot \sin^2(\Delta\theta)\right)\right)$$

$$T_{\theta s'} = 10 \cdot \log\left(\exp\left(-\frac{1}{2} \cdot \frac{\left(k_0 \cdot n_z \cdot \sin(\Delta\theta) \cdot w_{02s}\right)^2}{w_{02s}^2 \cdot \left(1 + \mu_s^2\right)}\right)\right)$$

$$T_{\theta s'} = 10 \cdot \log\left(\exp\left(-2 \cdot w_{01}^2 \cdot \left(\frac{\left(\pi \cdot n_z \cdot \sin(\Delta\theta)\right)^2}{\lambda}\right)^2 \cdot \frac{\mu_s^2}{1 + \mu_s^2}\right)\right)$$

The separation of the axes prompts the following intermediate step,

$$T_\theta = \sqrt{T_{\theta s'}} \cdot \sqrt{T_{\theta s'}} = T_{\theta s} \cdot T_{\theta s} = T_{\theta x} \cdot T_{\theta y}$$

whereby the **individual efficiency of the angular misalignment and inclusion of the mode-field mismatch** for each coordinate axis results in the following equation:

$$T_{\theta s} = 10 \cdot \log\left(\frac{2 \cdot \mu_s}{1 + \mu_s^2}\right) + 10 \cdot \log\left(\exp\left(-w_{01}^2 \cdot \left(\frac{(\pi \cdot n_z \cdot \sin(\Delta\theta))^2}{\lambda}\right) \cdot \frac{\mu_s^2}{1 + \mu_s^2}\right)\right) \quad (3.30)$$

With this equation, there is now a connection, in which the asymmetry of the field distribution can relatively easy be taken into consideration. It is only necessary that individual efficiencies according to Eq. (3.15) merge. In the case, when mode-field mismatch is not present because of simplification or symmetry, the front term can be omitted.

After the discussion about the loss mechanisms in fiber–chip coupling, the adaptation of the fields and their effects in the fiber–chip coupling will be presented in Chap. 4.

References

Agrawal, G.P.: Fiber-Optic Communications Systems. Wiley, New York (1992)

Andersen, W.T.: Consistency of measurement methods for the mode field radius in a single-mode fiber. IEEE J. Lightwave Technol. **2**(2), 191–197 (1984)

Bludau, W.: Lichtwellenleiter in Sensorik und optischer Nachrichtentechnik. Springer, Berlin (1998)

G.650 I-TR (1994) Transmission media characteristics: definition and test methods for the relevant parameters of single-mode fibres

Geckeler, S.: Lichtwellenleiter für die optische Nachrichtenübertragung. Springer, Berlin (1990)

Green, P.E.: Fibre Optic Networks: Prentice Hall. Englewood Cliffs, New Jersey 07632 (1993)

He, Y., Shi, F.: Beam Propagation Method and Microlens Design for Optical Coupling. VDM Verlag Dr. Müller, Saarbrücken (2010)

Ishii, M., Hibino, Y., Hanawa, F., Nakagome, H., Kato, K.: Packaging and environmental stability of thermally controlled arrayed-waveguide grating multiplexer module with thermoelectric device. J. Lightwave Technol. **16**(2), 258 (1998)

Kawano, K., Saruwatari, M.: A new confocal combination lens methods for a laser-diode module using a single-mode-coupler. J. Lightwave Technol. **LT-3**(4), 739–745 (1985)

Kogelnik, H.: Coupling and conversion coefficients for optical modes. Proc. Symp. Quasi-Opt **14**, 333–347 (1964)

Kogelnik, H.: On the propagation of Gaussian beams of light through lenslike media including those with a loss or gain variation. Appl. Opt. **4**(12), 1562–1569 (1965)

Kogelnik, H.: Theory of Optical Waveguides. Springer, Berlin (1975)

Kogelnik, H., Li, T.: Laser beams and resonators. Proc. IEEE **54**, 1312–1329 (1966)

Kuhmann, J.: Untersuchung von Flip-Chip-Bondprozessen zur selbstjustierenden, flussmittelfreien Montage von OEICs. Ph.D. thesis/dissertation, Heinrich-Hertz-Institute, Berlin (1998)
Mahlke, G., Gössing, P.: Lichtwellenleiterkabel. 4. Auflage edn. Publicis Corporate Publishing (1995)
Marcuse, D.: Light Transmission Optics. Van Nostrand Reinhold, New York (1972)
Marcuse, D.: Theory of Dielectric Waveguides. Academic Press, New York (1974)
Marcuse, D.: Loss analysis of single-mode fiber splices. Bell Syst. Tech. J. **56**, 15 (1977)
Marcuse, D.: Gaussian approximation of the fundamental modes of graded-index fibers. J. Opt. Soc. AM **68**(1), 103–109 (1978)
März, R.: Integrated Optics: Design and Modeling. Artech House, Boston (1995)
Maxwell, J.C.: A Dynamical Theory of the Electromagnetic Field. Scottish Academy Press, Edinburgh (1982)
Maxwell, J.C.: Über Faradays Kraftlinien (1855/1856). Ostwalds Klassiker der exakten Wissenschaften, vol. 69. Reprint [der Ausg. Leipzig, Akad. Verl.-Ges., 1898], 3. Aufl. edn. Deutsch, Frankfurt am Main (2001)
Maxwell, J.C., Boltzmann, L.: Über Faradays Kraftlinien (1855/1856). Ostwalds Klassiker der exakten Wissenschaften, vol. 69, 44., erw. Aufl. edn. Deutsch, Frankfurt (2008)
Maxwell, J.C., Boltzmann, L.: Über Faradays Kraftlinien (Trans. t. Camb. Phil. Soc., vol. 10, p. 27, gelesen am 10. Dec. 1–855 u. 11. Feb. 1856, Maxw. Scient. Pap., vol. 1, p. 155). Ostwald's Klassiker d exakten Wiss, vol Nr 69, 2.Aufl.,unveränd.Nachdr. edn. Akad. Verl.-Ges., Leipzig (1895)
Neumann, E.G.: Single-Mode Fiber: Fundamentals. Springer, Berlin (1998)
Opielka, D.: Optische Nachrichtentechnik. Vieweg Verlag (1995)
Petermann, K.: Microbending loss in monomode fibres. Electron. Lett. **12**(4), 107–109 (1976)
Petermann, K.: Constraints for fundamental-mode spot size for broadband dispersion-compensated single-mode fibres. Electron. Lett. **19**(18), 712–714 (1983)
Reider, G.: Photonik—Eine Einführung in die Grundlagen. Springer, Heidelberg (1997)
Saruwatari, M., Kawate, K.: Semiconductor laser to single mode fiber coupler. Appl. Opt. **18**(11), 1847–1856 (1979). doi:10.1364/AO.18.001847
Saruwatari, M., Nawata, K.: Semiconductor laser to single-mode fiber coupler. Appl. Opt. **18**, 9 (1979)
Saruwatari, M., Sugie, T.: Efficient laser diode to single-mode fiber coupling using a combination of two lenses in confocal condition. IEEE J. Quantum Electron. **6**, 7 (1981)
Snyder, A.W., Love, J.D.: Optical Waveguide Theory. Chapman and Hall, New York (1983)
Streckert, J., Brinkmeyer, E.: Characteristic parameters of monomode fibers. Appl. Opt. **21**(11), 1910–1915 (1982)
Wengelink, J.: Photolithographie mit semitransparenten Masken, Ph.D., Cuvillier Verlag, Göttingen (1996)

Chapter 4
Fiber-Optical Coupling

Abstract In modern optical communication systems, it is of the highest importance to transmit as much optical power from the transmitter to the receiver. It seems that future systems will not be that strongly dependent on good optical waveguides. Actually, even after 25 years of existence of low-loss glass fibers, the coupling efficiency remains the biggest concern of the system engineers. In this chapter, the most important principles of the optical coupling that are relevant for engineers working with this topic are discussed.

4.1 Adjusting Techniques

The adjustment techniques of photonic packaging for fiber–chip coupling can be divided into two fundamentally distinguishable functional groups:

1. Active techniques
2. Passive techniques.

4.1.1 Active Techniques

The active techniques use the ability to control the adjustment under real electrical excitation conditions of OEICs. By using electrically actuated elements, the adjustment of the coupling is carried out online by micromechanical actuators and fixed permanently after optimization of the optical coupling. This method is very time and labor intensive and therefore relatively expensive. The mechanically active coupling is normally performed by hand. Approaches to automation are tro find only in the area of PC-controlled maximum value search and subsequent fixation of the coupling, e.g., by laser welding techniques (see Chap. 8) or adhesion (see Chap. 9). Both, the stripping and preparation of fiber ends, the insertion of the fibers and the removal of the packages as well as the electromechanical preparation of the inner housing assembly to be completely carried out by hand or in individual production.

© Springer-Verlag Berlin Heidelberg 2015

U.H.P. Fischer-Hirchert, *Photonic Packaging Sourcebook*,
DOI 10.1007/978-3-642-25376-8_4

For smaller quantities or prototypes, this method has of course a privilege, but should be replaced for reasons of cost for high production lots by passive adjusting techniques.

4.1.2 Passive Techniques

Passive adjustment technique means a fiber–chip coupling by pre-alignment grooves that are especially suitable for mass production. The use of high-precision lithography techniques plays a crucial role. There are three methods in this area of application, but these are currently insufficiently used in the telecommunications industry due to the low production numbers:

1. LIGA—Lithography/electroplating/impression
2. Flip-chip (expensive basic technology)
3. Chemical- or laser-etched U-grooves

The LIGA technique (Brück 2001; Ehrfeld et al. 1994) was developed as an evolution of the macroscopic impression techniques from Karlsruhe Institute of Technology, briefly referred to as KIT, and was realized by using a precise preform to manufacture a master, which is able to produce PMMA components with micrometer accuracy. It provides a preform to produce optical waveguides and U-grooves for fiber alignment in the material within one production cycle. Waveguide core material and cladding must then be spun in further process steps on the chip. The glass fibers can then be inserted into the U-grooves and are passively aligned, which is accurately enough for multi-mode applications. For the mass market of the consumer market, this technique seems future-sent.

In flip-chip-bonding technology, the face-down devices are mounted (see Fig. 4.1). This type of contact allows an unpackaged chip assembly that are called "direct-chip-attach" calls (Goodwin et al. 1991; Lau 1995).

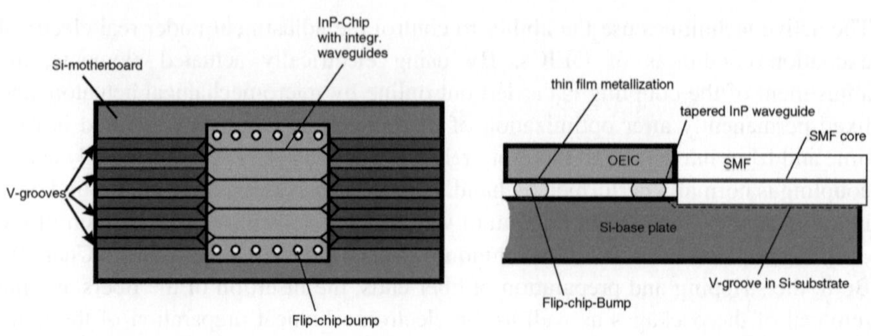

Fig. 4.1 Schematic representation of a fiber–chip coupling using the FC-bonding technology for contacting and self-adjustment Fischer (2002b)

By the self-adjustment during the soldering, the flip chip technology can be a way on implementing flip chip technology in photonics. Thus, single-mode fibers can be coupled with waveguides within the required alignment tolerances of submicron distances. Studies by Fischer and Kuhmann (Fischer 2002b) at the Heinrich-Hertz Institute, Berlin, have shown that the permitted tolerances could be met by ±1.0 μm in a single-mode fiber-optic coupling in the lateral direction. It seems possible to replace the complex and costly active coupling of fibers and waveguides with high-frequency optical signal processing by the low-cost mass-market flip chip technology and thus complements the LIGA technique in the passive adjustment range for RF-OEICs.

Small bond geometries provide excellent high-frequency characteristics in the range of frequencies greater than 20 GHz (Karpuzi 2009). Parasitic effects are significantly reduced, and thus, the flip-chip bonded devices are particularly well suited for the optical assembly and interconnection technology. In photonics, the focus is often on the low-loss transmission of high frequencies. The flip-chip-bonding technology is far superior to the conventional wire bonding techniques in the HF range, but also significantly more expensive to technology.

Soldering flux is added to prevent oxide layers so that good connections between the liquid solder and the pads are corresponding. Photonics, however, must be entirely dispensed with the use of fluxes, not to destroy the sensitive optical surfaces of the OEICs. Flux can wet the surface of a diode laser chips, thus changing its emission properties negatively. Therefore, one reduces, during the soldering process, the oxide layers with the addition of formic acid or Hydrogen gas (Kuhmann 1996).

The flux-free process is, in addition to the required positioning accuracy, the main difference between microelectronic and photonic application of this contacting method.

V-grooves can be produced by anisotropic wet etching of silicon. Microsystems technology by wet etching of the silicon wafers (Beyer and Eigler 1996; Menz and Mohr 1997; Steckenborn et al. 1991) is an established process and allows fixation of fibers, fiber arrays, and lenses with very high precision. The FC-bonding technology in combination with high-precision patterned V-grooves represents an inexpensive way to produce self-adjusting electrical and optical connections.

Furthermore, it is possible to structure plastics topologically in the submicron range by laser ablation using Excimer Lasers or Femtosecond Lasers. Instead of working with an expensive master form as in the LIGA process which for prototype devices (Brück 2001) is too expensive to produce, you can ablate the plastic with the help of short-wavelength laser radiation. Suitable laser sources like the excimer laser are working at UV-output wavelengths between 350 and 190 nm. Due to the short wavelengths, the polymer chaines are cracked and peeled off layer by layer, so that a U-groove can be milled. A batch production is possible by the use of masks, which cover the entire wafer. The parts to be removed remain open and can be simultaneously irradiated by excimer laser. This production method is suitable to produce quickly and inexpensively optical waveguides in single-mode technology and U-grooves for the reception of glass fibers (Fischer 2002a; Fischer and Graener 2005).

4.2 Fixation Techniques

Regarding the use of special fixation techniques, many different techniques have been patented by the most part of the packaging companies to prevent the access of outsiders because of secrecy. In the meantime, clamping techniques accomplish for prototyping and simple applications (see Chap. 6). Laser welding is used typically in different shapes for very long-term stable housing and gluing for easy multi-mode applications, but also occasionally for single-mode usage. The adhesives generally have the property of acting in conjunction with hygroscopic water. Therefore, more gastight modules are required for adhesive fixation. Together with a relatively low glass transition temperature, all the adhesives have problems with long-term stability. In the publications that have been summarized in this book, several newly developed different fixation techniques are presented.

In the fixation techniques, we further differentiate the area of single-fiber coupling and the multi-fiber coupling. The multi-fiber technology is significantly more difficult in construction and adjustment of the fiber–chip coupling. It must be simultaneously aligned with each other up to 32 fiber ports and then finally fixed. The fibers are usually present in array form as semi-finished Si with chemically etched V-grooves. Only the sum of all alignment tolerances of the array fibers often results in a deviation of greater than one micron. This complicates the adjustment and can normally produce no optimal coupling results. The alignment tolerances for the straight fiber–chip coupling are longitudinally ±2.5 μm and laterally ±25 μm. However, you can only reach a coupling efficiency of about 10 % for conventional OEICs of InP or GaAs. Only using silica components, coupling efficiencies of over 90 % can be achieved, since the optical fields of both waveguides fit very well with each other (see Chap. 5). The fixation is usually realized in spite of the disadvantages of adhesives in the optical path butt with ended fibers (see Chap. 11).

In order to improve the coupling efficiency to OEICs made of semiconductor materials, mode-field transformers are integrated into the waveguide of the OEICs. These devices are used to achieve a coupling efficiency of about 50 % (see Chaps. 10 and 11), or lens systems are introduced to match the optical fields of both waveguides in the installed optical path.

In modern optical communication systems, it is of crucial importance to send as much optical power of the transmitter to the receiver. It seems possible that future systems will no longer be so dependent on the fiber attenuation, but even after 35 years of existence of low-loss optical fiber, the coupling efficiency remains an utmost concern of system engineers. In this chapter, the basic principles for optical coupling are discussed, which are crucial for photonic engineers.

4.3 Characteristics of a Good Coupling

In an optical coupling system between two waveguides, the power transmitted from one system into another can be described as follows (see Fig. 4.2):

$$T = P_2/P_1 \tag{4.1}$$

P_1 Optical power in system 1
P_2 Optical power in system 2

The transmittance T is a number less than unity and is usually expressed as a percentage. Transmission losses occur because of the transition of light through the space between two subsystems. Thereby, the light can be attenuated by the following mechanisms:

- Absorption
- Conversion to phonons and photons, respectively
- Scattering

Sometimes, it is more useful to express transmittance T in dB. One defines then the insertion loss (IL) as follows:

$$\mathrm{IL} = -10\ \log_{10}T \text{ in dB} \tag{4.2}$$

Corresponding values for the IL between linear and logarithmic values are given in Table 4.1.

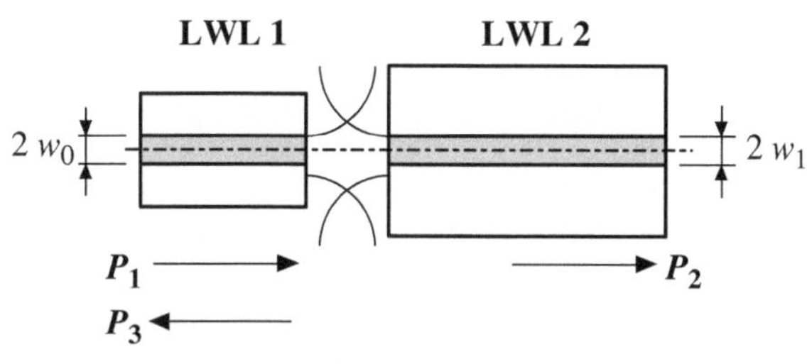

Fig. 4.2 Transmission and reflection between two waveguides

Table 4.1 Insertion loss

Transmission in %	Insertion loss in dB
100	0
80	1
50	3
10	10

Other T/IL pairs can be derived in the same way. In addition, the cascade of losses in a transmission system can be summed up to a total power loss, which includes all individual losses. The individual losses may appear discretely (at the interfaces or components) or continuously through the glass fiber section.

4.4 Reflections

Another important parameter in the optical coupling system is the power reflected back into the first component:

$$R = P_3/P_1 \tag{4.3}$$

P_1 Optical power in system
P_3 Back-reflected optical power in system

Like T, the reflectance R is also a number less than unity and is usually expressed as a percentage. Physically, it means that reflections are caused by the difference of the refractive indices between the optical components and the optical connection elements. The return loss (RL) can be defined as follows:

$$\mathrm{RL} = -10 \log_{10} R \text{ in dB} \tag{4.4}$$

The already shown principles valid for IL are also valid for RL. High IL is not desirable in fiber optics. It can be improved using the following mechanisms:

- More effective coupling mechanisms
- Stronger light source
- More sensitive detectors
- Optical amplification

On the other hand, it is not so easy to find solutions for reducing the reflections. In optical systems, single and multiple reflections can significantly increase the transmitter's noise, and degrade the signal quality at the receiver side. In optical amplifiers, the noise figure can be increased and the amplification can be significantly reduced. Some methods for reducing the reflections are as follows:

- Index matching with gel or oil
- Coatings for optical surfaces
- Angular surfaces
- Optical isolators

To achieve the bit error rate of 10^{-9} in digital systems, a reflection loss of 40 dB is required for all installed components. In amplitude-modulated analog systems, the target return loss is 55 dB. In wide area networks with bit rates higher than 10 Gbit/s, the value of 55 dB is also sometimes required.

4.5 Mode Fields in Waveguide Structures (Spot Size)

A term often used in coupling calculations is mode-field radius (w_0). Mode-field diameter ($2w_0$) is also known as spot size or focus diameter. The term originates from the calculations of light propagation of cylindrical gas lasers. In glass fibers, it refers to the intensity profile, which has a circular symmetry and can be well approximated by a Gaussian distribution along a radial coordinate r. The Gaussian power intensity profile has only a deviation of about one percent to the exact mathematical description, which is realized in the core area by Bessel functions of the first kind and nth order J_n and for the cladding region by an exponentially decaying Hankel function (see for more details Chap. 3), respectively. Therefore, in this chapter, there will be only the description of optical modes by using the Gaussian function.

$$p(r) = p(0) \cdot \exp\left\{ -2\left[\frac{r}{w_0}\right]^2 \right\} \tag{4.5}$$

w_0 mode-field radius (MFR).

The optical field of a step-index fiber is shown in Fig. 4.3. The normalized refractive index difference is 0.3 %, and the core radius a is 4.25 μm. As already discussed in the chapter about the optical waveguides, the same rules apply for all optical waveguides. The mode-field radius plays the key role in the coupling process between different components, even if the fields under consideration are elliptical or have deviations from the ideal Gaussian profile.

As already mentioned in Chap. 3, the normalized frequency V can take values between 1.9 $< V <$ 2.4 (Saruwatari and Nawata 1979). The mode-field radius is defined as the distance from the fiber axis where the power density has its maximum value down to the radius value where the power has dropped to $1/e^2$ (13.5 %). It can be noticed that a non-negligible part of the optical power of 13.5 % is guided through the cladding at the wavelength of 1.55 μm. The mathematical relationship between the mode-field radius and the core radius is as follows:

$$w_0 = a\left[0.65 + \frac{1.619}{V^{3/2}} + \frac{2.879}{V^6}\right] \tag{4.6}$$

with

$$V = \frac{2\pi}{\lambda} a \sqrt{n_1^2 - n_2^2} \tag{4.7}$$

a core radius of a waveguide, λ wavelength,n_1 core refractive index, and n_2 cladding refractive index.

The core radius, given by the change in the refractive index, is always lower than the mode-field radius (4.5 μm at 1300 nm and 5.25 μm at 1550 nm). Another fact is that the large portion of the guided wave is guided outside of the core. This portion can be described by the following mathematical expression:

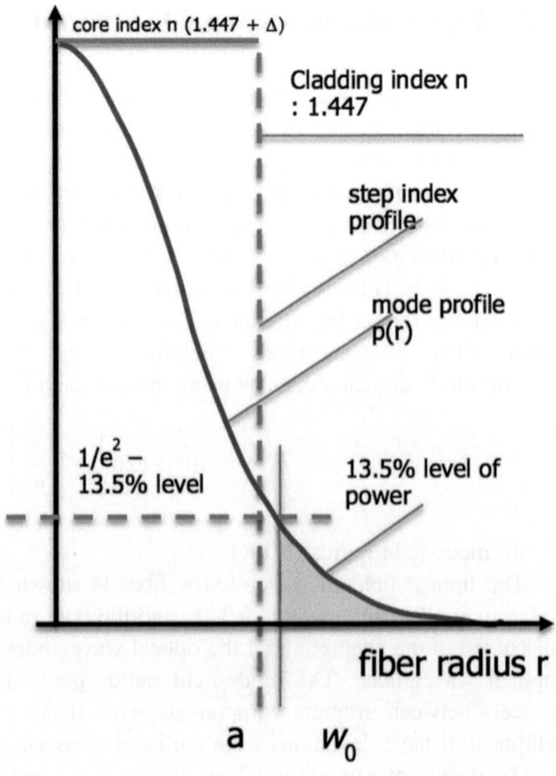

Fig. 4.3 Gaussian distribution of a wave guided through a glass fiber

$$F = \exp\left\{-2\left(\frac{a}{w_0}\right)^2\right\} \tag{4.8}$$

It amounts to about 17 % at 1300 nm and 25 % at 1550 nm. Since a SMF becomes multi-modal at 1240 nm, i.e., there are several modes of propagation, the broadening of the wave into the cladding will be much lower.

As the wave exits the end facet of the waveguide, the mode-field radius widens with the distance z according to the following formula, which is also graphically represented in Fig. 4.4:

$$w^2(z) = w_0^2\left[1 + \left(\frac{z \cdot \lambda}{2 \cdot n \cdot \pi \cdot w_0^2}\right)^2\right] \tag{4.9}$$

The spatial distribution of the mode field can also be expressed by the far-field angular distribution. The width of the near-field w_0 can be converted to the far-field angle (for $z > 100$ μm) by a simple transformation of Eq. (4.9). The Gaussian profile then yields:

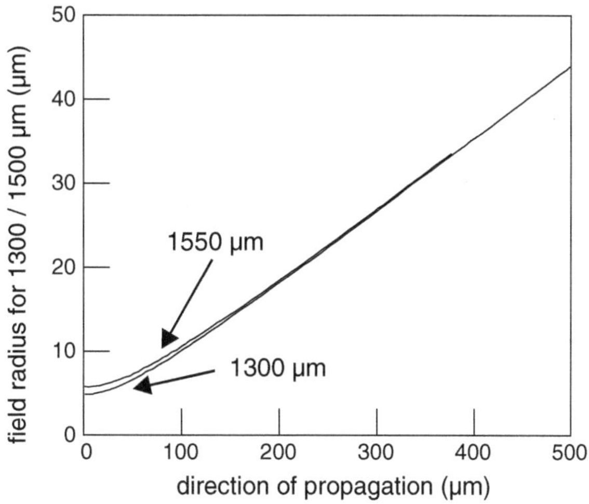

Fig. 4.4 Mode-field widening after exiting the waveguide

$$\tan\Theta = \frac{w(z)}{z} = \left[\frac{\frac{z\cdot\lambda}{\pi\cdot n\cdot w_0}}{z}\right] = \frac{\lambda}{\pi\cdot w_0} \tag{4.10}$$

$$w_0 = \frac{\lambda}{\pi\cdot n\cdot \tan\Theta} \tag{4.11}$$

for $w_0 \ll 1$ and $z \gg 100\,\mu\text{m}$

Equation (5.8) also gives us the dependency of the mode-field widening on the wavelength and on the refractive index n of the medium outside of the waveguide.

The increase in mode-field radius between 1300 and 1550 nm results in the broadening of the far-field angle by approximately 1.3 %.

The sine of the far-field angle θ derived from the Eq. (4.10) is identical to the definition of the numerical aperture NA given in Sect. 3.3.7 and, as already shown, can vary between 0.15 and 0.25 in the case of glass fibers. Typically, far-field angles of the semiconductor lasers result in the numerical apertures that are higher than those of the glass fibers (up to NA = 0.5 or 30°). This further means that it is not possible to achieve the coupling efficiency of more than 15 % from the laser into the glass fiber without mode-field adaptation.

4.6 Coupling Efficiencies

For efficient power transfer between the optical components, the largest possible overlapping of the mode profiles is necessary. The mode-field overlapping of two different Gaussian beams is shown in Fig. 4.5. For single-mode waveguides, the

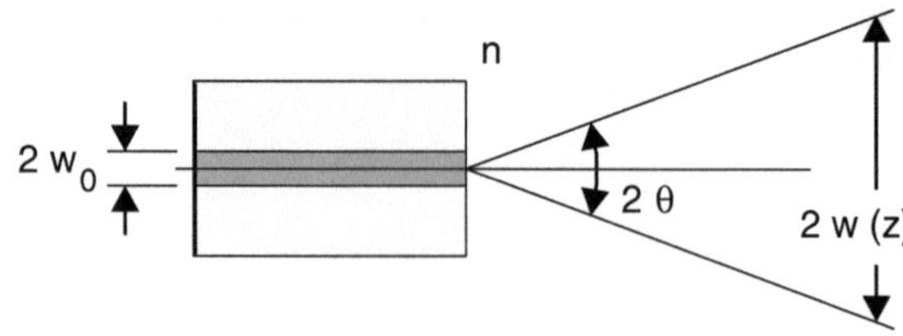

Fig. 4.5 Relationship between far and near field

extent of overlapping can be described with simple approximations. Let us define κ_0 as the relationship between mode-field radiuses w_0 and w_1 as follows:

$$\kappa_0 = \frac{4}{\left(\frac{w_1}{w_0} + \frac{w_0}{w_1}\right)^2} \tag{4.12}$$

The loss (in dB) at the interface between two fields can be described according to (Saruwatari and Sugie 1981) as mode-field mismatch (in dB) in the following way:

$$L(\kappa_0) = -10\log(\kappa_0) \tag{4.13}$$

The coupling efficiency between two optimally adjusted waveguide elements can be estimated with this simple equation, thereby, the waveguides can be of any type or shape. The only requirement is that the distribution of the optical fields in the vertical and horizontal plane of propagation is not significantly different. Up to the field ratio of 1:3, the Eq. (4.13) does not introduce a large error and can be reliably used. Otherwise, the overlapping must be separately estimated for both planes of propagation:

$$L(\kappa_{0\Sigma}) = \frac{L(\kappa_{0\perp} + \kappa_{0=})}{2} \tag{4.14}$$

If one wants to calculate not only the losses due to the unequal field distributions, but also the additional losses caused by the inaccurate alignment of the waveguides to one another, it is necessary to solve the overlap integral between the two beams with their fundamental modes by means of Hermitian–Gaussian functions (Fig. 4.6). According to Saruwatari (Saruwatari and Nawata 1979) which was in detail discussed in Chap. 3, the portion of coupled beam energy can be expressed as follows:

$$\begin{aligned} \eta &= \left|C_{x0,0}\right|^2 \left|C_{0,0}\right|^2 \\ C_{x0,0} &= C_{0,0}\exp\left(-\frac{jk\theta x_0}{2}\right) \cdot \exp\left[-\frac{x_0^2}{q}\left(\frac{1}{w_1^2} + \frac{jk}{2R_1} - \frac{jk\theta}{2x_0}\right) \cdot \left(\frac{1}{w_0^2} + \frac{jk}{2R_0} - \frac{jk\theta}{2x_0}\right)\right]z \end{aligned} \tag{4.15}$$

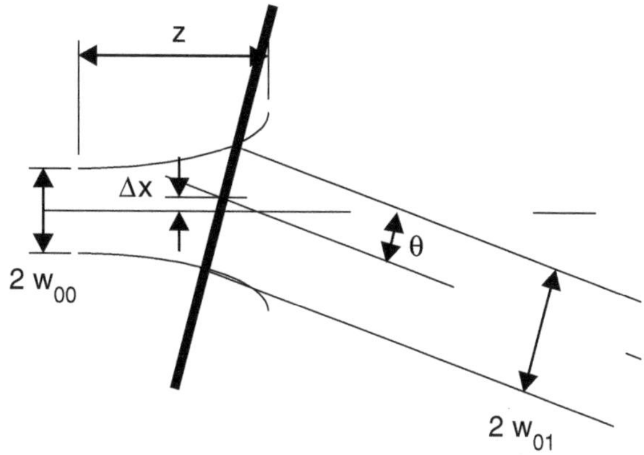

Fig. 4.6 Coupling between two Gaussian beams with lateral, longitudinal, and angular misalignment

η is then given by the following expression:

$$\eta = \kappa \exp\left(-\kappa\left\{\begin{array}{c} \frac{x_0}{2}\left(\frac{1}{w_1^2}+\frac{jk}{2R_1}\right)+\frac{x_0}{2}\left(\frac{1}{w_0^2}+\frac{jk}{2R_0}\right)+ \\ +\frac{\kappa^2}{8}\left[\left(\theta^2 w_1^2-\frac{2x_0\theta w_1^2}{R_1}\right)+\left(\theta^2 w_0^2-\frac{2x_0\theta w_0^2}{R_0}\right)\right]\end{array}\right\}\right)$$

where κ stands for:

$$\kappa = \frac{4}{\left(w_1^2 w_0^2\left[\frac{1}{4}\left(\frac{1}{w_1^2}+\frac{1}{w_0^2}\right)^2-\frac{k^2}{4}\left(\frac{1}{R_1}+\frac{1}{R_0}\right)^2\right]\right)} \tag{4.16}$$

Under the assumption that $R_0 = \infty$ represents the fiber end face and

$$R_1 = z\left\{1+\left[\frac{\pi w_1^2}{\lambda}\right]^2\right\}$$

The coupling efficiency can be simplified as follows:

$$\eta = \kappa \exp\left(-\kappa\left\{\frac{x_0}{2}\left[\frac{1}{w_1^2}+\frac{1}{w_0^2}\right]+\pi\theta^2\left[\frac{\left(w_1^2(z)+w_0^2\right)}{2\lambda^2}\right]-\frac{x_0\theta}{w_1}\right\}\right)$$

with

$$\kappa = \frac{4w_1^2 w_0^2}{\left[w_1^2+w_0^2\right]^2+\frac{\lambda^2 z^2}{n\pi^2}}$$

$$w_1^2(z) = w_1^2\left\{1+\left[\frac{\lambda z}{n\pi^2 w_1^2}\right]\right\}$$

where λ stands for the wavelength. Under the assumption that $z = 0$, the coupling efficiency is further simplified to:

$$\eta_0 = \kappa_0 \exp\left(-\frac{2x_0^2}{\left[w_1^2 + w_0^2\right]^2} - \frac{2\pi^2\theta^2 w_1^2 w_0^2}{\lambda^2 w_1^2 + w_0^2}\right) \tag{4.17}$$

with

$$\kappa_0 = \frac{4}{\left[\frac{w_0}{w_1} + \frac{w_1}{w_0}\right]^2} \tag{4.18}$$

Analogous to the Eq. (4.2), the transmission loss in dB can be calculated according to the Eq. (4.13).

$$L(\eta) = -10\log(\eta_0) \tag{4.19}$$

The exact calculation is relatively complex. Therefore, it is often easier to estimate transmission losses by approximate formulas. A good approximation, with the inaccuracy in the 5 % range, was offered by Mickelson and Basavanhally (1997). However, it is applicable only to the identical single-mode waveguides such as two SMF fibers. Then, the transmission loss (in dB) in case of the lateral misalignment in x- or y-direction can be approximated as follows:

$$L(x) \approx 4.343\left[\frac{x}{w_0}\right]^2 \tag{4.20}$$

The losses in the case of longitudinal or angular misalignment can also be estimated with the same approximation:

$$L(z) \approx 5.3\left[\frac{\lambda z}{10nw_0^2}\right]^2 \tag{4.21}$$

$$L(\alpha) \approx 2.7\left[\frac{w_0 n\alpha}{10\lambda}\right]^2 \tag{4.22}$$

Both possibilities presented here for calculating the transmission losses, which include lateral and angular misalignment, show that these losses are independent of each other and that individual losses (in dB) can be easily summed up to result in the total transmission loss. The only requirement is that the spot size of the incoming light beam is located at the fiber surface.

Comparison of Fig. 4.7a, b shows that for the same value of misalignment, the loss due to lateral misalignment is significantly higher than the loss caused by longitudinal misalignment. By comparing the approximate formulas given by Eqs. (4.19) and (4.20), it can be seen that the misalignment in the z-direction is

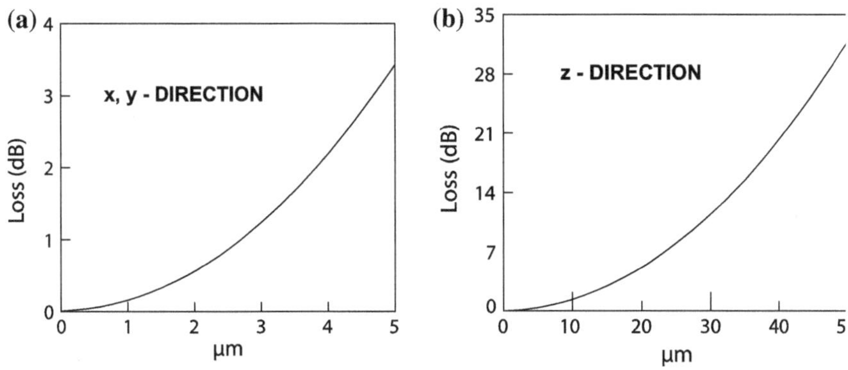

Fig. 4.7 **a** Transmission loss between two identical waveguides in the presence of lateral misalignment. **b** Transmission loss between two identical waveguides in the presence of longitudinal misalignment

Fig. 4.8 Transmission loss between two identical waveguides in the presence of angled misalignment

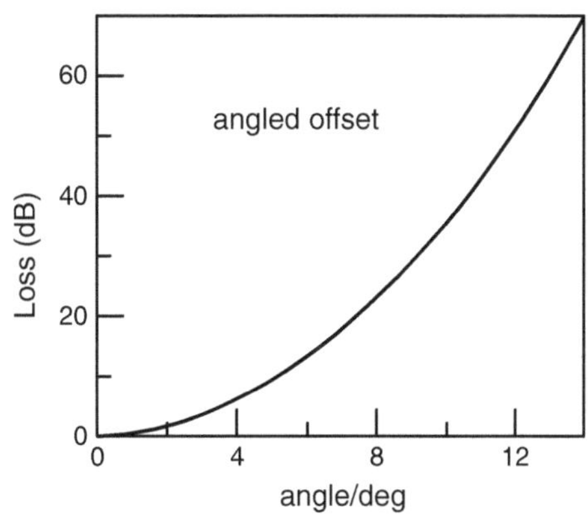

reduced by a factor of 10 in the denominator. This is not the case for the misalignment in the *x*-/*y*-direction.

The loss due to the angular misalignment (Fig. 4.8) rises very quickly for values above 2°. For 10° angular displacements, the additional attenuation is already 10 dB. The misalignment angle of 1.05° results in the additional loss of 1 dB. An overview of discussed loss mechanisms is given in Fig. 4.9. To give the reader the better feeling about the influence of different mismatch mechanisms, the numerical values of different mismatches that result in 1 dB transmission loss are given in Table 4.2. The calculations were performed for a wavelength of 1310 nm and a mode-field diameter of 9 µm for both waveguides (SMF). The losses are bidirectional and equal and can be easily summed up for small individual contributions.

In addition to the intrinsic losses, which are caused by mismatch of mode fields or misalignments, extrinsic losses introduce even more attenuation between two

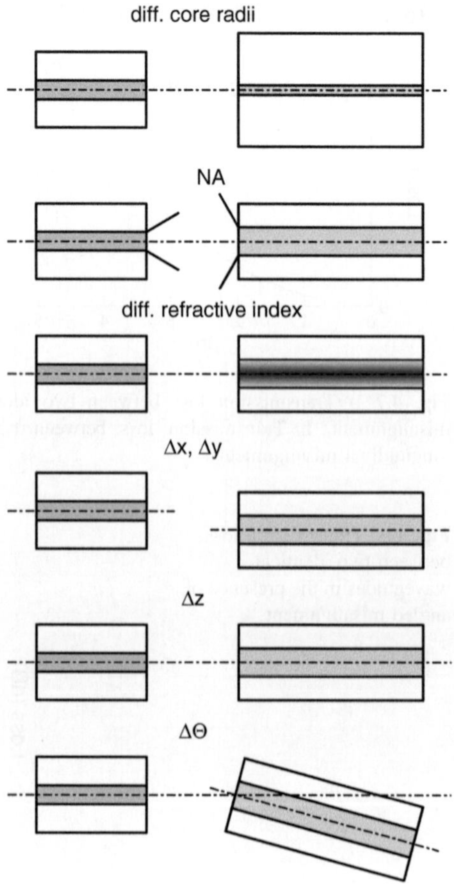

Fig. 4.9 Transmission loss mechanisms between waveguides (here two glass fibers)

Table 4.2 Coupling errors and mismatch/misalignment values (at 1.55 μm) for 1 dB excess loss

Coupling error type	Error value
Field mismatch	16 %
Different core radii	
Different numerical apertures	
Step index to graded index	
Lateral offset	2.7 μm
Longitudinal offset	28 μm
Angular offset	1.05°

waveguides. Dirt or scratches on fiber end faces, cleft or split exit windows have the absorbing or scattering effect on the exiting light, respectively, allowing only a portion of it to reach the opposite waveguide.

4.7 Laser–Fiber Coupling

The radiated light cone of a laser diode is elliptical and has a big aperture, from 20° to 60°, depending on the direction of propagation (Fig. 3.16). An obvious solution to minimize the coupling loss into the glass fiber would be to set two cylindrical lenses one behind the other. They should convert different aperture values in vertical and horizontal direction of propagation into the circular profile of a monomode glass fiber with the opening angle of 11.5°. The following imaging systems come into consideration in such optical design:

- **Lens systems**
  - Cylindrical lenses
  - Spherical lenses
- **Graded-index lenses** (Grin- or Selfoc®-lenses)
- **Fiber taper** (lensed fiber end faces)

Due to the small dimensions of the fiber-optic components, the goal of the design engineer is to minimize the amount of needed optical components. Only when the highest possible transmission values are requested, or a parallel light beam between the OEIC and the fiber is required, for example, to insert the optical isolator because of the reflection loss, a multi-lens system is selected. High costs of the adjustment process make these multi-lens systems very expensive.

In case of coupling a multi-mode fiber with a thin lens (Fig. 4.10), a magnification factor between the object and the image can be calculated in the following way with the conventional geometrical optics:

$$\beta = L'/L$$

The general thin lens equation sets the focal length in relation to the object and image distance:

$$1/f = 1/x + 1/x' \tag{4.23}$$

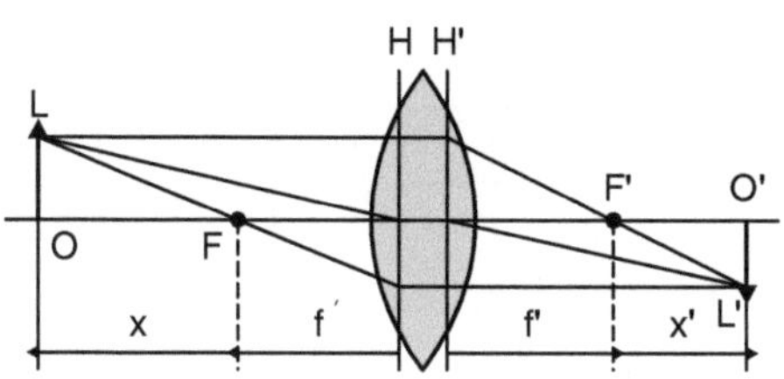

Fig. 4.10 Geometrical optics mapping

Under the assumption of the same media on both sides of the lens, the magnification factor can be calculated by the ratio of the screen distance to object distance:

$$\beta = x'/x$$

For the calculation of coupling between multi-mode waveguides, one can map the opening of the fiber core A to the opening of the laser A':

$$A\mathrm{NA}^2 = A'\mathrm{NA}'^2 \tag{4.24}$$

The losses in this type of imaging (mapping) can be calculated according to the Eq. (2.23). If the opening and the numerical aperture are known, the respective parameters of the coupled multi-mode waveguide can be calculated.

If the size of the focal point has the same order of magnitude as the wavelength, one cannot use simple geometrical optics imaging any more. Instead, the spreading of a Gaussian beam must be considered. This is the case when, for example, the picture of a single-mode waveguide should be mapped on the laser diode (Fig. 4.11).

The distance from the smallest beam diameter to the lens corresponds to the focus length and to the value $2w_0$ of the focus diameter. The focus point is defined as the area of the smallest beam diameter, in which the wave front is planar. This corresponds to the focal point at the corresponding focal length F. The illuminated diameter of the lens will be here denoted as D. The depth of focus (DOF) corresponding to the mode-field diameter can be easily calculated with the following formula if the wavelength and the F-number of the lens are known:

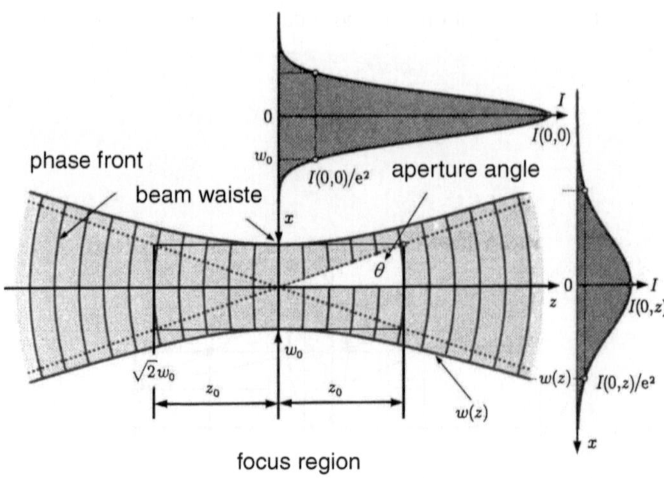

Fig. 4.11 Gaussian beam propagation outside of a waveguide

$$F - \text{number} = F\# = (F/D) = 2w_0/(4\lambda/\pi) = 1/(2\,\text{NA})$$
$$\text{DOF} = (8\lambda/\pi)(F/D)^2$$

Example:

- Lens diameter $D = 1$ mm,
- Focus length $F = 2$ mm
- Wavelength $\lambda = 1.55$ µm

$$\text{DOF} = \left(8/3.14 \cdot 1.55 \times 10^{-6}\,\text{m}\right)\left(2 \times 10^{-3}\,\text{m}/10^{-3}\,\text{m}\right)^2 = 16\,\mu\text{m}$$
$$2w_0 = \left(4/3.14 \cdot 1.55 \times 10^{-6}\,\text{m}\right)\left(2 \times 10^{-3}\,\text{m}/10^{-3}\,\text{m}\right)^2 = 4\,\mu\text{m}$$

With a 1-mm-diameter lens, it is possible to create a focus diameter of 4 µm and a mode-field diameter of 16 µm at the distance of 2 mm from the lens. For a more detailed calculation of lenses at fiber ends as well as for the Selfoc®-lenses, please refer to the (Ladany 1993).

An overview of the imaging techniques used in optical modules is shown in Fig. 4.12.

In the simplest case, a straight-cut fiber (butt-fiber) is positioned directly in front of the laser. The coupling efficiency that can be thereby achieved is 10–15 % (−10 dB). This solution is typically used in low-priced systems.

The efficiencies between 20 and 50 % can be achieved by means of ball lenses with radii between 0.3 and 3 mm. In the commercial laser modules with TO-3 housings, such lenses are usually already built in. This allows the fiber to be cost-effectively placed in the right position in front of the lens.

Selfoc®- or Grin-lenses are cylindrical parts, which are constructed similarly to glass fibers with core and cladding structure. In addition, the refractive index of the glass in the core of the lens is adjusted along the direction of propagation in such a way that the light takes a nonlinear path within the lens, as shown in Fig. 4.13. In order to produce a parallel beam between the laser diode and fiber, two identical Grin-lenses should be placed one behind the other, so that the parallel light beam is formed in the space between them.

In order to reduce the adjustment costs, the number of adjusting elements should be reduced. An effective way to achieve this is to realize an imaging element directly on the glass fiber's end face. This is known as a fiber taper. One solution can be made by means of casting resin, which has a refractive index adapted to the fiber ($n = 1.446$). The fiber is briefly dipped in the resin, so that a half-ball lens is formed on its end face once it is pulled out. The coupling efficiency of up to 30 % can be achieved. Another solution is to melt a ball of glass to the end ofthe fiber using a splicing machine with electrodes which heaten the fiber and the glass ball.

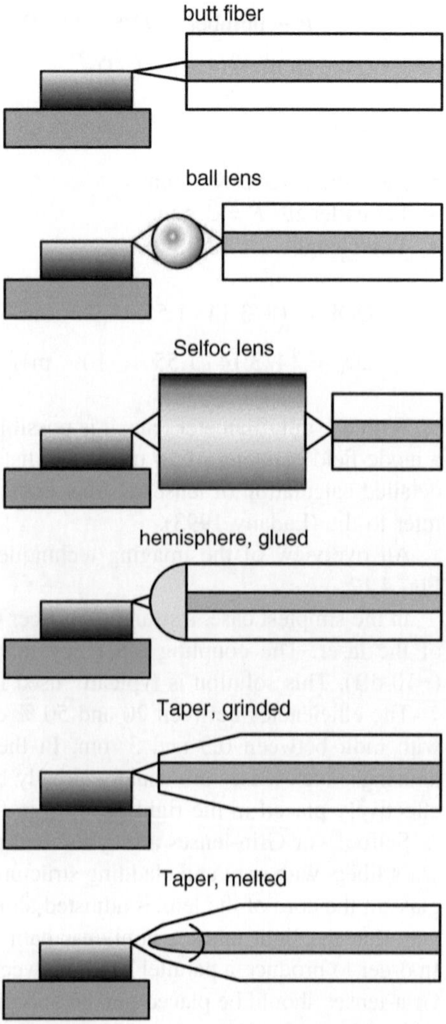

Fig. 4.12 Different coupling techniques for focusing the light of a semiconductor laser into the fiber

A more complex variant of the taper with a cone shape can be produced by wet etching, by mechanical grinding (sanding), from the fused material, or by laser ablation. Here, the cone shape acts as a lens. If the fiber end is warmed up to the melting point and then additionally pulled, a semi-spherical lens is formed (Fig. 4.14). The coupling efficiency of this most often used solution is between 30 and 70 %. If two different radii are polished on the vertical and horizontal plain on the fiber end face, the coupling efficiency can be increased to even more than 90 %. However, such tapers are very expensive (Fig. 4.15).

When coupling with a taper, in one longitudinal displacement point, the coupling efficiency from the laser into the fiber is the highest. In this case, the distance between the laser and fiber taper depends on the chosen radius of the taper.

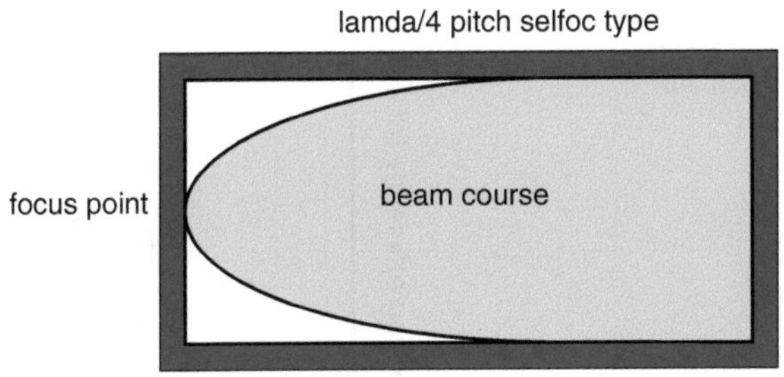

Fig. 4.13 Beam propagation within Selfoc® lens

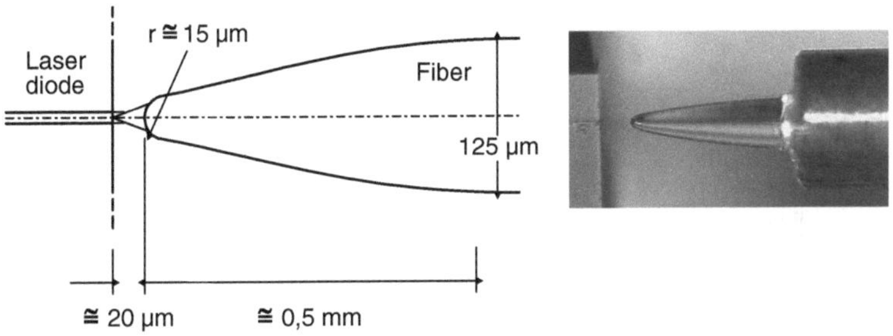

Fig. 4.14 Taper with core and rejuvenating cladding

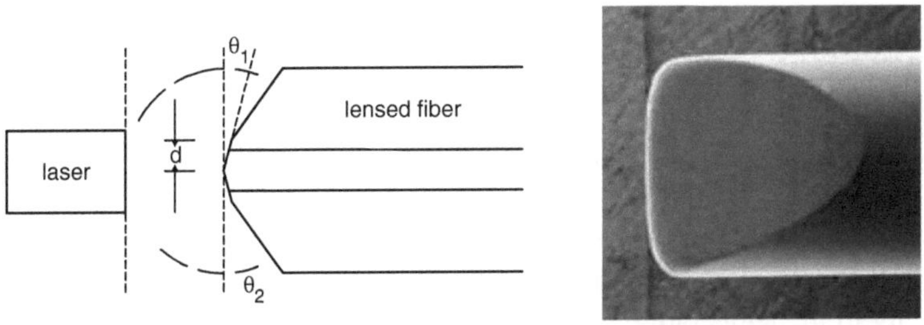

Fig. 4.15 Taper with different taper angles

Rule of thumb: The smaller the half-sphere radius of the taper, the smaller is the distance from the laser and the greater is the coupling efficiency. The coupling efficiencies of three different fiber ends [small lens, big lens, and straight-cut fiber (Butt-fiber)] are compared in Fig. 4.16. It can be seen that a taper with a 10 μm radius has the highest coupling efficiency. However, the danger of having a contact between the chip and the taper is huge and consequently the risk of damaging the

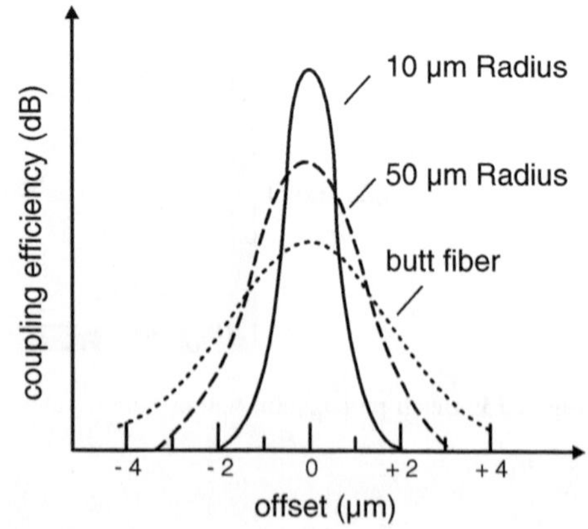

Fig. 4.16 Comparison of the coupling efficiency of tapers with different radii and butt coupling

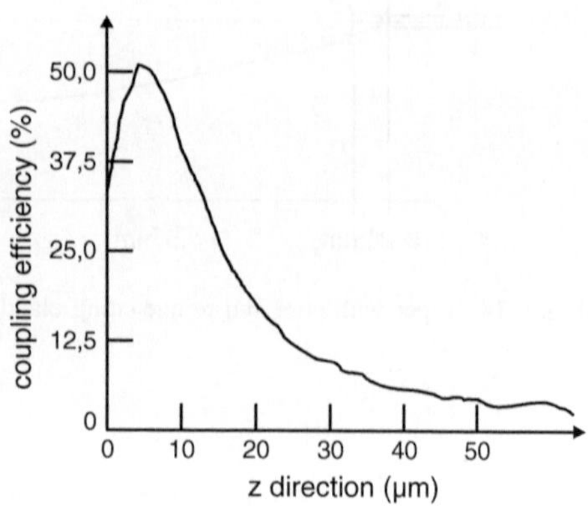

Fig. 4.17 Efficiency of the laser-to-fiber coupling for axial misalignment of a taper with 10 µm radius

laser diode with the fiber component. Distances between the OEIC and the fiber taper end are typical from 5 and 15 µm.

It should be additionally considered that the spatial sensitivity significantly increases with the higher coupling efficiency, as can be seen in Fig. 4.17. The sensitivity to the temperature changes and vibrations is in that case also increased. At high coupling efficiencies the mechanical stability of the whole structure is crucial to the long-term stability of the module.

Another important aspect is the avoidance of back reflections into the transmitter diode. Planar surfaces (e.g., fiber end faces) perpendicular to the optical axis behave very unfavorably in this respect with about 4 % reflection at the air-to-glass transitions. One could try to compensate the refractive index change between the OEIC (e.g., InP $n = 3.3$) and air ($n = 1$) by cutting the fiber end with an angle of about 7°, or by coating the fiber surface with a material similar to those used in camera lenses or glasses. In optical connector couplings, an index matching gel can be used. It has the refractive index of 1.5 and is applied on the connectors' end faces.

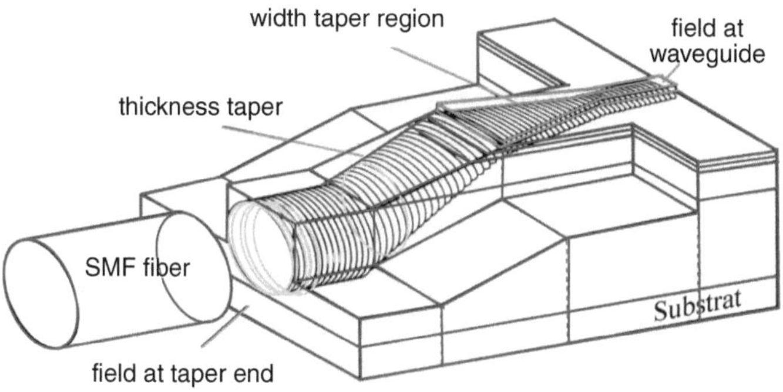

Fig. 4.18 Vertical and horizontal waveguide taper

Further on, it is of crucial importance that the system is mechanically stable with respect to vibrations and thermal influences. In industrial applications, both the fiber tapers and lens systems are used for fiber–chip coupling.

4.8 Waveguide Taper

A way to realize mechanical insensitive coupling modules is the insertion of waveguide structures in the semiconductor material to adjust the mode fields of the laser to the glass fiber. These structures are called waveguide tapers or mode-field expanders. Some realization options are listed below. One can make a difference between a lateral (thick) taper and a vertical (wide) taper (Honecker et al. 2002). The combination of two of these would be ideal to optimally adjust the field of a single-mode glass fiber.

For the simplicity, only a single vertical taper is typically realized during the production of a semiconductor. The lateral taper is technologically more difficult to produce and is still rarely commercially available. Anyway, the coupling efficiency significantly increases in comparison with the butt-fiber coupling. The comparison of coupling efficiencies between butt and taper coupling is shown in Fig. 4.18. The coupling efficiency would be different for horizontally and vertically polarized light. A waveguide taper has a gain of more than 7 dB over butt-fiber coupling. Therefore, a coupling efficiency of more than 50 % can be expected.

4.9 Mode-Field Measurement Methods

To predict an effective optical coupling between various optical components, it is necessary to exactly determine the mode fields of the OEICs, as shown in Sect. 3.3. If the mode field or the spot size is known, the overlap integral of the Gaussian beams can be used to predict the coupling efficiency. Here the comparison of the fields of active OEICs (such as lasers) with the fields of tapered single-mode fibers is of particular interest, since such coupling arrangements are frequently used in real systems. During that time, two methods for measuring the spot size have been established. These are as follows:

- Near-field method,
- Far-field method.

Other methods have also been developed (Andersen 1984), ITU-T G652 (Keil 1984). These are however beyond the scope of this book.

There are several methods to characterize the optical mode field of photonic ICs. The most used method is the so-called far-field method. This method provides a good resolution of the field in a large distance in comparison with the wavelength of the used light. Additionally, the setup is easy to realize with standard electromechanical parts such as rotating stages. In this case, the field is scanned by a small photodiode, rotating around the output side of the waveguide. The EIA standard, RS-455-47, describes the measurement procedure to qualify optical fibers. The resolution of the method has its limits by the scanning steps and the active field of the photodiode. You must keep in mind that in the far field, a smearing of the fine contour of the near field will appear. The setup needs a large room around the DUT, because the detector motion takes wide space.

To get more detailed information about the field distribution at the end facet of the waveguide, an additional method is to use a microscope objective to magnify the spot to a camera system. This technique is called in literature near-field observation. The resolution is limited by the diffraction limit of the used light of the exciting optical source, which is fed through the tested waveguide. At 1550 nm, the resolution is limited to 2 μm in both lateral dimensions. For typical optical fields of waveguides of laser diodes, the resolution is insufficient to get a detailed knowledge of the optical field parameters. The resolution is also adulterated by the lens due to aberrations, which broaden the focus on the focus plane, and the DUT must be fixed in micrometer precision in front of the setup. This is a time-consuming procedure limiting the near-field technique to laboratory use. Both techniques are wasteful due to time-consuming measurement and positioning times. These facts are focusing the field analysis on random sample survey and disqualify these techniques basically for high-output industrial use.

4.9.1 Near-Field Method

The near-field measurement method can be used for high-accuracy measurement of the mode-field radii w_x and w_y of a waveguide under investigation. This method is described in DIN standards (DIN EN 60793-1-45:2004-07 and DIN 58002:2001-12). The advantage of this measurement technique is high measurement resolution.

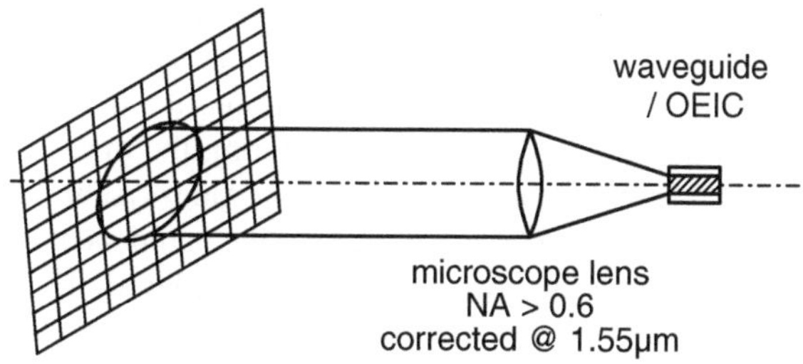

Fig. 4.19 Near-field measurement setup

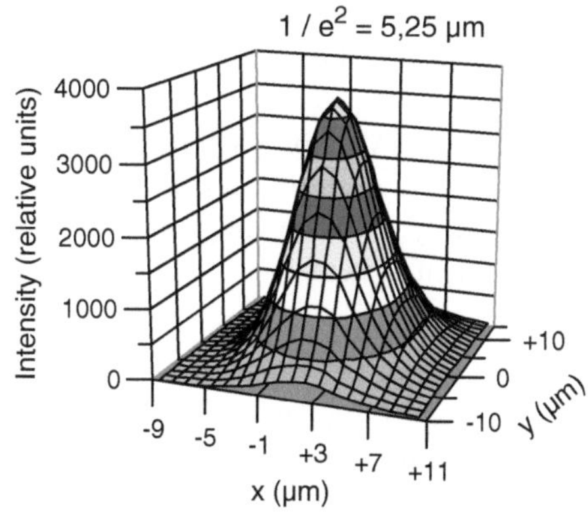

Fig. 4.20 Near-field of a single-mode fiber (SMF) measured at 1550 nm

However, high precision is typically followed by high equipment costs. As shown in Fig. 4.19, the near-field intensity distribution exiting the waveguide is enlarged by means of an objective and displayed on the image plane. The coordinates X and Y have been shown here in accordance with ISO/CD 11807-1. On the image plane, the intensity distribution is spatially resolved and measured by means of a photodetector. High magnification allows very precise estimation of the light distribution in the waveguide of OEICs. An example of such a measurement on a SMF fiber is shown in Fig. 4.19.

The measuring system consists of a light source used for the excitation of OEICs, an imaging system (preferably a microscope objective), and a receiver system, which is usually a Vidicon video camera tube with electronic image processing (Fig. 4.20).

The complete measurement setup should be installed so that all components are aligned on the same optical axis. The intensity of the light source must be adjusted so that the full dynamic range of the camera system can be used, typically with 8-bit resolution.

If the optical attenuators are used in order not to overexcite the camera, they must attenuate the light homogenously over the whole scanning area. When characterizing the passive optical components, the waveguide can be illuminated externally, so that leaky modes, scattered light, or irradiated light are avoided, and the waveguide is excited by a single mode. The center wavelength and the spectral bandwidth of the light source must thereby be known. If the component is polarization dependent, the angle of polarization must be kept constant during the whole measurement process. In addition, the temperature of the device under test should be stabilized and recorded during the measurement.

The numerical aperture of the imaging system should be at least three times larger than the numerical aperture of the component to be measured. Otherwise, the measurement accuracy will be severely reduced. The magnification factors of used lenses should be known. In addition, the lenses should have the anti-reflective coating to reduce the reflections and interferences. It is also recommended that the lenses are corrected for the wavelength range of interest in order to avoid the distortion of the near-field image. The lateral extent of the detector should be chosen so that the image of the waveguide corresponds to at least four times the $1/e^2$ spot size.

An infrared Vidicon photodetector system is commonly used in the praxis. It is known to show strong deviations of local sensitivity and intensity dynamics. Consequently, the following points are of particular significance when installing the measurement system:

- Isotropy, spatial invariance
- Nonlinearity lower than 5 %
- Variation of sensitivity of less than 2 %
- Background variations of less than 1 %
- Signal-to-noise ratio of less than 20
- Spatial resolution greater then 10 times $1/e^2$ spot size
- Digital analysis with more than 6 bits

The measurement should be only performed when the light source with the coupling system, the objective and the detector system have reached the steady-state operation over time. This should prevent the additional intensity fluctuations to reduce the measurement accuracy.

4.9.2 Median-Field Method

The drawbacks of the near- and far-field techniques are evidently. Therefore, at the Harz University, a new setup to measure the optical field without any additional optics or micrometer-precise adjustment of the DUT was developed. The new technique is called median-field method. Here, an easy optical setup is combined with a mass production compatible measurement cycle. A closer description of this conversion and the measurement setup can be found in Fischer and Windel (2004).

The intensity is recorded metrologically by a small photodiode in a 2-axis scanning setup, using a distance r in the range of 2–3 mm from DUT to photodiode, which is sketched in Fig. 4.21. The spacing r can be seized on use of a 3-axis

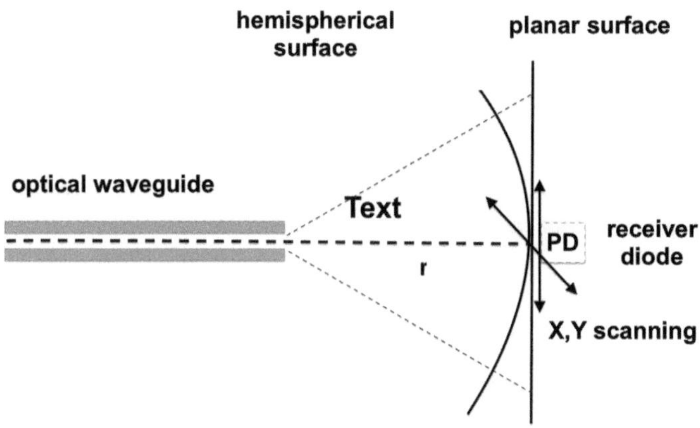

Fig. 4.21 Median-field measurement setup

system, starting from zero spacing at the DUT to 3 mm longitudinal traverse path. It is constant during the measurement. To get comparative data to the far-field method, a conversion of the measured intensities must be made. The spacing of the photodiode from the longitudinal axis no longer directly aligns the photodiode to the test object. It would have to be turned, in order to keep the effective receipt surface constant, hereby the distance of the photodiode to the DUT is changing. Both factors affect the effective surface area of the measuring receiver device, and therefore, a distance coefficient must be introduced.

The distance coefficient is specifically calculated for each detection point and stored in an online table. The surface of the receiver photodiode is assumed as rectangular. In order to achieve the conversion from Cartesian coordinates to hemispherical plane, the corner points are first considered as vectors, as shown in Fig. 4.21.

Fig. 4.22 Reception plane and DUT

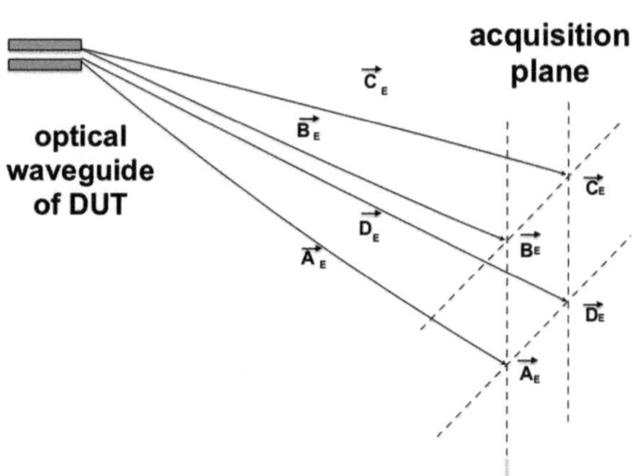

By normalizing the vectors on the length of the beam that reaches the center of the rectangle, one obtains the effective area of the illuminated surface, which is depicted in Fig. 4.22.

The smaller distance of the photodiode reduces the plane of the surface at the Z-axis. It must also considered that the surface has to be rotated, which becomes an irregular square in the aquisition plane, and the surface is calculated according to the equation for the general square:

$$A = \frac{1}{2} * d_1 * d_2 * \sin\angle(d_1, d_2) \tag{4.25}$$

$d_{1/2}$ are the diagonals between the corner points of the measuring plane. Now, the surface can be calculated by:

$$A_{\mathrm{EN}} = \frac{1}{2} * \left|\overrightarrow{A_{\mathrm{EN}}C_{\mathrm{EN}}}\right| * \left|\overrightarrow{B_{\mathrm{EN}}D_{\mathrm{EN}}}\right| * \sin\angle\left(\overrightarrow{A_{\mathrm{EN}}C_{\mathrm{EN}}}, \overrightarrow{B_{\mathrm{EN}}D_{\mathrm{EN}}}\right) \tag{4.26}$$

The difference of a plane area and the semispherical one is shown in Fig. 4.23. To calculate the area of the surface on the spherical surface, the vectors of the corner points of the receiving diode are standardized on the radius of the sphere. The radius corresponds direct to the distance of the receiver diode from the DUT on the z-axis. It stretches to an anomalous square, whose surface can be determined as follows:

$$A_K = \frac{1}{2} * \left|\overrightarrow{A_K C_K}\right| * \left|\overrightarrow{B_K D_K}\right| * \sin\angle\left(\overrightarrow{A_K C_K}, \overrightarrow{B_K D_K}\right) \tag{4.27}$$

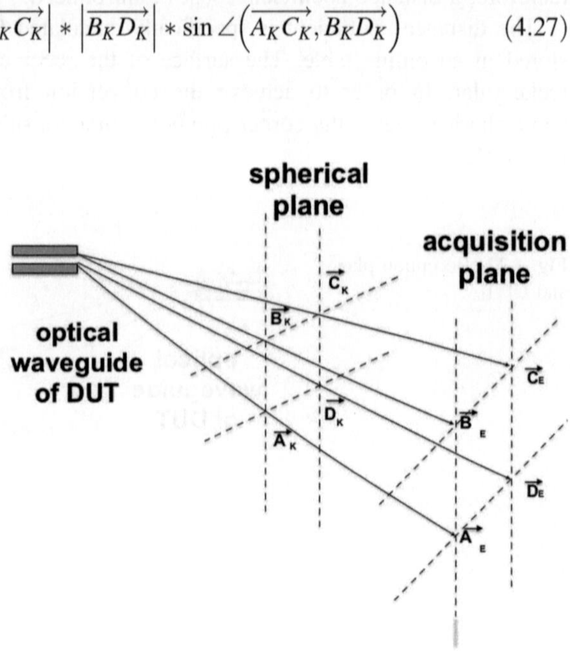

Fig. 4.23 Difference of plane areas of flat measurement surface and spherical surface

For the determination of the distance coefficient, both surfaces must be kept in relationship. The distance factor κ depends on the coordinates, where the receiver diode is working in the acquisition plane

$$\kappa(\text{distance\_faktor}) = \frac{A_{\text{EN}}}{A_K} \tag{4.28}$$

The intensities collected at the respective coordinates must be calibrated using the distance factor κ, in order to get the proper intensity values.

4.9.2.1 Automated Acquisition

Automation of the median-field acquisition is realized using a 6-axis-motion system from Physic Instruments (PI) F-604. The motion of the photodiode is controlled by a Labview program. It also calculates the distance coefficient κ. A virtual instrument (VI) was developed, which controls the PI system and processes the measuring data. The user front end is depicted in Fig. 4.24 (Windel 2006).

One can see at the red-framed part the user interface to print in the file name for the storage of the measuring data. Further on, the user is asked to type the incrementation steps size in μm and the count of measurement steps per line ("steps").

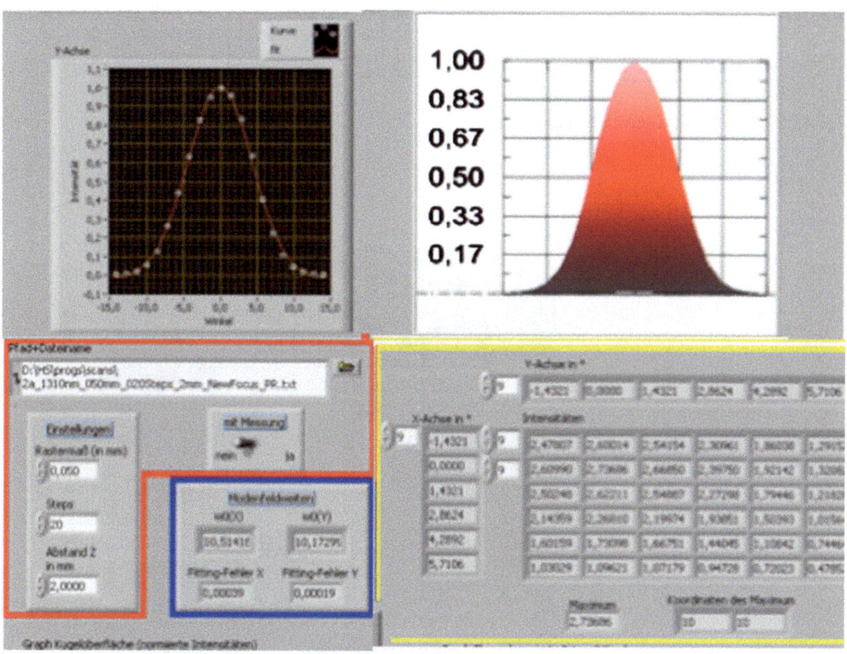

Fig. 4.24 Labview VI user front end for automated field acquisition

These inputs are necessary for the automated acquisition of the intensity in connection of the VI drivers developed by PI.

Within the yellow-bordered part of the surface on the right upper side, one can see the measured intensities converted with the distance coefficient, the maximum intensity and their coordinates are indicated in the table. In the blue frame, the optical field widths in *x*- and *y*- direction, determined by the fitting process, are indicated. The intensity field and the result of the fitting process are shown in the two graphs.

4.9.2.2 Measurement Setup

To check the accuracy of the method, a SMF manufactured after ITU-G.652 was measured. This reference measurement was realized with the following measurement setup, shown in Fig. 4.25.

As optical source, the Anritsu MU951001A integrated laser set was used. The module consists of two laser diodes with at 1310 and 1550 nm, which are alternatively changeable. The detector was a photodiode from RCA with 50 μm active surface. The devices were installed on a motion controller system of Physics

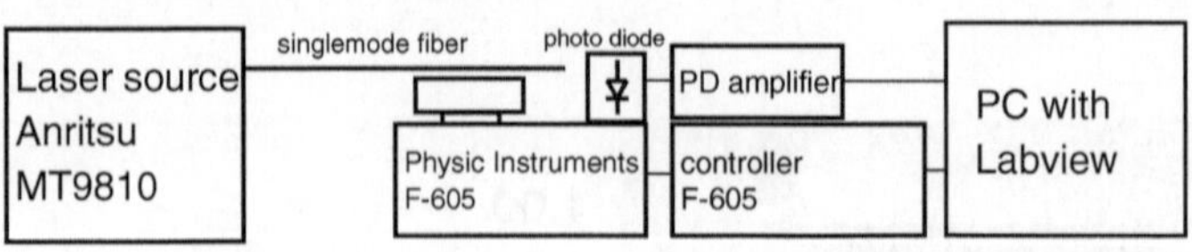

Fig. 4.25 Schematic measurement setup

Fig. 4.26 Photograph of setup

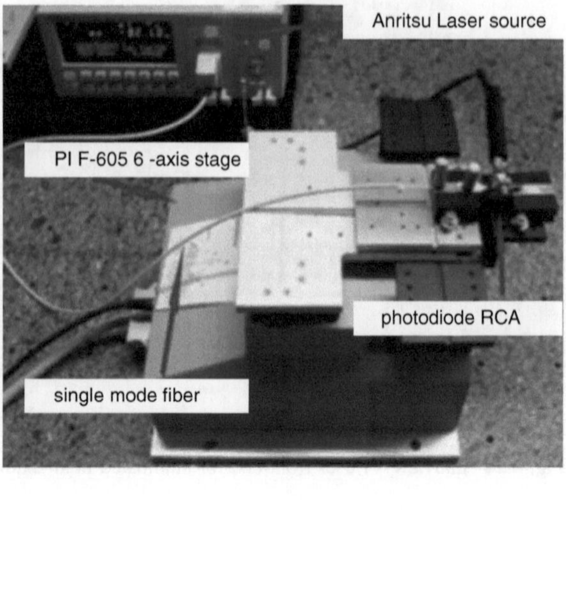

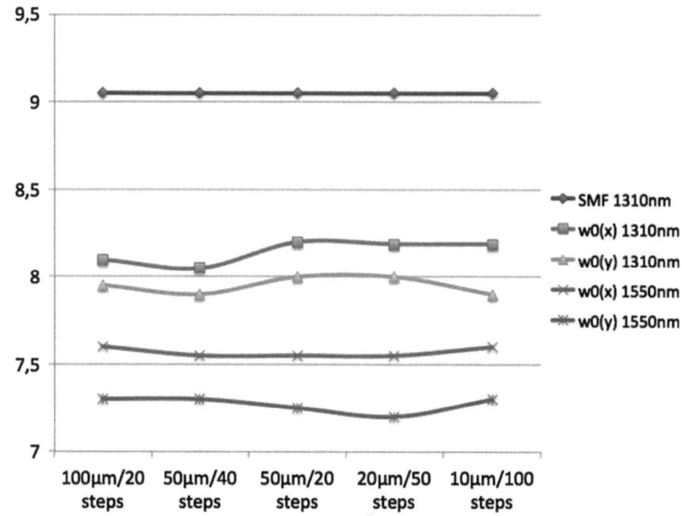

Fig. 4.27 Comparison of mode fields with different detectors and wavelengths with and without distance calibration

Instruments and controlled by an additional PC, which serves only as interface for the PC where the recorded data processing and calculating are performed (Fig. 4.26).

4.9.2.3 Results

A standard single-mode fiber was subsequently measured to analyze with different measurement points and measuring point intervals and the accuracy of the method at two wavelengths (1550 and 1310 nm). No significant discrepancy was detected comparing all variations. In Fig. 4.27, the optical mode-field diameters determined are depicted. A value of 9 µm was expected according to the ITU-G.652 standard. From the results, it can be seen that the values are mainly too small without distance factor. The results show values at 1310 nm of $2w_0 \approx 7.4$ µm and $2w_0 \approx 8$ µm were at 1550 nm excitation wavelength. The deviation based on the fact that the distance between DUT and sensor is not accurate enough measured.

In order to use this new median-field method, it is important to calibrate the whole measuring setup on the distance from the DUT to the photoreceiver. The help of highly accurate distance sensors could realize this determination. An easier approach is to make use of standardized mode-field width of a SMF at 1310 nm. The mode field can be measured at the beginning of the measurements using 1310 nm excitation source fed into the SMF. The calibration factor for the distance can easily be calculated. To get a highly accurate distance factor F_K, a mode-field calibrated fiber from the National Institute of Standardization should be used conforming with the ITU-Standard at $\lambda = 1310$ nm:

$$F_K = 1.21945$$

The whole time for the complete scan of the field was less than one minute, which is a substantial improvement in relation to traditional far- and near-field measuring procedures. In order to reduce the absolute measurement times further, the use of a CCD camera is necessary. This setup has the advantage that the measurement of the mode field can be realized in only one step and measurement times will reduce less than one second.

The median-field method is ideal to measure the spot size of single-mode optical components. In comparison with the classical far-field and near-field method, the new method is ideal for use in automated systems, e.g., at the production of lasers, where the lasers must be characterized very quickly on bar. The measurement time can be reduced in combination with additionally current-power characterizations.

Using automated concepts with Labview programming, the median-field method shows high potential for automated high-speed and high precise mode-field measurement in agreement with ITU Recommendation G.652.

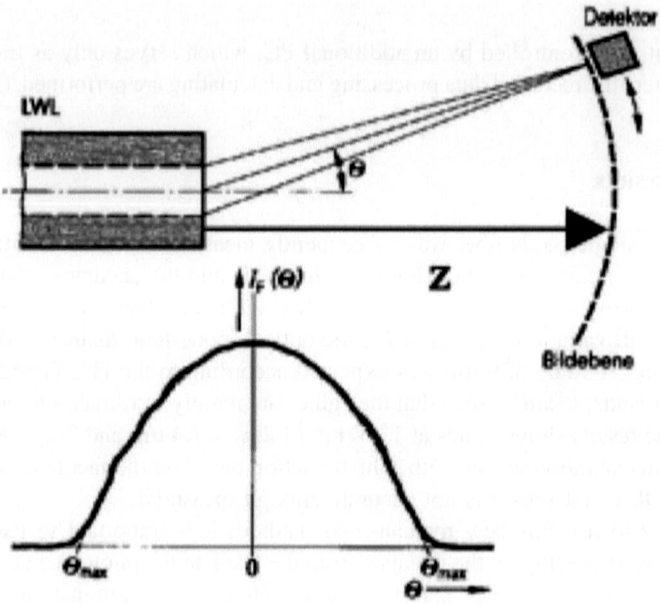

Fig. 4.28 Far-field measurement setup

4.9.3 Far-Field Method

As just shown, it is relatively expensive to build up the near-field measurement system correctly. A more simple solution is offered by the far-field measurement method (Fig. 4.28). As already shown in Sect. 4.3, Eq. (4.8), it is possible to estimate the near-field distribution of light based on the angular distribution of the far field by means of a simple approximation. There is a direct relationship between the spot size and the tangent of a far-field angle. Analogous to the mode-field diameter of the near field, the far-field angle is defined as the angle from the axis where the intensity of light has dropped to $1/e^2$ of its maximal value.

The measurement setup is similar to the one for the near-field, except of the measured OEIC/waveguide and the photodetector which can be circularly moved across two axes around the OEIC. The detector first circularly scans across the first angular axis. After that, the light intensity of the chip is scanned across the second angular axis. This is how the intensity profile of the emitted light is obtained. The angular distribution of light can be converted to the mode-field diameter of the waveguide by means of the following formula:

$$w_0 = \frac{\lambda}{\pi \cdot n \cdot \tan \Theta} \tag{4.29}$$

Like in the case of the near-field measurement, the difference between maximum and minimum measured intensities must be at least 64 digital levels. Since commercially available photodiodes are typically used as detectors, it can be assumed that they are very homogenous and linear. The actuator should have the angular resolution of less than 0.5°, in order to provide high resolution of the measurement. Scattered light is avoided by means of a phase-sensitive detection with a lock-in amplifier.

Comparison of the two measurement systems (near- and far-field) shows only a small difference of less than 10 % because the near-field measurement has better

Fig. 4.29 Far-field measurement curve and Gaussian approximation

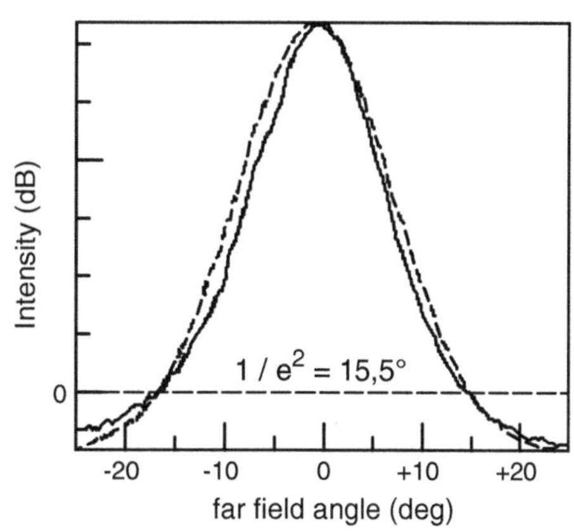

resolution than the direct calculation if the spot size based on the BPM program. A typical far-field distribution of a drawn fiber taper is shown in Fig. 4.29. A Gaussian approximation is also shown in the figure with a dashed line. It can be noticed that the measured angular distribution is relatively noisy.

It can also be seen that both curves match quite well at the value of $1/e^2$. Since the measured curve is slightly asymmetrical, the mean of the two values has to be used in this case. The far-field angle equals 17° on the left side and 14° on the right side, resulting in a mean value of 31°/2 = 15.5°. By substituting this value in the Eq. (4.27), a corresponding mode-field radius is obtained:

$$w_0 - 1.55\ \mu\text{m}/(3.14 \cdot 1 \cdot \tan(15,5°)) = 1.78\ \mu\text{m}$$

This is a typical value for fiber tapers with radii around 10 μm.

4.10 Summary

The optical coupling between different optical components requires low coupling losses and low reflections. In most cases, the geometrical optics cannot be used. Instead, the wave analysis should be applied. The optical modes of the components (laser, fiber, waveguide) are usually described with a Gaussian distribution, and the coupling efficiencies can be calculated by means of the overlap integrals. The biggest problem represents the mechanical alignment of the components.

References

Andersen, W.T.: Consistency of measurement methods for the mode field radius in a single-mode fiber. IEEE J. Lightwave Technol. **2**(2), 191–197 (1984)

Beyer, W., Eigler, H.: Moderne Produktionsprozesse der Elektrotechnik. Expert-Verlag, Renningen-Malsheim, Elektronik und Mikrosystemtechnik (1996)

Brück, R.S.: Angewandte Mikrotechnik. Hanser Verlag (2001)

Ehrfeld, W., Abraham, M., Ehrfeld, U., Lacher, M., Lehr, H.: Materials for LIGA products. In: Proceedings of IEEE Workshop on Micro Electro Mechanical Systems, MEMS '94, pp. 86–90 (1994)

Fischer, U.H.P.: Optoelectronic Packaging. Berlin/Offenbach, VDE-Verlag GmbH, pp. 125 (2002a)

Fischer, U.: Optoelectronic Packaging. VDE Verlag, Berlin (2002b)

Fischer, U.H.P., Graener, H.: Bauteile der optischen Nachrichtentechnik in Gläsern und deren Packaging. In: Innovationsforum Strukturierung von Gläsern Sachsen-Anhalt, 14/15. February Barleben, Germany (2005)

Fischer, U.H.P, Windel, T.: Medianfield-methode—a planar method to obtain the spot-size of single mode optical components. In: Laser Assisted Net Shape Eng, Bamberg, Meisenbach-Verlag, pp. 1249–1256 (2004)

Goodwin, M.J., Meseley, A.J., Kearly, M.Q., Morris, R.C., Kirkby, C.J.G., Thomson, J., Goodfellow, R.C., Bennion, I.: Optoelectronic component arrays for optical interconnection of circuits and subsystems. J. Lightwave Technol. **9**(12), 1639–1645 (1991)

Honecker, J., Umbach, A., Trommer, D., Eckhardt, T., Fischer, U.H.P.: High-speed photo diode modules with up to 45 GHz modulation bandwidth for optical communication systems. In: Conference on Optical Fiber Communication, Technical Digest Series, Optical Fiber Communication Conference and Exhibit. Anaheim, CA, pp 724–725 (2004)

Karpuzi, Oz: 40 GHz bis 100 GHz Bonding und Packaging. Paper presented at the ITG-Fachtagung Photonische Aufbau- und Verbindungstechnik, Wernigerode (2009)

Keil, R.: Experimental investigation of the beam spot-size radius in single-mode fiber tapers. Electron. Lett. **20**(15), 621–622 (1984)

Kuhmann, J.F.: Untersuchungen zu einer flußmittelfreien und selbstjustierenden Flipchip-Bondtechnologie für photonische Komponenten. Technische Universität, Berlin (1996)

Ladany, I.: Laser to single-mode fiber coupling in the laboratory. Appl. Opt. **32**, 3233–3236 (1993)

Lau, J.H.: Flip Chip Technologies. McGraw-Hill, New York (1995)

Menz, W., Mohr, J.: Mikrosystemtechnik für Ingenieure. VCH Verlagsgesellschaft mbH, Weinheim (1997)

Optoelectronic Packaging. Wiley Series in Microwave and Optical Engineering. Wiley, New York (1997)

Saruwatari, M., Nawata, K.: Semiconductor laser to single-mode fiber coupler. Appl. Opt. **18**(11), 1847–1856 (1979)

Saruwatari, M., Sugie, T.: Efficient laser diode to single-mode fiber coupling using a combination of two lenses in confocal condition. IEEE J. Quant Electron. **6**, 7 (1981)

Steckenborn, A., Winkler, T., Jantke, G., Arndt, F., Schlaak, H.F.: High precision wafer orientation for micromachining. Microsyst. Technol. 467–471 (1991)

Windel, T.: Entwicklung einer planaren Messmethode zur Bestimmung von optischen Modenfeldern. http://www.cuvillier.de/flycms/de/html/30/-UickI3zKPS,yc0g=/Buchdetails.html (2006)

Chapter 5
RF Lines

Abstract The wide use of integrated microwave and millimeter-wave circuits requires a packaging and interconnection technology, which keeps in mind on the one hand the specific circumstances of the case materials and on the other hand meets the special requirements of high-frequency connection to the chip. The necessary structures are typically implemented as microstrip line or coplanar line. For particularly high-frequency connections to more than 100-GHz semirigid coaxial cables are used. In this chapter, the basic principles of the coaxial and coplanar lines are represented.

5.1 Maxwell's Equations

The entire field of electrical, magnetic, and optical phenomena can be completely described by four Eqs. (5.1)–(5.4), which were published by James Clerk Maxwell in 1864:

$$\mathrm{rot}\, E = -\frac{\mathrm{d}B}{\mathrm{d}t} \tag{5.1}$$

$$\mathrm{rot}\, H = \frac{\mathrm{d}D}{\mathrm{d}t} \tag{5.2}$$

$$\mathrm{div}\, B = 0 \tag{5.3}$$

$$\mathrm{div}\, B = 0 \tag{5.4}$$

$$D = \varepsilon \cdot \varepsilon_0 E \tag{5.5}$$

$$B = \mu \cdot \mu_0 H \tag{5.6}$$

$$j = \sigma E \tag{5.7}$$

© Springer-Verlag Berlin Heidelberg 2015

U.H.P. Fischer-Hirchert, *Photonic Packaging Sourcebook*,
DOI 10.1007/978-3-642-25376-8_5

These equations and the still-relevant material Eqs. (5.5) and (5.7) contain everything you need to know about electric and magnetic fields. Bold characters are identified as vectors. The parameter $\mu_0 = 1.256 \times 10^{-6}$ (V s/A m) is called the magnetic constant or induction constant. μ is a dimensionless material constant and is called permeability. Its amount is of the order of one. The permittivity is $e_0 = 8.85 \times 10^{-12}$ (F/m), and the relative dielectric constant $e = c/c_{\text{vak}}$ is defined by the ratio of the speed of light in the medium to the speed of light in vacuum. This reminds us of the refractive index of light in the media and is the electrical analog of this. Thus, high-frequency waves behave in their propagation in conduits increasingly similar to light in optical fibers, the higher the frequency of the wave. The structure of Maxwell's equations is now made clear by means of simple examples to bring the reader closer to the propagation of electromagnetic waves into electrical RF lines. It is recommended to study the latter at Appendix literature for a deeper understanding.

First, it is important to look at the induction law of Eq. (5.1). In the environment of current carrying conductors, always forces occur, which are usually much larger than the electrostatic, and differ also with respect to the direction of the force action. They are called magnetic forces.

A straight wire with a current flow in straight direction surrounds himself with a magnetic field whose field lines include the wire as closed circles and extend in planes which cut through the wire vertically. If you think of the current flow in the wire by a time-varying electric field ***E*** replaced, we obtain a magnetic field ***H***, the ring develops around the time-varying electric field ***D***, as in Fig. 5.1 can be seen. On the other hand, a current flow in an electrically conductive wire could be achieved by the change of a magnetic field, or by reversal of the direction or intensity change.

Generalizing this idea to the pure fields, time-varying magnetic fields ***H*** produce electric fields, which also set up in a ring around the magnetic field, as in the case of the changing electric field.

This is the meaning of Eq. (5.2), which is also referred as Ampere's law. Currents are the causes of magnetic fields, and they are vortex of the magnetic field. We call $j + \mathrm{d}\boldsymbol{D}/\mathrm{d}t$ as the total current density, resulting from the conduction current density j and the displacement current density $\mathrm{d}\boldsymbol{D}/\mathrm{d}t$.

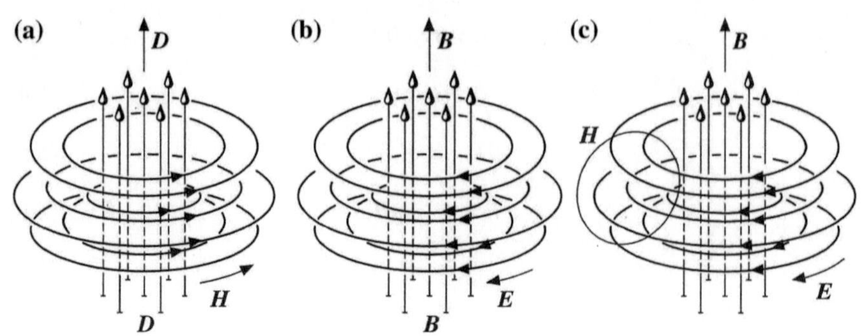

Fig. 5.1 Changes in the magnetic and electric fields

If you connect up both statements of Eqs. (5.1) and (5.2), then the fields complement each other in their temporal change in such a way that they always mutually generate itself, which is shown in Fig. 5.1c.

Equation (5.3) implies that the field lines of the magnetic induction are free of sources and sinks, i.e., there are no "magnetic monopoles," in which the field lines begin and end as in the electric field.

The last of Maxwell's equations (5.4) states that electric field lines have a source, namely the electrical charge carriers $s = \mathrm{d}Q/\mathrm{d}V$, i.e., the charge present per unit volume.

The symmetry of Maxwell's equations is particularly clear when one considers the case of the vacuum without charges and electric currents:

$$\mathrm{rot}\, H = \frac{\mathrm{d}D}{\mathrm{d}t} \quad \mathrm{rot}\, E = -\frac{\mathrm{d}B}{\mathrm{d}t}$$

$$B = \mu_0 H \qquad D = \varepsilon_0 E$$

$$\mathrm{div}\, B = 0 \qquad \mathrm{div}\, D = 0$$

It is clearly seen that a changing electric field ($\mathrm{d}\boldsymbol{D}/\mathrm{d}t$) generates a magnetic vortex field (rot $\boldsymbol{H}$). Conversely, ($\mathrm{d}\boldsymbol{B}/\mathrm{d}t$) causes a varying magnetic field, an electric vortex field (rot E) out. This describes the mechanism for the generation and propagation of electromagnetic waves in vacuum.

5.2 Wave Types

For the propagation of electromagnetic waves, there are basically two possibilities to be distinguished:

- Spread in space or
- Spread along lines.

As lines in the classic sense refer to a double conductor, isolated from each other, there are elongated conductors. Such simple conductors are usually used for transmission of low frequencies. In the high frequencies, coaxial lines, strip lines, and waveguides are the most common cable types. The frequency range of the transmitted waves extends over the microwave range between 1 and 300 GHz. The line types are also divided into

- Lecher or TEM lines and
- Waveguides.

Some examples are shown in more detail in the following. In Lecher lines, the electric and magnetic fields are polarized perpendicular to the propagation direction (Fig. 5.2). That is why they are also called transversal electro magnetic (TEM) lines. The line inside is completely field-free in TEM lines, also the electric field lines are

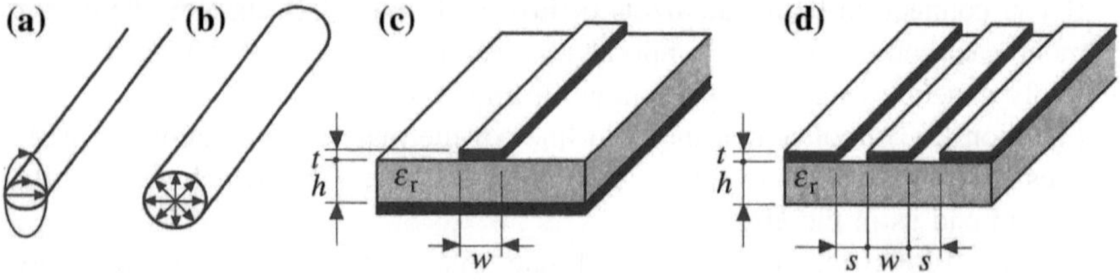

Fig. 5.2 TEM-conductor lines a two-wire line, b coaxial, c microstrip line, d coplanar

Fig. 5.3 Circular waveguide

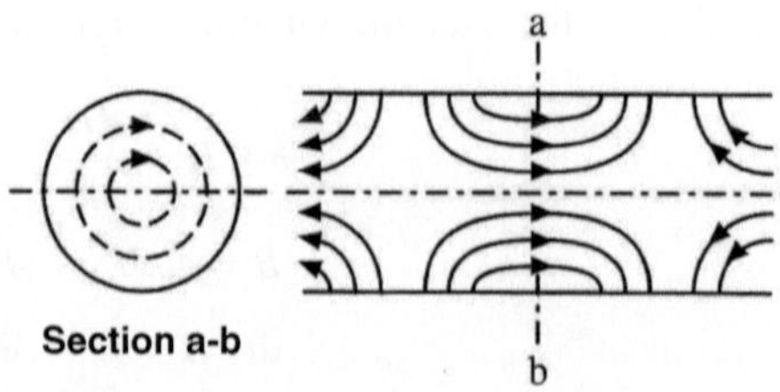

perpendicular to the surface, while the magnetic field lines are only tangential to the surface. In waveguides (Fig. 5.3), however, electric or magnetic field components are polarized in the direction of propagation, which will not discussed in further detail.

Once lines are composed of different layers, one finds wave propagation similar to the TEM waves, but has some differences. We know such behavior in microstrip lines and coplanar lines, but not in coaxial lines. As a rough basis for the cross-sectional dimensions of the conductor, it can be started from the assumption that the geometric dimensions of the conductor compared to one-quarter of the wavelength of the field in the dielectric should be small.

In the TEM lines, no fields in the propagation direction are present when the conductor material is completely lossless and the dielectric located between the conductors is absolutely homogeneous. The conductor materials such as gold, copper, or silver used in practice one can assume that the propagation of pure TEM waves is present with high probability.

The simplest TEM line is the two-wire line (Fig. 5.2a). Because of their quite modest high-frequency characteristics, this cable type should be mentioned for completeness only. The coaxial line has much better RF performance and is in the broadband transmission system, the cable type, which was the very common one in the wide-area technology until the end of the 1980s. Relatively small losses of only 10 dB/km at 2.5 GHz were standard, but every 10 km to the signal refresh a station had to be erected. In photonic packaging, this cable type has been proven in the coupling to high-frequency components up to 100 GHz. This type is used in research and in applications such as network analyzers up to several hundred GHz. The microstrip line (Fig. 5.2c) is, however, only up to 20 GHz used in the module construction. The coplanar line can be used well above 100 GHz (Fig. 5.2d).

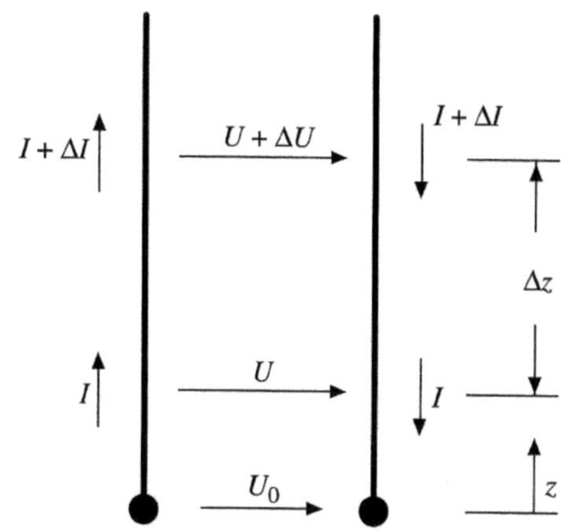

Fig. 5.4 Transmission line equations

5.3 Conduction Equations

In the following, the propagation of TEM waves in simple two-wire lines should be considered in more detail. The line equations are to be derived qualitatively and give an introduction to the RF characteristics, since the elementary wave propagation in double lines, the simplest form of wave propagation constitutes (Fig. 5.4). When the construction of the line is known, one can specify for such a line, the self-induction, capacity, and line resistance. It relates the four-line parameters to a unit length and referred to as the "inductance," the self-inductance per unit length, as "capacitance," i.e., the capacitance between the two conductors per unit length, and thus the ohmic line resistance per unit length or "resistance coating." In addition, it introduces the "conductance value," i.e., the conductance of the cross-derivative between the two conductors per unit length. From Ohm's law, $U = RI$, and then, the line equations results the following:

$$\frac{\mathrm{d}U}{\mathrm{d}z} = -(R' + \mathrm{j}\omega L') \cdot I \tag{5.8}$$

$$\frac{\mathrm{d}I}{\mathrm{d}z} = -(G' + \mathrm{j}\omega C') \cdot U \tag{5.9}$$

To determine the profile of the voltage along the line, one differentiates Eq. (5.8) and sets the size of Eq. (5.9) to

$$\frac{\mathrm{d}^2U}{\mathrm{d}z^2} = (R' + \mathrm{j}\omega L') \cdot (G' + \mathrm{j}\omega C') \cdot U = \frac{\mathrm{d}^2U}{\mathrm{d}z^2}\gamma_z^2 \cdot U = 0 \tag{5.10}$$

$$\text{mit}\, \gamma_z = \pm\sqrt{(R' + \mathrm{j}\omega L') \cdot (G' + \mathrm{j}\omega C')} \tag{5.11}$$

The size γ_z is called "propagation constant" because any electrical properties of the propagating wave is implemented in it. Since the expression for γ_z is complex, it consists of a real part and an imaginary part:

$$\gamma_z = \pm(a + \mathrm{j}b).$$

The real part is a measure of the attenuation and is called "attenuation coefficient," and the imaginary basis of the phase-rotating property of the line and is called a "phase coefficient." Solving the differential equation (5.10) by U, one can calculate the following:

$$\begin{aligned} &= U \cdot e^{\mathrm{j}\omega t} = U_0 \cdot e^{-\gamma_z} e^{\mathrm{j}\omega t} + U_1 \cdot e^{-\gamma_z} e^{\mathrm{j}\omega t} \\ &= U_0 \cdot e^{-a_z} e^{\mathrm{j}\omega\left(t\frac{\beta_z}{\omega^z}\right)} + U_1 \cdot e^{-a_z} e^{\mathrm{j}\omega\left(t\frac{\beta_z}{\omega^z}\right)} \\ &= U_0 \cdot e^{-a} e^{\mathrm{j}\omega\left(t\frac{z}{v}\right)} + U_1 \cdot e^{-a_z} e^{\mathrm{j}\omega\left(t\frac{z}{v}\right)} \end{aligned} \tag{5.12}$$

Assuming a wave which spreads only in the z-direction. Since the first portion having an amplitude which decreases more and more by increasing in the z-direction, this means, at $z = 0$, therefore the voltage at the cable beginning $U_1 = 0$. By substituting Eq. (5.12) in Eq. (5.9), one obtains the following:

$$I = I_0 \cdot \mathrm{e}^{-a_z} \mathrm{e}^{\mathrm{j}\omega\left(t - \frac{z}{v}\right)} + I_1 \cdot \mathrm{e}^{-a_z} \mathrm{e}^{\mathrm{j}\omega\left(t - \frac{z}{v}\right)} = I_0 \cdot \mathrm{e}^{-\gamma_z z}$$

mit

$$\begin{aligned} I_0 &= \frac{U_0}{Z_e} \\ I_1 &= \frac{U_1}{Z_e} \\ Z_e &= \sqrt{\frac{R' + \mathrm{j}\omega L'}{G' + \mathrm{j}\omega C'}} \end{aligned} \tag{5.13}$$

Z_e is referred to as "impedance of the line." This results using Eq. (5.11) to the following propagation constant:

$$\gamma_z = \sqrt{-\omega^2 \cdot L'C'} + \frac{R'}{2} \cdot \sqrt{\frac{C'}{L'}} + \frac{G'}{2} \cdot \sqrt{\frac{L'}{C'}} = \alpha_z + \mathrm{j}\beta_v \tag{5.14}$$

β_v is the phase constant of the lossless line ($\beta = (2\pi/\lambda_z)$) with λ_z as wavelength, and α_z is as already indicated above, the attenuation constant:

$$\alpha_z = \frac{R}{2 \cdot Z_e} + \frac{G \cdot Z_e}{2} \tag{5.15}$$

$$\beta_v = \frac{\omega}{v} = \frac{2\pi}{\lambda} \approx \omega\sqrt{L'C'} \tag{5.16}$$

At a constant phase, this corresponds to the condition of

$$t - z/v = \text{const} \tag{5.17}$$

The traveling speed of the phase is dz/dt and Eq. (5.17) identical with v. The second term of Eq. (5.12) is attenuated in the negative direction of propagation. This is called the returning or reflected wave. And the current I consists of two parts, a forwarding and a reflected wave. According to Eq. (5.13), each partial wave consists of the ratio of voltage and current through the line impedance, regardless of the terminating resistor. Magnitude and phase of the terminating resistor determine the ratio U_0/U_1, I_0/I_1 directly.

5.4 Skin Effect

At high frequencies, the current is no longer distributed over the entire conduit cross section, but forces itself to the surface. The reason is the internal inductance. Let us assume that a current flows through a surface element dr/ds in a conductor (Fig. 5.5), with a magnetic field $\boldsymbol{B}$.

In turn, its change induces an electric vortex field $\boldsymbol{E}_{\text{ind}}$. It is on the side facing the axis opposite to the applied field $\boldsymbol{E}$. Therefore, the resultant field from the central axis to the outside must decrease, as well as the generated current $\boldsymbol{I}_{\text{ind}}$. At high frequencies, the current is almost entirely pushed to the surface. At a depth of d [according to Eq. (5.18)], it has already dropped to $1/e$ (ρ = specific resistance):

$$d = \sqrt{\frac{\rho}{\pi \cdot \mu \cdot \mu_0 \omega}} \tag{5.18}$$

Fig. 5.5 Skin effect

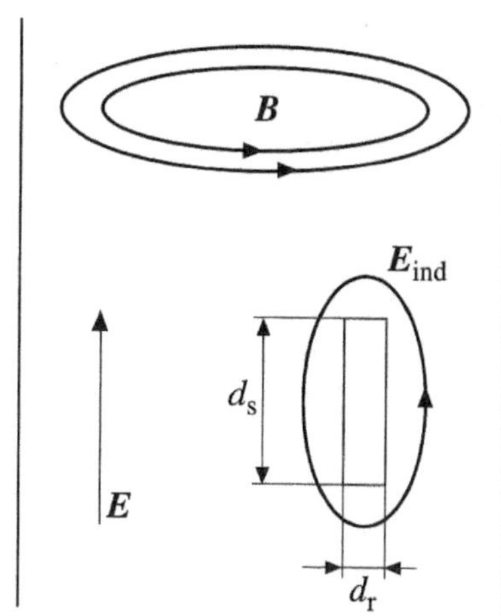

In addition, the skin effect obtained a phase shift between current and voltage, so that the same wire for high frequencies having a higher electrical resistance. This results in that in the RF technique, thin-walled, highly conductive coated with gold lines are used and the quality of the surface roughness and plays a crucial role in the RF transmission characteristics.

5.5 Coaxial Cables

The coaxial line is an often used cable type for the transmission of high-frequency signals. It consists of a cylindrical inner conductor and an outer conductor coaxially surrounding it. If the inner conductor is formed of only one wire, one speaks of a pure coaxial line. The intermediate space is filled with a dielectric that is adapted with its specific dielectric constant to be attained to the cutoff frequency of the transmission bandwidth of the cable. The electromagnetic field remains completely inside the cable, and the electric field lines, such as the radii of the magnetic field lines, such as concentric circles about the longitudinal axis of the cable. The characteristic impedance of commercially available RF cable is internationally standardized to be 50 W. Table 5.1 lists some of the cables with the cutoff frequencies and diameters.

The cables commonly used in the RF construction techniques are shown in Figs. 5.6, 5.7, and 5.8. A distinction is made in the cables for the flexible and semirigid (semirigid) versions. The latter are particularly interesting for frequencies above 40 GHz, since the electrical data due to the rigid cabling remain constant in them, they provide a lower attenuation and higher transmission performance compared with the flexible lines.

Table 5.1 Types of coaxial cables

Cable	CA50020[a]	CA50034[a]	RG178	RG58
Inner conductor (mm)	0.111	0.2	0.3 (7 × 0.1)	0.9 (190.18)
Outer conductor (mm)	0.5	0.86	1.8	4.95
Dielectric material	Polytetrafluorethylene (PTFE)	PTFE	PTFE	PTFE
Conductor material inside	StCuAg	StCuAg	StCu	StCu
Conductor material exterior	Cu silvered	Cu silvered	Cu silvered	Cu silvered
Cutoff frequency (GHz)	270	160	5	3
Attenuation (dB/m)	7	5	3	1.4
Dielectric constant	2	2	2	2

[a]Precision tube Co.

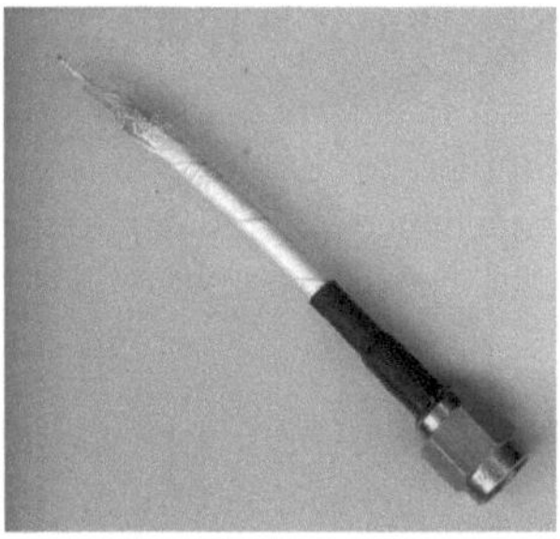

Fig. 5.6 Coaxial cable SMA-connector and RG178-cabel

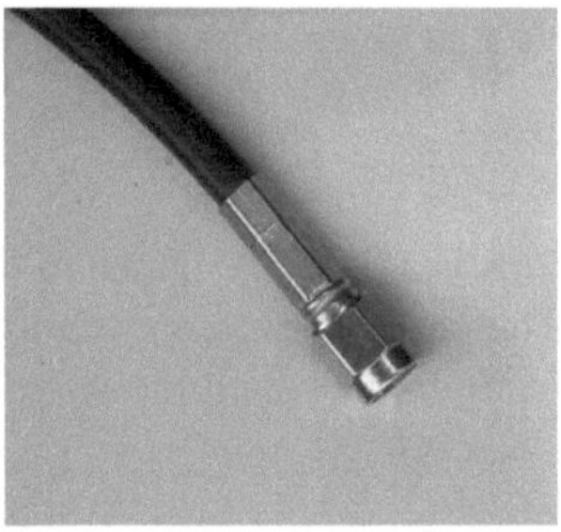

Fig. 5.7 Coaxial cable with SMA-connector RG58-cable

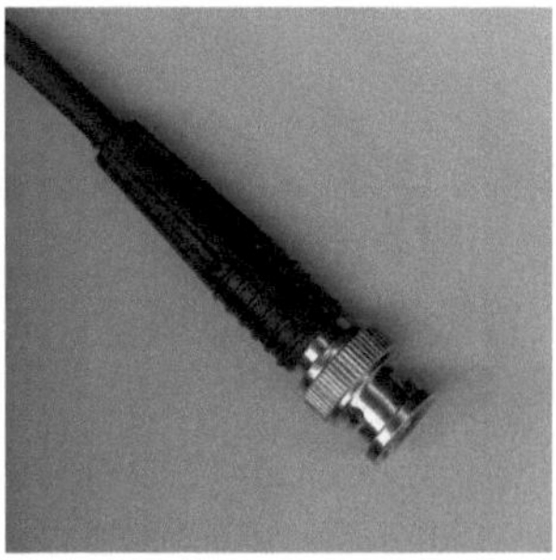

Fig. 5.8 Coaxial cable with BNC-connector and RG58-cable

The types of plugs used are divided according to the area of the cutoff frequency. The most common connector types are listed in Table 5.2. The type of SMA is the most widespread and has up to 18-GHz bandwidth, which covers a wide range of applications. In the applications of optical design technology, the V-connector variants are used in most modules now.

Table 5.2 RF connectors and their bandwidth

Connector types	Bandwidth (GHz)
BNC	0–2
SMA	0–18
K	0–45
V	0–65
W	0–110

5.6 Wave Impedance

Similar to the previously presented line resistance $\boldsymbol{Z_e}$, the concept of wave impedance $\boldsymbol{Z_f}$ is understood as the ratio of the transverse components of $\boldsymbol{E}$ and $\boldsymbol{H}$:

$$Z_f = E/H \tag{5.19}$$

Accordingly, only the forward propagating or reflected TEM wave must be taken into account by the definition of the standing wave:

$$Z_f = E_p/H_p = -E_r/H_r \tag{5.20}$$

At a given cross section along the cable, the field resistance is obtained as below:

$$Z_{f1} = (E_p + E_r)/(H_p + H_r) \tag{5.21}$$

This is a field in contrast to the characteristic impedance of location-dependent size. The characteristic impedance can be calculated from the material constants of the medium, in which the wave is passing:

$$\begin{aligned} \frac{E_p}{H_p} &= Z_f = \sqrt{\frac{\mu_0 \cdot \mu}{\varepsilon_0 \cdot \varepsilon}} = Z_0\sqrt{\frac{\mu}{\varepsilon}} \\ Z_0 &= \sqrt{\frac{\mu_0}{\varepsilon_0}} = 120\pi\Omega = 377\ \Omega \end{aligned} \tag{5.22}$$

Here, $\boldsymbol{Z_0}$ is the wave impedance in vacuum and in good approximation in air (ε = 1,006).

The line resistance $\boldsymbol{Z_e}$ can be also calculated. The speed of light is

$$c = 3 \times 10^{-8}\ \mathrm{m/s}$$

we obtain the following expression:

$$Z_e = \frac{\sqrt{\varepsilon}}{C'} \cdot \frac{1}{c} = 33.3\ \Omega \cdot \sqrt{\varepsilon}\frac{\mathrm{pF/cm}}{C'} \tag{5.23}$$

One must therefore only measure the capacitance at the end of the line to determine the cable resistance to the attention of Z_e.

5.7 Coplanar Lines

A coplanar line is a special variant of the coaxial line with three conductors. The two ground conductors are at the same level as the signal line and two coupled slot lines are corresponding. The structure of a coplanar line is shown in more detail in Fig. 5.9.

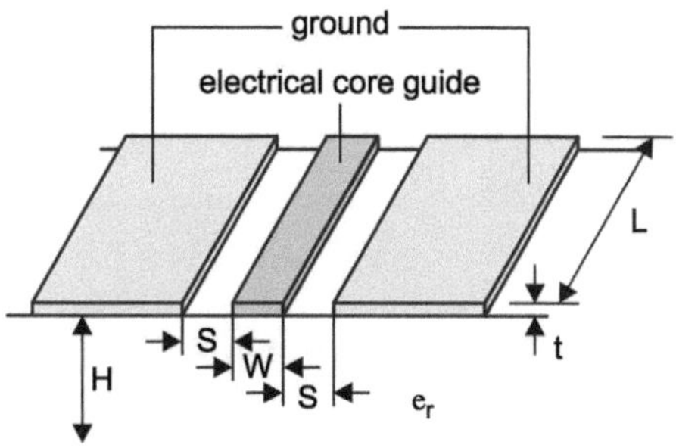

Fig. 5.9 Geometric structure of a coplanar line

Like the coaxial cable, the coplanar lines have large differences in the transmission bandwidth due to the selection of the line parameters. To determine the line parameters exactly and thus the bandwidth limit of the line at a given substrate material is an important factor in the use of this cable type. The course of the field lines in the wave propagation as a common shaft and the push–pull wave is shown in Fig. 5.10.

It can be seen that the field lines are very crowded at the slot between the ducts, and the wave is greatly affected by inaccuracies in the manufacture of the conduit and the material used. The distances and widths of the conductive strips must be adapted to the substrate material used.

At the same time unwanted modes must be suppressed in order to achieve the high cutoff frequencies, which can be expected from this conductor type. An important factor is to fix the ground lines to a common potential, so that the in-phase mode is suppressed.

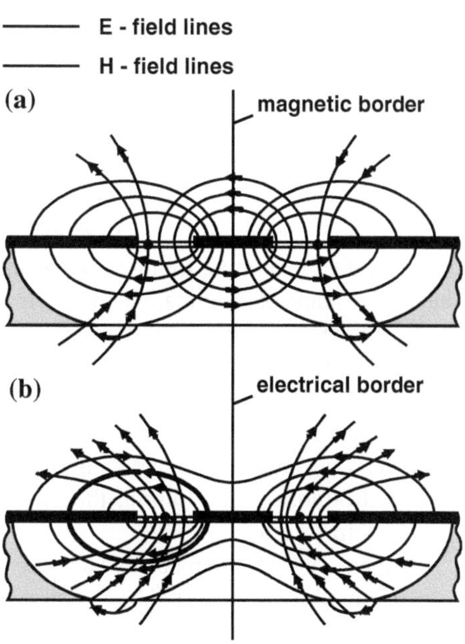

Fig. 5.10 Field lines in the coplanar line **a** push–pull wave, **b** common-mode wave

Here, reference is made to the existing literature in the Annex, how these lines are to be dimensioned. There are examples shown for CPW structures, and their measured transmission bandwidths are compared (Schauwecker 1996; Janssen 1992; Zinke 1995).

5.8 Substrate Materials

The die with all of the additional components and microwave terminals is usually introduced in a housing, which consists of gold-plated stainless steel or brass. The chip itself is bonded to a substrate or carrier attached with solder (see Chap. 6) and by planar or coaxial RF lines to the corresponding sockets in the housing connected to the outside world. In addition, the chip having short electric wires (bonding wires) is connected to the intermediate carrier. A variant of the intermediate carrier are prepared in a planar form. The material to be processed should be mechanically stable and RF-capable. This is focused in the choice of the dielectric constant of the substrate material, which assures coatability with the conductor material and the bondability of the metal surface. For this purpose, the substrate height should be about the chip height to avoid long bonds. A bonding wire always forces an inductance that acts as a low pass and clearly restricts the possible range to high frequencies. As a guide, it is believed that 1-mm bond wire length has an inductance of about 0.8 nH/mm. With respect to the cutoff frequency f_c, the serial impedance is important to determine. Most of these are chosen to 50 Ω. The cutoff frequency is calculated by the following:

$$f_c = R/(2\pi L) \tag{5.24}$$

Now, when the same bond wire from a low-impedance source supplies a laser with a dynamic internal resistance of 5 Ω, the cutoff frequency is only 1/10 of the lowest possible value. PTFE substrate (polytetrafluoroethylene, Teflon) is today the most common material for use in microwave technology. Next is the RT/Duroid, while for mass applications low-cost RO3000 is used. These substrates are a PTFE resin compound with ceramic filling to adjust the dielectric constant. Another very stable material is made of alumina, which is already one of the ceramic materials. Its high-frequency characteristics are very stable to more than 100 GHz, and it is often used in expensive circuit designs. Recently, also silicon wafers are used as an intermediate carrier, since they are ideal for optical motherboard technology (see Chap. 10). The advantage is based in the very precise fiber adjustment provided in the silicon substrate V-grooves. These are realized by anisotropic wet chemical etching process with hydrofluoric acid. Herein, the glass fibers can be implemented prior to the optical chip (see Sect. 4.1.2).

Further materials are mechanically easy to work like duroplastic TMM substrates manufactured by Rogers, USA, with good dielectric parameters. Different dielectric

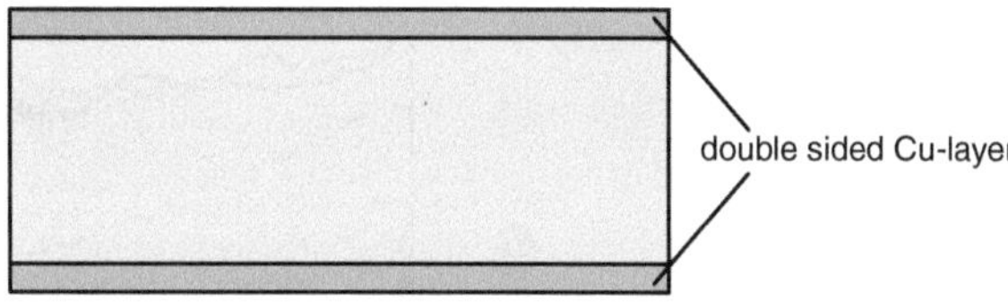

Fig. 5.11 Construction of a TMM substrate

values between $2 < \varepsilon < 10$ give a lot of freedom for the engineer. The resin, which embeds the ceramic, is a highly cross-networked carbon (Fig. 5.11).

All substrates can be conventionally processed like printed circuits up to 300 °C. Additional, these substrate materials are storage stable and can be can be provided with electrical vias.

As can be seen from Table 5.3, the individual RF properties are dependent of the respective dielectric constants. The thermal expansion coefficient of TMM fits very well to many metals, such as brass. The expansion coefficient of Al_2O_3 and silicon fits very well with stainless steel, the preferred housing material at Butterfly and DIL packages.

To give an insight into the transmission properties of the materials, the attenuation curves of the substrates discussed are shown in Figs. 5.12 and 5.13.

This can be also seen in the measured curves of Schauwecker (1996) that the transmission bandwidth for the TMM3 substrate is the best choice. With increasing dielectric constant, the bandwidth decreases. The substrate TMM10 has a strong initial attenuation of about 0.5 dB at 0.1 GHz to about 3 dB at 45 GHz, and this substrate can be used only with other line parameters.

The printed circuit parameters are the following:

$W = 200$ μm, $H = 380$ μm, $t = 17$ μm, $L = 5$ mm and titanium-gold metallization (Ti/Au) mit 20/300 nm thickness

A comparison of the results with TMM substrates of alumina ceramics (Al) and silicon substrates is depicted in (Fig. 5.13). One can recognize in the case of coated silicon substrates, very unattractive initial losses of up to 1.2 dB, which even more strongly increase at high frequencies. A suitable linearity of the transmission can be

Table 5.3 Characteristics of RF substrate materials

Properties	Si	Al_2O_3	TMM3	TMM10
ε_r	11.8	9.8	3.27	9.2
Resistivity (Ω/cm)	10	10^6	3×10^9	2×10^7
Water absorption (%)	0	0	0.12	0.2
Thermal conductivity (W/(m K) 25 °C	140	24	0.68	0.73
Thermal expansion (10^6/°C)	2.44	4.4	16	16
Surface roughness (μm)	<1	24	2	2
Specific gravity	2.33	3.75	1.78	2.77

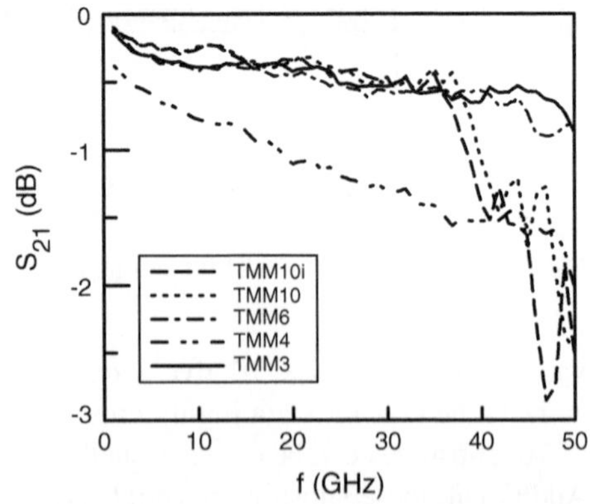

Fig. 5.12 Attenuation curves of coplanar lines on RF substrates with different dielectric constants

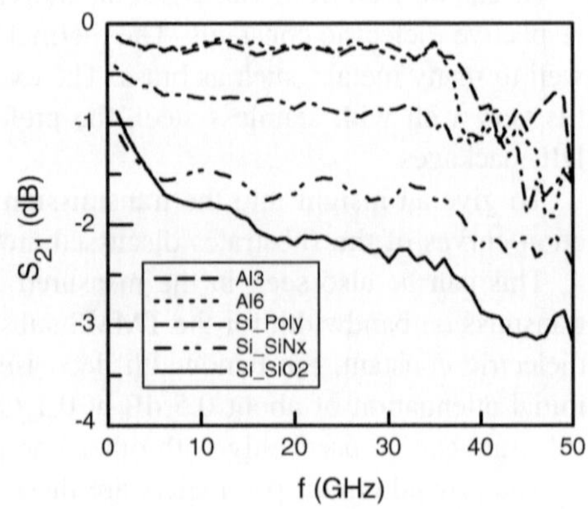

Fig. 5.13 Attenuation curves of Al-coated ceramics and silicon wafers

seen at the Al substrates. However, its range is limited sharply at about 35 GHz. This shows that the parameters of the Al substrates are not ideally adapted to high frequencies in this specific case.

The manufacture of the conductor tracks used here is discussed in more detail in Sect. 5.10. Here, it should only be noted that the measured lines have a length of 50 and 10 mm. Table 5.3 contains a summary of the properties of some substrate materials often used in the mounting technology.

5.9 High-Frequency Connection of OEICs

The transition from a planar microwave transmission line to a coaxial duct is of great importance. This transition, shown in Fig. 5.14, is the ultimate arbiter of up to which frequency limit the high-frequency circuit is used in practice.

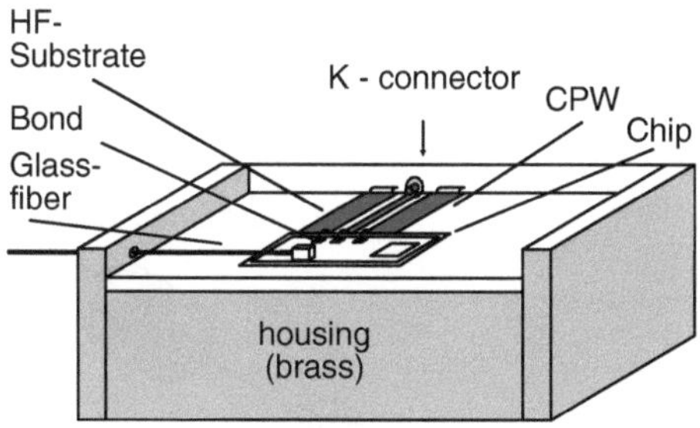

Fig. 5.14 Schematic diagram of a photodiode with optical fiber coupling and high-frequency plug connection

5.9.1 K-plug Connection

The scope of the commercially available connectors ranges from the DC area to very high frequencies in the GHz range. According to manufacturer's specifications, the K-connector (the letter representing the corresponding frequency band) is working well from the DC range up to 46 GHz. In Fig. 5.16, the modules of one of the most common types of K-connectors are shown in an exploded view in Figs. 5.15 and 5.16.

The K-connector consists essentially of three parts:

- Glass feedthrough
  The component is soldered into the housing wall (Fig. 5.15) and consists of Corning 7070 glass. The inner pin is coated with gold and has a gold-plated sheathing of Kovar. The pin diameter is 0.012 in. or 305 μm, while the sheath diameter is 0.078 in. or 1.981 mm.
- Inner conductor
  The inner conductor (center conductor) is placed over the glass bead and connecting the duct with the connector to be mounted later. The center pin is in better stability of a copper–beryllium alloy.

Fig. 5.15 Schematic representation of the connection of a coplanar line to a K-connector with glass-bead installation

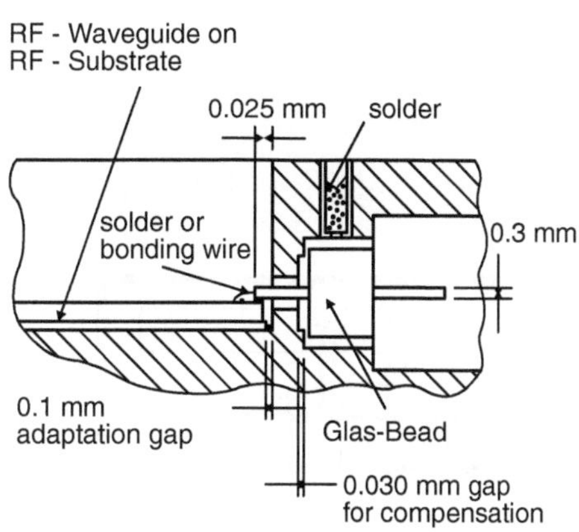

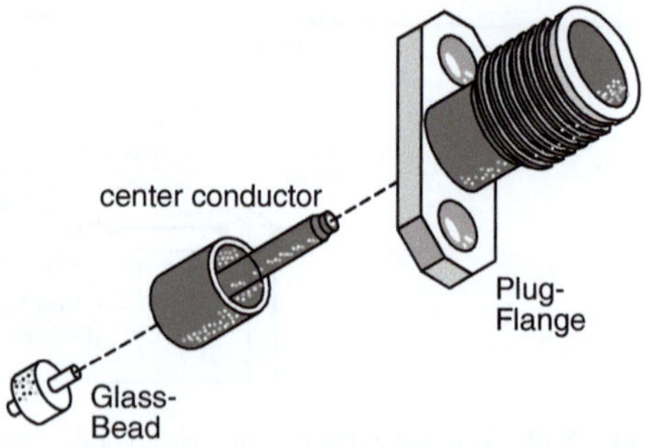

Fig. 5.16 Structure of a K-connector

- Outer conductor
  This component is mounted over the housing and is used with the connector with the mating connector (Fig. 5.16).

Other connector types are available for the range up to 65 GHz (V-plugs) and up to 200 GHz (W-plug). Their structure is not very different from that of the K-connector except that the dimensions decrease greatly. Beyond the frequency range of 200 GHz, to date, only special waveguide connections like hollow conductors are known.

5.9.2 RF Supply from the Plug to the OEIC

In Fig. 5.17, a laser module with a high-frequency terminal (RF) with a frequency range of more than 50 GHz is shown. The RF connection to the OEIC is realized by the so-called V-plug, semirigid lines, and conventional bond wires.

The module consists of a miniature optical bench, which supports the chip, the fiber optics, and the adjustment mechanism. The platform is mounted on a Peltier

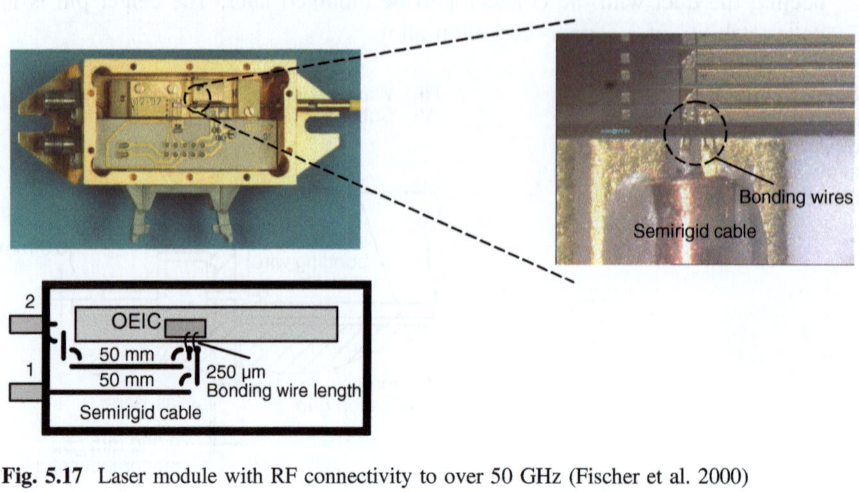

Fig. 5.17 Laser module with RF connectivity to over 50 GHz (Fischer et al. 2000)

cooling element, which provides the thermal stabilization of the OEIC. All low-frequency supplies (NF) are supplied with a standard flat plug and a thick-film circuitry. The high frequency is passed via two independent lines of semirigid cables manufactured by Suhner (EZ34 with 0.9 mm diameter and 200-μm inner conductor) via the Wiltron-V jacks directly to the OEIC without using a glass bead. This realization has the opportunity to reduce the reflection. This guarantees a very good RF connection without irregularities in the reflection characteristics.

At the interface between semirigid cable and OEIC, it is necessary to connect this interface with bonds, which must be fixed as short as possible. In the example shown, the end of the cable is fixed to an intermediate ceramic with an electric coplanar line. The coplanar RF line should be brought as close as possible to the chip. The bonding wires to the chip are about 200 μm long in the present example. In an additional module, the uporting optical bench was miniaturized (Eckhardt et al. 2000).

The results in reflectance measurements show a smooth characteristic of 45 MHz to 50 GHz: −30 dB at 1 GHz and −13 dB at 50 GHz. The attenuation also results in a smooth curve in which the curve is primarily determined by the inherent cable attenuation of 50 dB/m. At 50 GHz, the loss amounts to 3 dB. It can be expected that this mounting technology is also suitable for much higher frequencies up to 100 GHz.

5.10 Production of Coplanar Lines

The production of coplanar lines on different substrates usually follows the same process scheme. This is illustrated in Fig. 5.18 as an example for TMM3 substrates and for coated substrates (silicon substrates). The pre-treated substrate (here TMM) is spin coated with an isolating Polymide in a first lithography step and exposed, so that the interconnections with Ti/Au are passivated. In a second lithography step, the conductors are patterned for etching the copper layer. The exact sequence of processes is explained in the following sections. If you are dealing with a substrate to be coated material such as silicon, so first the insulation must be applied. This is absolutely necessary, since the silicon is not an ideal insulator, therefore a short-circuit current would flow between the leads. The protective layer may, for example, be applied by the spin coater. After then, the further structuring and vapor deposition of the substrate must be carried out.

5.10.1 TMM Substrate

Some properties of the TMM have been described in more precisely previously in this chapter, but still a few more details of the processing of this substrate material must be added to completion (Schauwecker 1996).

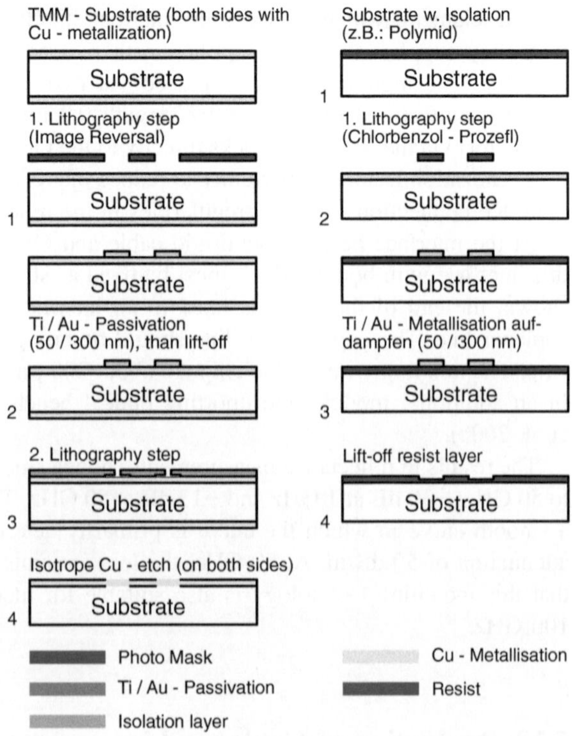

Fig. 5.18 Manufacturing process sequence for the preparation of coplanar lines on different substrates

The copper coating of the TMM substrate has typically a surface roughness of ±1.3 μm over a length of the substrate by more than 2 mm. Therefore, it is necessary to smooth the surface for further processing in the thin-film technology. For the production of coplanar lines, further processing steps are necessary in the following, which will be discussed in more detail now.

A galvanic reinforcement of gold metallization of the CPW lines must be routed. Only with these steps, a good workability of the wires with wire bonds to connect multiple RF networks can be ensured.

Process flow

It begins with a cleaning and subsequent polishing of the substrate surface. Then, the copper layer is etched off of the back and the resist mask must be removed on the polished side with acetone.

On top of the the remaining copper layer of the upper side, a thin, 50/300-nm titanium-gold layer, the "plating base," must be evaporated. Then, the first lithography step with the outgassing of the resist and subsequent structure exposure begins. A typical mask pattern is shown in Fig. 5.19. By means of masks of different width for the CPW conductor tracks, an optimum adjustment of the RF transmission function can be determined experimentally. In the present example, the inner track widths *W* were varied between 70 and 300 μm. A tapering (tapered

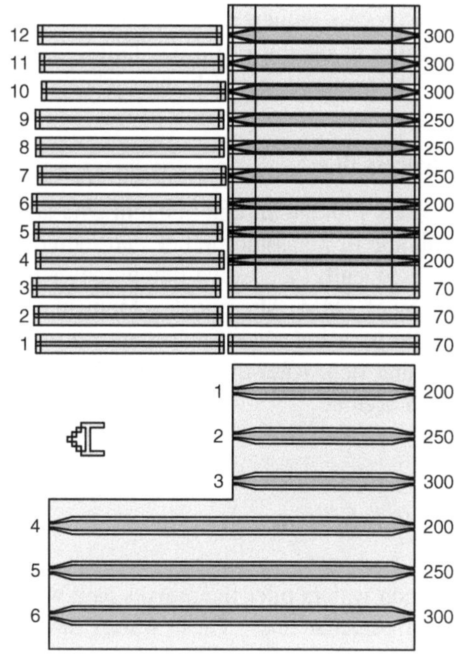

Fig. 5.19 Overview of masks and appearance of each coplanar structure. Representation of coplanar lines on the mask for TMM10 in the first lithography step

electrical waveguide) at the end of the CPW line for better adaptation to the transition to the RF connector is also clearly visible.

The exposed resist is then cured to stabilize the process on the hot plate, after that step it is exposed and at the finishing step put into the developer bath.

Afterward, the structured samples will receive a galvanically reinforced, 2-micron-thick gold layer. Thereafter, the resist mask is removed with acetone. In the following step, the Ti/Au layer will be etched over the entire surface.

In the subsequent second lithography step, the coplanar lines are fixed using a standard positive resist process including annealing and exposure followed by the development of the structure. After developing a varnish curing is introduced, and the wet chemical etching of the RF structures with a fine etching solution is carried out. After the etching, the resist layer is removed with acetone, and the resulting structures can be measured electrically and optically.

5.10.2 Alumina Ceramic

The commercially most popular RF carrier material is an alumina ceramic (Al_2O_3 ceramic). It is available as standard with substrate heights ranging from 100 to 635 μm. Their advantages lie in its excellent mechanical stability and workability. In addition, the surface quality is sufficient for the production of the RF lines.

The process sequence do not need costly galvanic reinforcements of the gold layer, because of the thick film possibilities of the process type. Therefore the processing of this material is much more cheaper than the thin-film process on TMM and silicon.

Process flow

It starts with the imprint of a thin gold layer with a height of 5 μm. Using an etching solution of potassium iodide and iodine, this layer is thinned to about 3 μm thick. Subsequently, a positive lithography step is carried out analogously to the lithography of TMM. After the development of the resist layer, the resulting resist mask is cured on a hot plate for 3 min at a heat treatment of 120 °C for the subsequent etching step. After that, the gold layer is etched and removed after the resist mask with acetone.

5.10.3 Silicon

Silicon wafers offer themselves as a well-known material for intermediate support with coplanar lines. These wafers have an excellent surface quality for thin-film applications, and they are very inexpensive commercially available. They also enable microstructuring by simple anisotropic etching processes. Thus, it is possible to introduce the material into the V-grooves in which the optical fiber can be guided and positioned in front of a same OEIC. More detailed information on this topic is listed in Chap. 10.

However, strong limitations of the RF capability over 20 GHz also result from the high dielectric constant ε of 11.8. This value forces very narrow inner conductor width W and a very small spacing S of the coplanar lines. Errors in the manufacturing accuracy and additional edge roughness on the conductor edges result in a disruptive influence and thus an additional attenuation of the RF transfer function. Therefore, a very complex process control is necessary that drives the cost of such devices heavily.

Furthermore, it is required of RF substrates that the surface resistance should be in the MΩ range per cm. High-purity silicon is only up to 10 kΩ per cm available on the market. Therefore, an additional process step is necessary. Between conductor and substrate, a non-conductive layer has to be added, which may consist of polyimide, SiO_x, and SiN_x.

Process flow

Pyralin, polyimide, and the SiNx insulating layers are applied in a PCVD process, by chemical vapor deposition. The SiO_x layers can be deposited sufficiently thick in the simpler sputtering. Subsequently, the resist mask for the positive lift-off process can be applied to the real dimensions of the coplanar lines, vapor-deposited Ti/Au metallization, and lifted the resist mask.

In another manufacturing variant, the insulation layer is a thin, approximately 150-nm-thick Ti/Au layer sputtered onto the whole area and reinforced with a gold plating of about 2 μm in thickness. Thereafter, an isotropic etching process in a solution of potassium iodide and iodine can remove the resist.

References

Eckhardt, T., Fischer, U.H.P., Ziegler, R.: DIL-size reusable modules with up to 50 GHz modulation bandwidth for optical communications systems. In: Faulkner, D.W., Harmer, A.L. (eds.) WDM and Photonic Networks, pp. 181–184. IOS Press, Netherlands (2000)

Fischer, U.H.P., Peters, K., Ziegler, R., Pech, D., Kilk, A., Eckhardt, T., Mekonnen, G.G., Jacumeit, G.: New reconfigurable fiber-chip coupling method for multipurpose packaging with up to 50 GHz modulation bandwidth. Opt. Fiber Technol. **6**, 68–73 (2000)

Janssen, W.: Stufenleiter und Hohlleiter, p. 268. Hüthig-Verlag, Heidelberg (1992)

Schauwecker, B.: Realisierung, Konzept und Optoelektronischen. Technical University of Berlin, Aufbau-und Verbindungstechnik eines höchstfrequenten Empfängerschaltkreises (1996)

Zinke, B.: Hochfrequenztechnik I, p. 492. Springer, Berlin, Heidelberg (1995)

Chapter 6
Soldering, Adhesive Bonding, and Bonding

Abstract This chapter discusses and analyzes the bonding of OEIC and its electrical and thermal environment. First, the bonding interface between the OEIC and the heat sink would be investigated. In the following, failure mechanisms and reliability tests are discussed. Adhesive bonding takes more and more place in the electrical packaging especially for flexfoil and thermal sensitive chips. An overview of different gluing mechanisms is given. Finally, the different types of wire bonding techniques are depicted and compared in conjunction with practical examples for the application with RF connections.

6.1 Die Bonding

To build up laser on heat sinks, several factors makes it difficult due to

- Mechanical,
- Thermal,
- Optical, and
- Electrical problems.

The laser threshold depends on the laser temperature:

$$I_{\text{th}} = I_{\text{th0}} \cdot \mathrm{e}^{\left[\frac{T - T_{\text{ref}}}{T_0}\right]} \tag{6.1}$$

T_0 characteristic temperature

To prevent the threshold from increasing, the thermal load of the laser has to be minimized. Also, a good thermal contact between the heat sink and the OEIC is needed to achieve a heat flow of 4 kW/cm^2.

Different types of solder pastes are in use, which ensure a good thermal contact and mechanical stability as well as the ability to absorb most of the thermo-mechanical stress. The thermo-mechanical stress occurs between different materials due to its different coefficient of thermal expansion. The following solder materials are used:

© Springer-Verlag Berlin Heidelberg 2015

U.H.P. Fischer-Hirchert, *Photonic Packaging Sourcebook*,
DOI 10.1007/978-3-642-25376-8_6

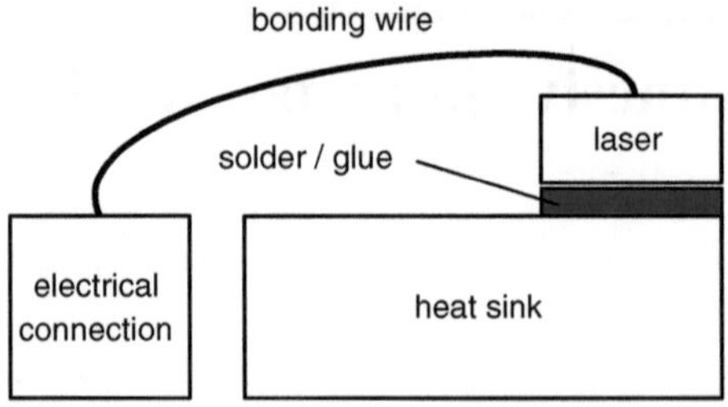

Fig. 6.1 Principle setup for die bonding

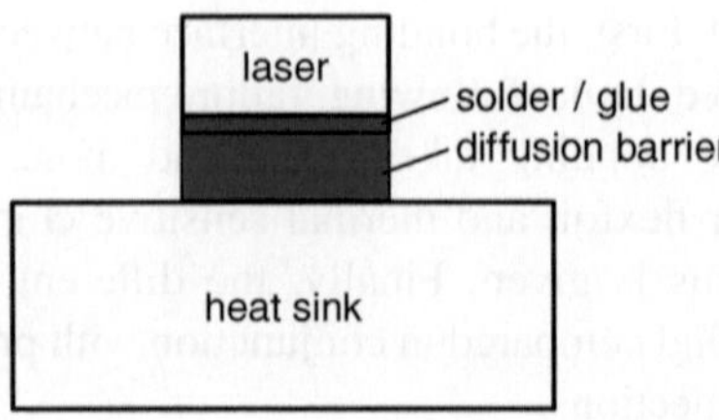

Fig. 6.2 Diffusion barrier

- Indium (In)
- Tin (Sn)
- Tin–lead alloys (Sn/Pb)
- Gold–tin alloys (Au/Sn) in eutectic composition, which is currently the best solution

To separate the die and the bonding material, a diffusion barrier layer is applied. These prevent the die of diffused solder material, which is a common property of the solder pastes (Fig. 6.1).

Gold diffuses at room temperature relatively quick into the heat sink or the InP layer. Figure 6.2 shows the schematic system of the diffusion barrier. Therefore, nickel and platinum are used as diffusion barrier layers. A thin layer of gold is used to avoid the oxide growing. Otherwise, common methods to achieve a good solder flow are the use of formic acid, flux, or hydrogen reduction at temperature above 300 K.

6.2 Heat Sinks

For heat sinks, it is necessary to use materials with a similar CTE (coefficient of thermal expansion) compared to the die material. Materials are, e.g., copper, silicon, Kovar, and diamond in case of InP, which can be seen in Table 6.1. To ensure a good thermal contact, the surface of the heat sink should be lapped (thinned). The roughness of the surface should be less than 100 nm. A popular heat sink in

Table 6.1 Material coefficients in comparison for die bonding

Material	Thermal conductance in W/(mK)	Expansion coefficient in 10^{-6}/K
InP	67	4.56
InGaAs	66	5.66
GaAs (doped)	44	6.4
GaAs (undoped)	44	5.8
AlGaAs	11	5.8
AlGaInP	6	–
GaInP	5	–
In	81.8–86	29–33
Sn	64–73	19.9–23.5
80Au20Sn	57.3	16
77.2Sn20In2.8Ag	54	28
60Sn40Pb	44–50.6	24.7
88Au20Ge	44.4	12.9–13.3
52In48Sn	34	20
97Au3Si	27.2	12.3
5Sn95Pb	23–35	28.4–29.8
Heat sink materials		
Diamond (Type I)	2000	0.8
Diamond (Type IIa)	4500 bei 200 K	–
CVD diamond	1000–1600	2
Ag	427	19
Cu	398	16.5
Au	315	14.4
Silicon carbonate	193–250	2.3–3.7
15Cu85W	240	7.5
4Cu6Ni90W	230	5.4
BeO	220–260	6.5–7.3
30Cu70W	201	10.8
Al_2NO_3	170–200	4.3
W	178	4.5
Silvar	153	6.5
10Cu90W	147–209	6.5
Si	125–150	2.6–4.1
Mo	115–140	5.4
Ni	90	13
Kovar	16	5.9
Stainless steel	16	17.3
Invar	12	0.9
SiO_2	1.2	0.6

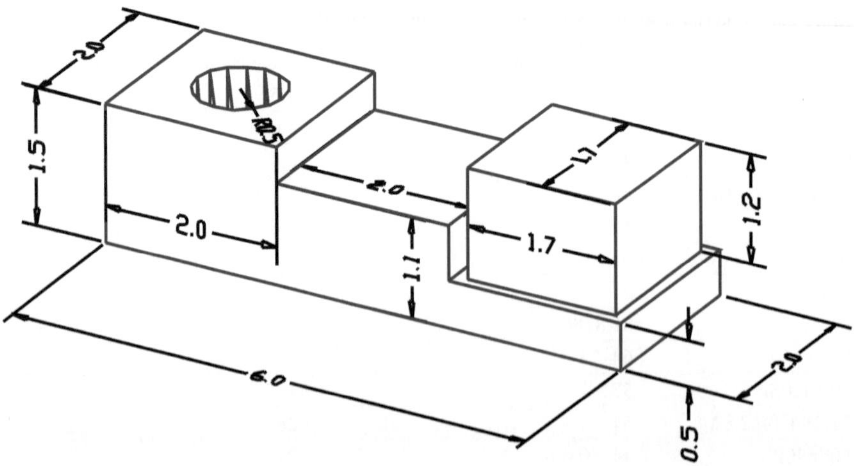

Fig. 6.3 Kyocera heat sink

industry and science is made by the company Kyocera (Fig. 6.3). It has the following dimensions: 2 mm × 2 mm × 4 mm, with a gold-plated screw hole.

Table 6.2 shows the common heat sinks and its appropriate solder systems. It is important that highly thermally stressable OEICs such as lasers should be lapped very thin and soldered on a heat spreader like diamond. This procedure leads to a better heat transfer. The OEIC has to be solderable, which can be achieved by vapor deposit of one or more contact layers.

The suitable combination for metallization of In/P and Ga/As material systems can be looked up in Mickelsen et al. (1997) and Kuhmann (1996). In particular, the metallization layer of Ti/Pt/Au turned out to be suitable for quaternary In/P systems.

Table 6.2 Materials for heat sink in combination with metallization

Heat sink	Adhesion layer	Diffusion barrier	Intermediate layer	Solder layer
SiO_2, ceramics or diamond	100 nm Ti	300 nm Ni oder W or Cr		
Cu	100 nm Ti	50 nm W	0.2–2 μm Ni, Sn, W	Ni, Sn
Cu	100 nm W			50 nm Au
Cu	100 nm Ni			50 nm Au
Semiconductor component				
InP	300 nm Ti	500 nm Pt		300 nm Au
InGaAsP–InP	50–100 nm Ti	50–100 nm Pt		50 nm Au
GaAs		20–100 nm W	50 nm Au oder Pt	
AlGaAs–GaAs	Cr–Au			2.5 nm In

6.3 Failure Mechanisms

Soft solders are preferably used to reduce thermo-mechanical stress. Inside the solder, a slow deformation called creeping occurs which causes a lift-off failure of the die and thus severely reduced the lifetime of the system. Also, Tin can creep on the laser facet and lead to whisker barriers. Hard solders show a lower susceptibility to whisker growth and creeping. The main disadvantage is the high solder temperature, which leads to higher thermo-mechanical stress inside the die and therefore to early failures.

Figures 6.4 and 6.5 show a common packaging method. It uses two different solders. Solder 1 has a low melting point and is applied to the die. Solder 2 has a high melting point and is deposited on the heat sink. It starts with melting the solder # 1. Then, the liquid solder 1 is applied to the solder # 2. The diffusion leads to a nearly stressless bond between the die and the heat sink. In addition, it is not necessary to polish the surface, which reduces the cost of production.

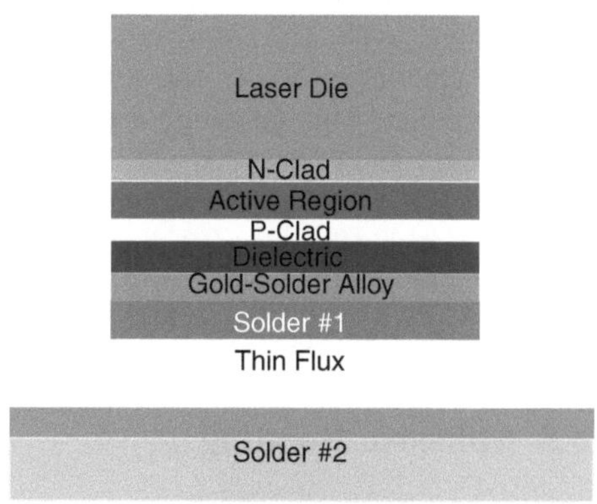

Fig. 6.4 Diffusion barrier

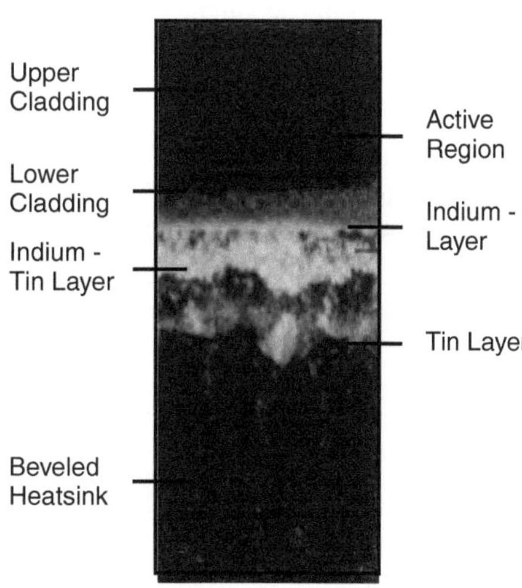

Fig. 6.5 Diffusion barrier (photograph)

6.4 Reliability Tests

The lifetime of OEICs is an important value, which has to be determined for every packaging solution. For example, it is necessary for the safety and reliable installation of undersea cable. The main variable to predict the lifetime is the mean time to failure (MTTF, which describe the average time to failure). The failure rate can be described mathematically by an Arrhenius plot:

$$L = C \cdot e^{\left[\frac{E_a}{kT}\right]} \tag{6.2}$$

E_a activation energy
k Boltzmann constant
T absolute temperature in Kelvin

The activation energy can be extracted by plotting the logarithm of the measured value against the inverse temperature. The activation energy is equal to the slope of the resulting graph. The failure criterion for laser is the raise of the threshold current over the amount of 50 %. The tests are accelerated by the use of high operation temperature and high operation current (e.g., $T = 100$ °C, $I = 20 \cdot I_{th}$).

Figure 6.6 shows a related test setup. Several OEICs are attached to a single heat sink, which is heated to a certain temperature. The results are depicted in Fig. 6.7. It can be seen that the strongest changes occur in the first few hours of the accelerating life test.

6.5 Adhesive Bonding

All OEICs without thermal load such as photodiodes can be fixed with adhesive. It results in a easier handling but leads to a shorter lifetime due to absorption of humidity.

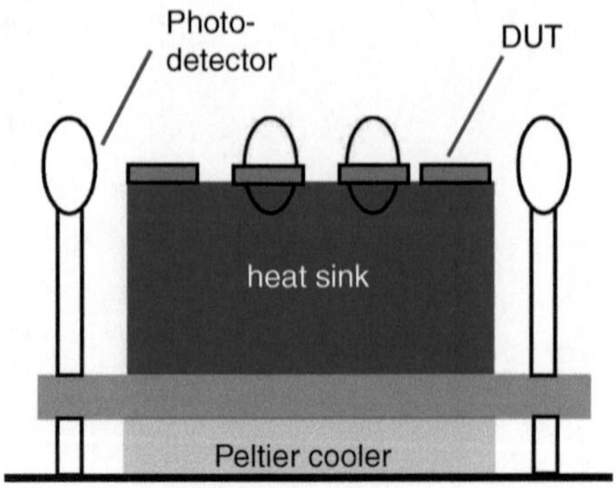

Fig. 6.6 Setup for a reliability test station for laser diodes

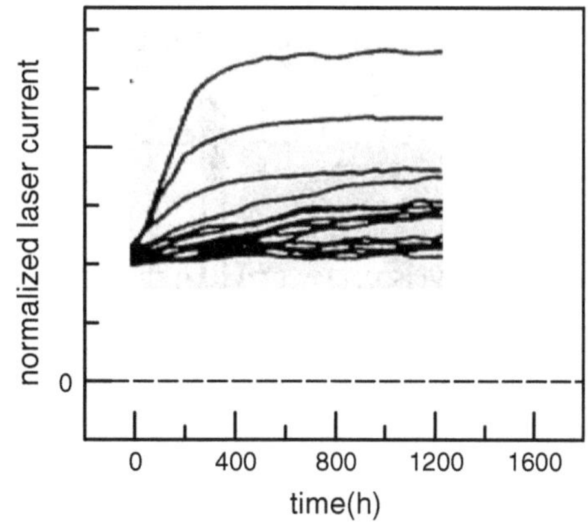

Fig. 6.7 Degradation behavior or typical laser diodes

Table 6.3 Application areas of different glue types

	Metal	Glass	Plastic
Metal	Cyanoacrylate	Two components	Acrylate
Glass	Two components	Two components	Two components
Plastic	Acrylate	Two components	Cyanoacrylate

It is necessary to choose the right kind of adhesive because they are addressed to different material combinations (Table 6.3). The adhesives have to fulfill certain conditions to obtain a successful bonding:

- Tempering
- Cleanliness of the surface
- Roughness of the surface
- Shrinking of the curing

Types of adhesives: (Table 6.4)

- Electrically conductive (with silver particles)
- Electrically non-conductive
- Thermally conductive for bonding of Peltier elements, heat sinks
- Underfiller/globetop for environmental protection
- Isotropic conductive adhesives (ICA)
- Optically transparent

6.6 Wire Bonding

Wire bonding (Reichel 1999) is one of the oldest techniques to electrically bond the semiconductor components. A thin wire bridge made out of aluminum or gold is welded at its ends to the bonding pads of the chip and the chip carrier. The bond is

Table 6.4 Different types of adhesives; TG glass temperature, RT room temperature, η viscosity, n refractive index, R electrical resistance

Designation	Application	TG (°C)	Pot life by RT	Hardening (°C/min)	η at 25 °C (cps)	n	R (W cm)	Specification particle size: Ø/max.
Polytec: H 20 E	Derive chip bonding electrically	85	2–4 days	150 °C/15 80 °C/12 h	30,000		1×10^{-4} to 4×10^{-4}	Filler: Ag, pasty, 20/50 µm
H 70 E	Chip substances thermally conductive bond	85	2–4 days	150 °C/15 80 °C/2 h	20,000		7×10^{-4}	Filler: Al_2O_3, pasty, 19/60 µm
E-2036	SMT calves electrically conductive	80	3–4 days	150 °C/30 125 °C/60	7500		0.8×10^{-4} bis 3×10^{-4}	Filler: Ag, pasty, 3/19 µm
415 G	SMT calves electrically conductive	54	30 min	RT/12 h	200,000		7×10^{-4}	Filler: Ag, pasty, 20/50 µm
Epo-Tek 302	Optical	50	5 min	RT/1 h	4000	1.566		Adheres to glass, quartz, ceramic, plastic
Epo-Tek 310	Optical	<RT	90 min	65 °C/120 RT/24 h	880	1.5071		Stick to glass and metal, vacuum-proof, flexible
Epo-Tek 905	General casting technique	65–70	1.5 h	60 °C/120 RT/24 h	150,000			Low shrinking filler: Al_2O_3, 19/60 µm
Kulzer: Technovit 4000	Cold-polymerized plastic in the form of syrup and powder	4 min	RT/8					Adheres to metal, shrink: 0.2 %
Panacol: Vitralit 6181	Optical			UV-A, 2 60 mW/cm²	5000	1.51		8.5 % shrink, −40 °C to +150 °C UV: 50 mm distance D = 150 µm

(continued)

Table 6.4 (continued)

Designation	Application	TG (°C)	Pot life by RT	Hardening (°C/min)	η at 25 °C (cps)	n	R (W cm)	Specification particle size: Ø/max.
Vitralit 6129	Optical			UV-A: 15 s 120 °C: 30 min	30,000			2.2 % shrink, −50 °C to +180 °C for glass, plastic, Al, ceramic
Cyanolit 101			10 s (Cu)	60 s (Al) 6 or 12 (5 to −25 °C)				Add gap <100 μm
Cyanolit 102			20 s (Cu) 80 s (Al)	6				
Cyanolit 202			13 s 40 s (PA) (PVC)	6				
Cyanolit 303			15 s	6				$T > 150°$
Doduco Auromal K1 A + B	Conductive adhesive		3 h	RT/30 h 80 °C/3 h	100,000		$<10^3$	$T > 150°$

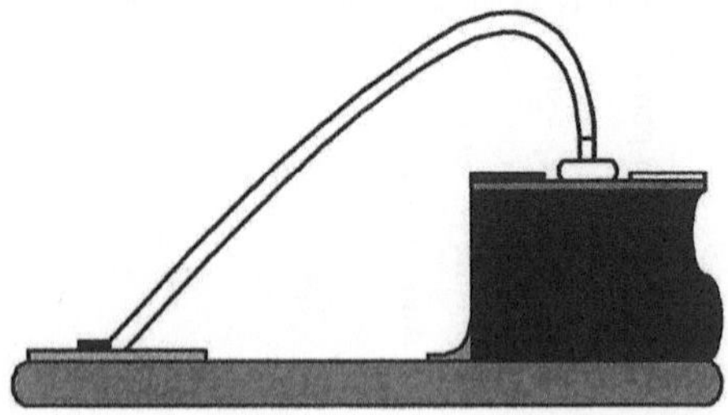

Fig. 6.8 Ball–wedge bond

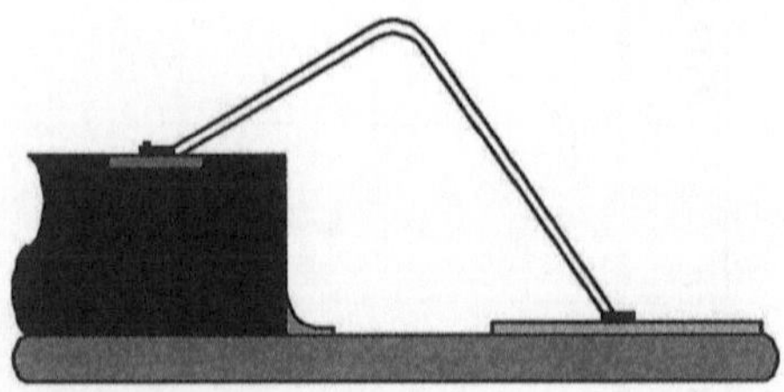

Fig. 6.9 Wedge–wedge bond

done without melting of the materials; that is why wire bonding belongs to the cold welding processes. Wire bonding methods can be classified according to the following criteria:

- Wire guiding (Figs. 6.8 and 6.9)
  - Wedge–Wedge
  - Ball–Wedge
- Kind of energy input
  - Thermo-compression bonding (TC, temperature and pressure)
  - Ultrasonic (US, Ultrasonic and pressure)
  - Thermosonic bonding (TS, temperature, ultrasonic, and pressure)
- Wire gauge and shape
  - Fine pitch wire bonding (∅ < 20 μm)
  - Standard pitch wire bonding (∅ 25.50 μm)
  - Thick wire bonding (∅ 100.625 μm)
  - Ribbon wire bonding (rectangular 50 μm × 30 μm)

6.7 Thermo-compression Bonding

The process is depicted in Fig. 6.10 (clockwise). The wire is guided to the end of tube of the bonding tool. An electrical discharge heated the bond tool, and thus, the wire melted and formed to a ball. Then, the ball is lowered and pressed to the bond pad.

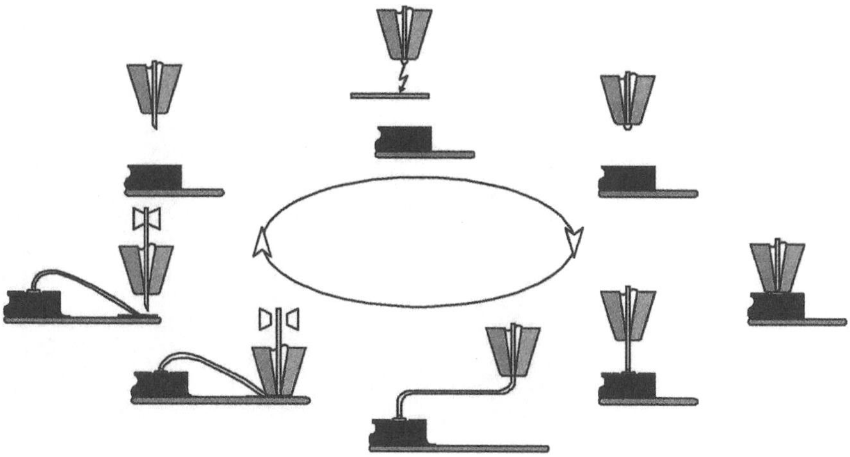

Fig. 6.10 Cycle of thermo-compression bonding

The high temperature of the bond tool and the pressure leads to a cold welding. In the next step, the bond tool is moved to the next bond pad and pressed to it as well. Last step is the shear of the wire. The result is an electrical bond between two pads.

Process parameter (Gold wire with 25 μm diameter)

- Bond temperature: 280–350 °C
- Bond pressure: 30–90 cN (higher pressure for the ball)
- Bond time: 20–200 ms

An example for a thermo-compression bonding of GaAs semiconductor component is shown in Figs. 6.11 and 6.12. The typical flat-pressed ball with the hemispherical shape can be seen in the figure.

Fig. 6.11 Ball bond

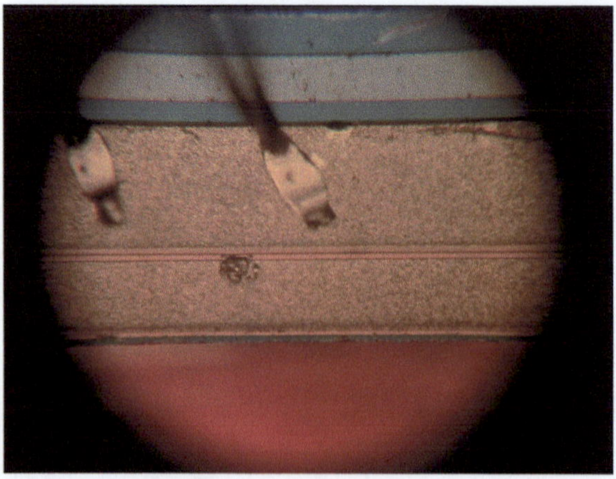

Fig. 6.12 Wedge bond

6.8 Ultrasonic Bonding

Figure 6.13 shows (clockwise) the process of the ultrasonic bonding technique. The bond wire is fed out obliquely at the end of the bonding tool tube. In the next step the bonding wire is guided downward and pressed down to the bond pad. The energy transmitted by the induced ultrasonic and the pressure leads to a cold welding between the wire and the bond pad metallization.

Then, the bond tool is moved to the next bond pad, and the same procedure is done to achieve a cold welding. To finish the electrical bond, the wire is sheared by the bonding tool.

This method enables the possibility of shorter bonding distance compared to the thermo-compression bonding method. That is why only a small curve of the bond

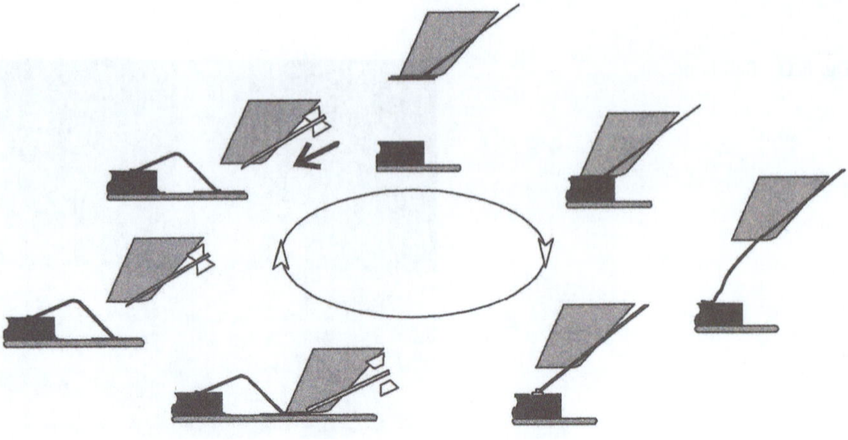

Fig. 6.13 Cycle of ultrasonic bonding

wire is necessary. A typical wedge bond is depicted in Fig. 6.12. The flat pressing of the wire and the small curvature of the bond can be seen.

Process parameter for a 25 μm wire:

- Bond temperature: Room temperature
- Bond pressure: 30–90 cN (higher pressure for the ball)
- Ultrasonic power: 100–500 mW
- Bond time: 20–200 ms

6.9 Thermo-sonic Bonding

Applying thermal and sonic energy does the cold welding during the thermo-sonic bonding. Therefore, similar process parameters compared to the other described bonding methods are used.

Process parameter (Gold wire with 25 μm diameter)

- Bond temperature: 0–250 °C
- Bond pressure: 30–90 cN
- Ultrasonic power: 100–500 mW
- Bond time: 20–100 ms

6.10 Bonding Tools

Table 6.5 shows the characteristics of commercially available automatic wire bond machines. A complete Wedge–Wedge wire bonder from the company Mech-EL-Industries can be seen in Fig. 6.14. It is an ultrasonic wire bonder for aluminum wires with 25 μm diameter. Additional, aluminum wires with the dimensions of 25 μm × 75 μm can be used.

The advantage of this wire bonder is low operating temperature of the die. This reduces the thermal stress. Sensitive optoelectronic components should not be exposed to temperatures above 100 °C.

Table 6.5 Characteristic data of for automated bonding machines

x/*y*-resolution	5 μm
Position accuracy	±5 μm
Rotational angular resolution	0.2° per step
Image storage	Up to 200 picture
Naming speed	About 60 ms
Input	About 2 kW
Throughput (ball–wedge)	6–8 Loops/s
Throughput (wedge–wedge)	Approximately 2 Loops/s

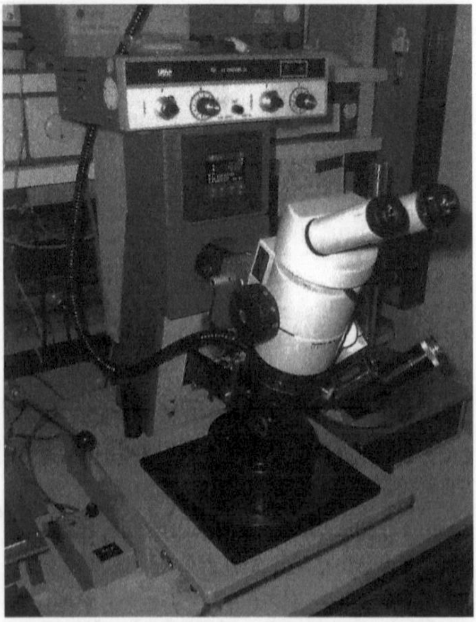

Fig. 6.14 Ultra sonic bond machine

The bond tool guides the bonding wire and then connects via high pressure and sonic or temperature the bonding wire to the bonding pad, which is depicted in Fig. 6.15. It is a masterpiece of precision mechanic engineering. The 15- to 25-µm-thick gold wire is guided nearly frictionless through the bond capillary tube.

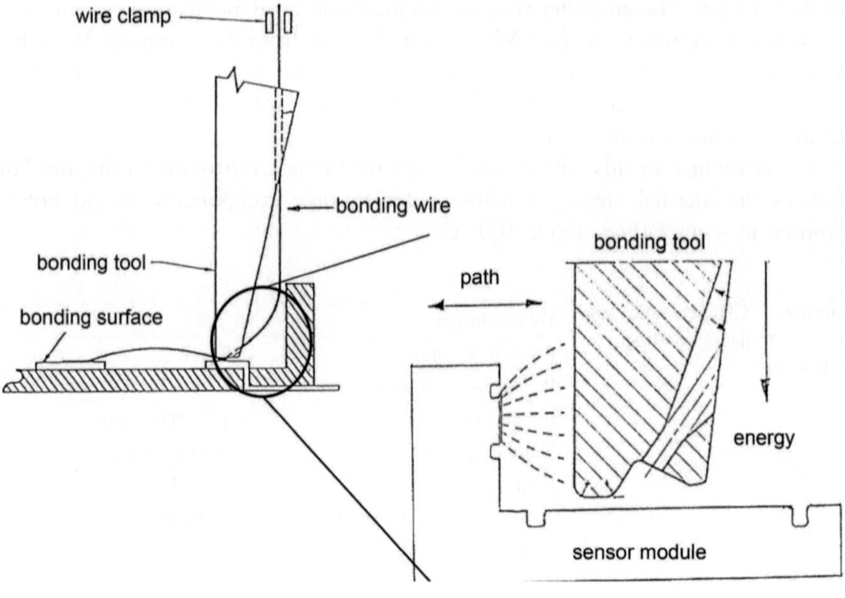

Fig. 6.15 Bonder head

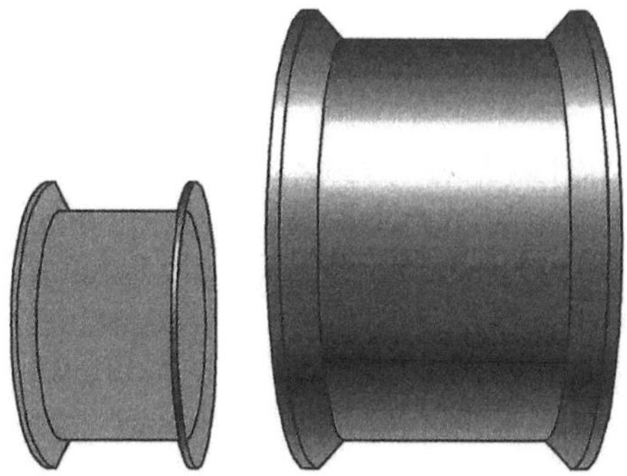

Fig. 6.16 Spooler for bond wires

The capillary tube also centered the wire. The energy to establish a bond has to be transmitted to the tip of the bonding tool. The amount of energy can be implimented by temperature and ultrasonic power in the range of several milliwatts. At the same time, a force sensor detects the applied pressure during the bond process. Therefore, it is necessary to measure the correct amount of ultrasonic energy to ensure a robust cold welding between the wire and the bond pad.

The appropriate wires are delivered on spools. The diameter for gold wires is between 17.5 and 50 μm by default. To enhance the hardness and the temperature resistance of the gold wire, it is alloyed with beryllium. Aluminum wires have a common diameter between 20 and 100 μm. The fatigue strengths can be improved by alloying the aluminum with magnesium. These wires are available up to 500 μm thickness. For some applications, e.g., signal transistors, a special bonding process with the use of palladium wires can be more economical. Compared with gold, palladium shows a higher mechanical strength but also a higher hardness and higher electrical resistance. For cheap productions, copper wires are a possible alternative to gold wires. Besides low material costs and a good electrical conductivity, copper possess a high temperature resistance. It is possible to use pure copper, thus no intermetallic phases occur which could reduce the mechanical properties.

The delivered wires are coiled upon spoons (Fig. 6.16)that have a diameter of 10–100 mm and a width between 20 and 120 mm. The length of the wire is between a few meters up to 100 m.

References

Reichel, H.: Lecture script Microsystems Technology II, p. 120. Technical University of Berlin, Berlin (1999)

Kuhmann, J.F.: Untersuchungen zu einer flußmittelfreien und selbstjustierenden Flip-Bondtechnologie für photonische Komponenten. Ph.D. Thesis, TU-Berlin (1996)

Mickelsen, A.R., Basavanhally, N., Lee, Y.-C.: Optoelectronic Packaging, p. 2612. Wiley, New York (1997)

Chapter 7
Optical Connection Technology

Abstract Optical connectors are used in a widespread application range, from low-cost automotive networks with polymeric fibers to multi-fiber single-mode connectors in optical glass fiber core networks. In this chapter, either single-fiber or multi-fiber connectors are described. Different types of single-mode and multi-mode connectors are currently available on the market. A description about the used types is included. In the second section of the chapter, the production and quality control of optical fiber tapers are represented.

7.1 Single-Fiber Connectors

For single-fiber connector, a distinction is made between straight- and helical-cut end facets of the connectors and between different connector types for specific applications:

- Polarization-maintaining fibers
- Matching gel/films
- Physical contact
- Air gap
- Automatic light obscuration
- Duplex connector

All losses between two optical waveguides may occur in the direct fiber-to-fiber coupling in a plug connection (Fig. 7.1). They can be largely avoided through precise guides and special surface treatment. Intrinsic losses can almost entirely be excluded by an appropriate choice of the fibers to each other. Extrinsic losses due to misalignment can be largely avoided by careful management elements. Nowadays, guide elements are commercially supplied with a precision of 2 μm. Absorption and scattering on the surface of the connector can be avoided by cleaning and polishing of the surface.

© Springer-Verlag Berlin Heidelberg 2015

U.H.P. Fischer-Hirchert, *Photonic Packaging Sourcebook*,
DOI 10.1007/978-3-642-25376-8_7

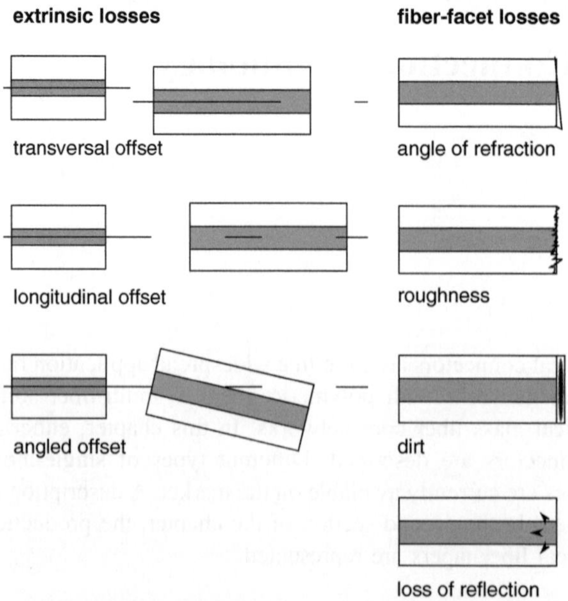

Fig. 7.1 Losses at the plug-to-plug coupling

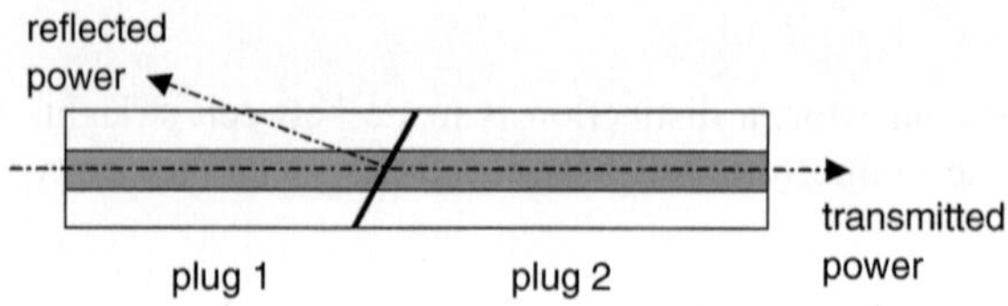

Fig. 7.2 Principle of an angled connector

Reflections of the connector end faces are avoided by the application of anti-reflective coatings or by tilting the end faces to each other at an angle of 5°–7° as generally applied in the industry (see Fig. 7.2). Tilting is applied to reduce the back reflections to the transmitter side down to −55 dB. The setup of the connectors is very similar in all solutions: The fiber is exactly centered and glued in a ceramic ferule. Almost all plugs of this ferrule have an outer diameter of 2.5 mm. Figure 7.3 illustrates the sectional view of a plug inner life. The nowadays most commonly used connector types are shown in Table 7.1.

For a more detailed study of the inner connector setup, I recommend the book written by Eberlein (2000). The German Telekom prefers the connector E2000 for the field use (Figs. 7.8 and 7.9), because of the high return loss. Another reason for the preferred use of the E2000 is the mechanical design, which leads to an automatically isolation of the light from the end of the connector when the plug connection is separated and provides a reliable dust cover.

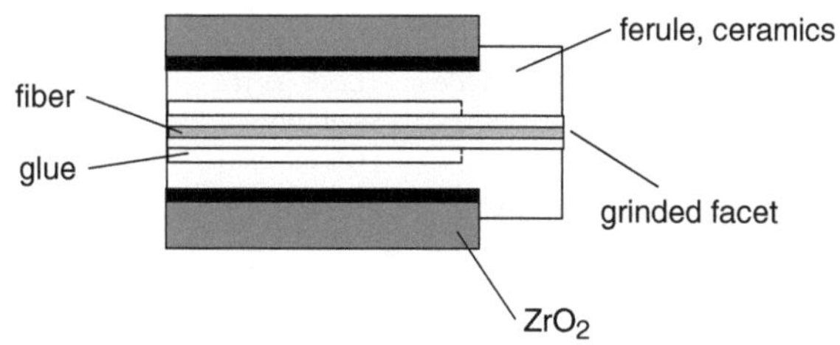

Fig. 7.3 Inner setup of a connector

Table 7.1 Single-fiber connectors

Type	Losses (dB)	Straight cut (dB)	Angular cut (dB)	Single(SM) Multi(MM)-mode	Polarization maintaining	Long-term stability (insertions)	Price
Mini BNC	0.21	–20		MM		>500	Low
ST	0.28	–20		SM, MM		>500	Medium
FC/PC	0.35	–30		SM, MM		>500	Low
FC/APC	0.4		–55	SM		>500	Medium
SMA	0.38	–20		MM		>500	Low
Radiall VFO		–30		SM	yes	>250	High
	0.7		–55	SM		>250	High
Radiall EC	0.2		–50	MM		>250	Medium
	0.5		–60	SM	yes	>250	Medium
Diamond E2000	0.18	–30		SM		>1000	Low
HRL-10	0.18		–55	SM	yes	>1000	Low
	0.3		–60	SM		>1000	High

Moreover, the price of the plastic connector is low, so you can assume that it will prevail in the coming years compared with the more conventional FC/PC connectors on the market. FC/PC connectors are mainly supplied by Japanese manufacturers as standard equipment, but have the disadvantage of being very quickly polluted and also to ensure only low back reflection loss.

7.2 Multi-mode and Multi-fiber Connectors

The multi-mode connector (see Table 7.2) is only common in the area of transmission technology for FDDI and computer linkage for Ethernet. Here, the data are transferred in duplex mode (one fiber for round-trip transportation). The

Table 7.2 Multi-mode connectors (SM Single-mode, MM Multi-mode)

Type	Losses (dB)	Straight cut (dB)	Mode type	Long-term stability (insertions)	Users	Price
ESCON	0.5	−20	MM	>500	LAN, ethernet	Low
FDDI M	0.5	−20	MM	>500	WDM	Medium
	0.5	−20	SM	>500		
MTConnect	0.3	−25	MM	>200	WDM	High
	0.5	−25	SM	>200		

commercially available connectors are listed in the Table 7.2. The MTConnector (Fig. 7.19) is available with 4, 8, 12, and 24 fibers in its housing. It consists of a silicon base plate, in which engraved V-grooves are situated, realized by chemical wet etching at intervals of 250 μm. The high precision of the etching technology ensures an alignment of the fibers to each other, which is for the complete fiber connector better than 1 μm.

7.3 Examples of Optical Connectors

For a better overview, all common connectors and associated couplings are shown in Figs. 7.4, 7.5, 7.6, 7.7, 7.8, 7.9, 7.10, 7.11, 7.12, 7.13, 7.14, 7.15, 7.16, 7.17 and 7.18. The figure shows the connectors together with the corresponding coupling.

The representations are complemented by photographs of bare fiber adapters for the connection of an untreated connector end with a meter or another connector.

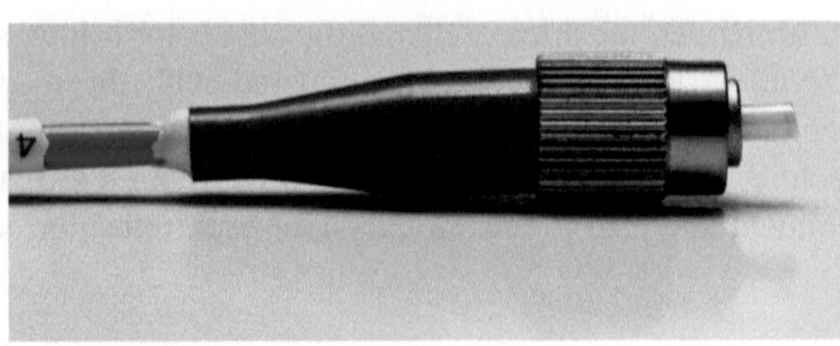

Fig. 7.4 FC/PC plug

Fig. 7.5 FC/PC coupling

Fig. 7.6 Radiall EC plug

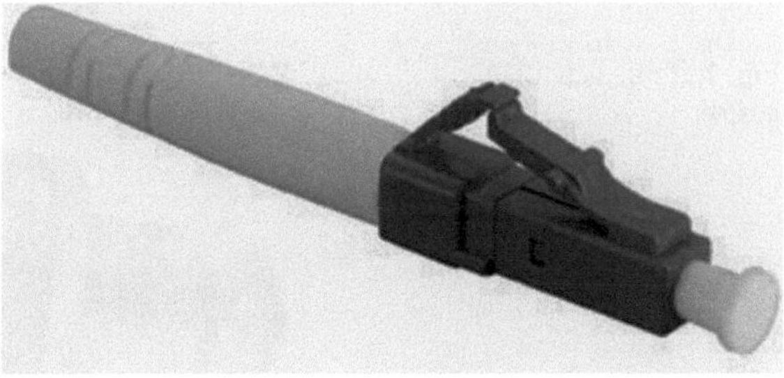

Fig. 7.7 LC-plug

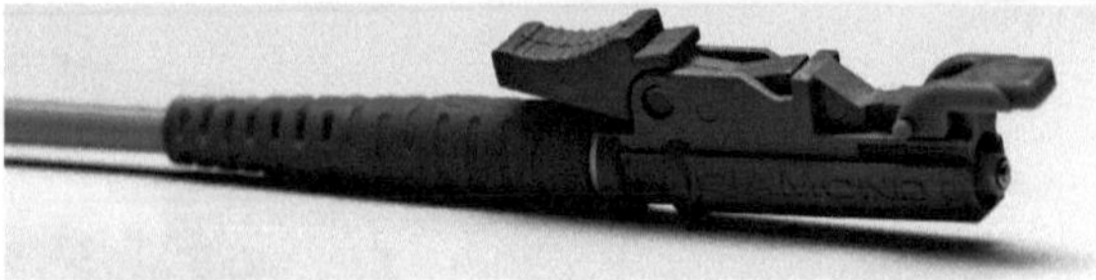

Fig. 7.8 Diamond E2000-connector

Fig. 7.9 Diamond E2000-plug

Fig. 7.10 Radiall VFO-plug

Fig. 7.11 Radiall VFO adapter

Fig. 7.12 DIN plug

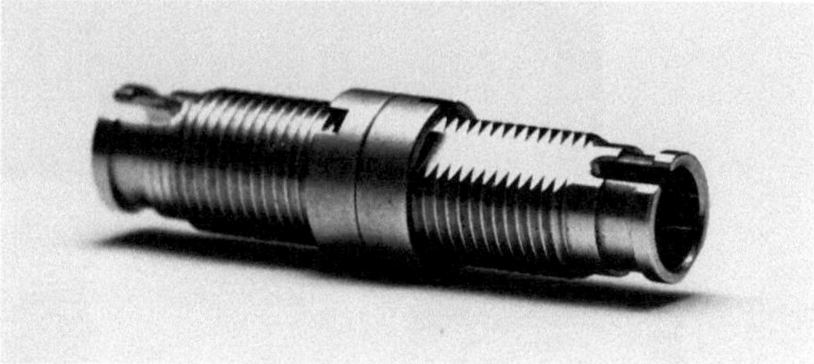

Fig. 7.13 DIN adapter

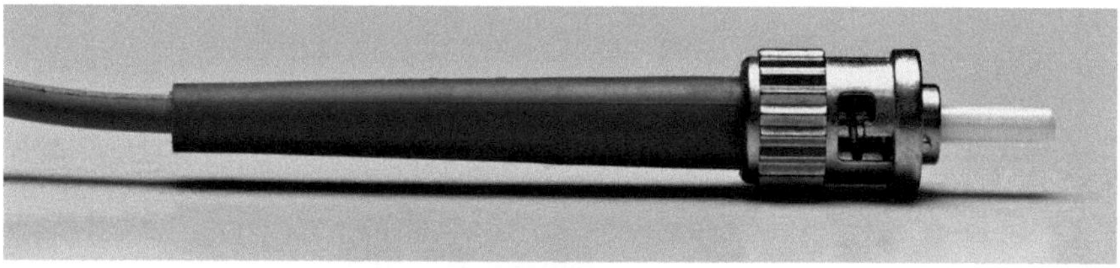

Fig. 7.14 ST plug

Fig. 7.15 ST jack

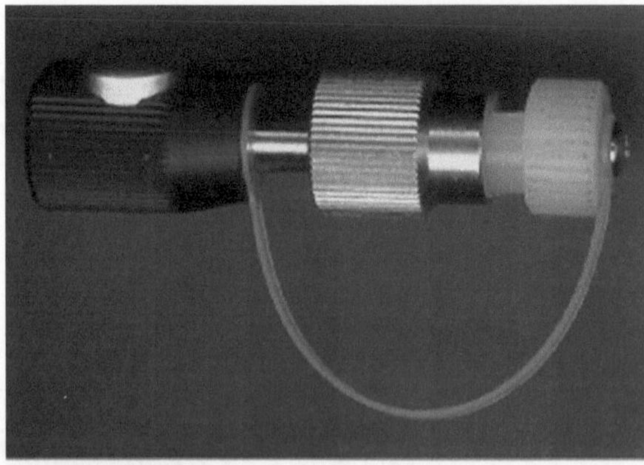

Fig. 7.16 Bare fiber adapter

Fig. 7.17 SC plug

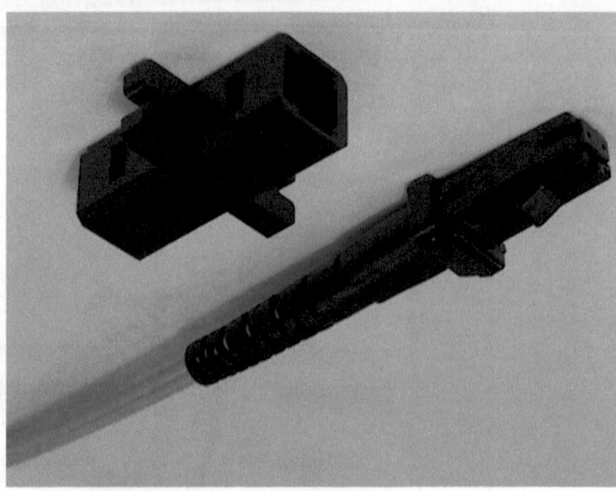

Fig. 7.18 MTConnector

7.4 Optical Fiber Tapers

7.4.1 Manufacturing of Fiber Tapers

This section provides detailed information about the production of glass fiber tapers from the melt. As already mentioned in Chap. 5, the fiber taper is a lenticular ending at the end face of a glass fiber. This type of lens has the advantage toward the discrete lenses, that only one component needs to be adjusted. Therefore, the manufacturing costs are low, and the mechanical construction of a chip-to-fiber coupling system remains very simple.

Optical fiber tapers can be produced in different ways:

- Melting
- Chemical etching
- Mechanical grinding
- Splicing of several fiber types
- Bonding with lens

A setup with a fiber taper is schematically shown in Fig. 7.19. The following properties are characteristic of a fiber-taper coupling:

- Losses: 3–5 dB
- Adjusting tolerance (1 dB additional loss): 0.5 μm
- Disadvantage: precise adjusting and fixation tools
- Advantage: only one device

The manufacturing of the lenses at the end of an optical fiber is typically performed in a drawing device, which is in our case presented in Fig. 7.20a, b.

A tension fiber is clamped in the lower part of an *x*, *y* adjustment device. A second optical fiber at which the lens is to be formed on is clamped at an appropriate length in the upper part of the coupling device after the complete stripping of the fiber. In the following, the fiber, which should be tapered, is heated with an arc of light on the lower part whereby the surface forces form a drop-shaped curve. The tension fiber is rounded in the same manner at the end face. In the next step, the tension fiber and the lens fiber are aligned in the *x*-, *y*-axes with each other so that both tips of the resulting curves at the end surfaces of the fibers are centered on one another. Now, the taper is reheated and the arc of light optimized so that the tapered fiber is heated at the top. The heating is controlled by means of the color of the glass melt. The intensity of the arc of light is

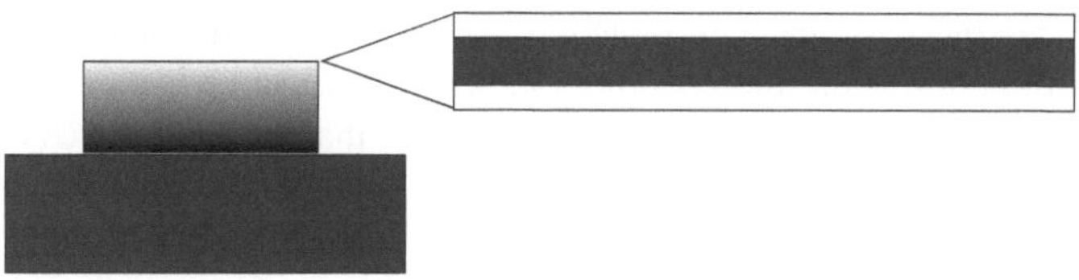

Fig. 7.19 Fiber-taper-chip coupling

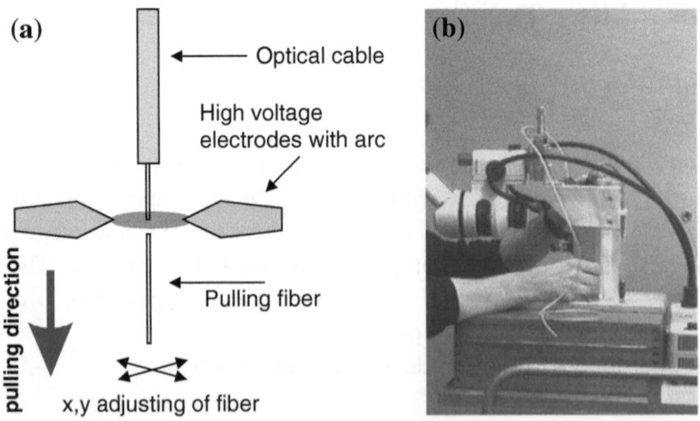

Fig. 7.20 **a** Taper-pulling process. **b** Taper-drawing machine

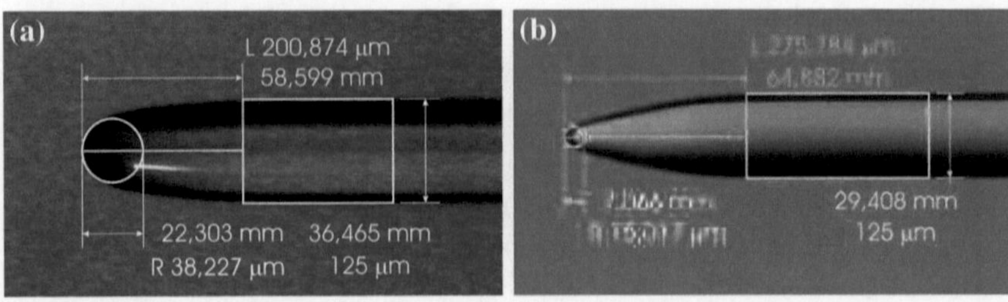

Fig. 7.21 **a** MMF taper radius 15 μm and **b** SMF taper with 80 μm radius

regulated by multiple activation and deactivation. In the following, the fibers are drawn down into the bottom of the melt in the taper. Thereby, the tension fiber coalesces with the taper. Thereafter, the fused fibers are heated and separated by a fast tug while maintaining the arc of light above the joint. Right after the tearing apart of the fibers, it builds a semispherical end at the lens fiber. The heat of the arc of light and the drawing speed can control the radius of curvature. The result is shown in Fig. 7.21, which represents the lenticular end of the fiber. Typical radii for the coupling of waveguides in Indium-OEICs are between 9 and 15 μm. As it can be noticed in Fig. 7.26, fabrication with low standard deviation $\sigma = 0.16$ μm is achieved. The measurement error is estimated to 0.1 μm.

The radius of the lens can be determined (Fischer and Windel 2006) by means of image processing systems. Unfortunately, the so-measured value provides only limited information about the coupling efficiency that can be achieved. Other factors play also an important role. It is often observed that the emission characteristics of the lens are not exactly parallel to the longitudinal axis of light propagation.

This deviation is called squint angle or banana-shaped form of the lens. If the angle is greater than 5°, the taper is not to be used because of insufficient coupling efficiency. To avoid the junk due to too large squint angles, it is very important;

both fibers are aligned parallel to each other, since during drawing, no readjustment is possible. A detailed statement on the coupling efficiency can be expected if the taper is measured on a test site with a standardized light source which viewing angle and intensity is exactly known. During the manufacturing process as well, the core and cladding media are mixed. Within the top of the taper arises an indefinable mixture of the two indices of refraction. If this process is moreover realized manually, the manufacturing parameters and thus the final product spread considerably. If an exact mathematical model describes the taper, you come very quickly to the limitations of this method. Each taper, due to its production, is unique to a specific coupling behavior. Statements about the coupling behavior, particularly when used in experimental structures, can only be made by a close examination of each taper.

7.4.2 Taper Measurement Setup

This workstation (Honecker 2000) allows the measurement of the coupling losses, which occur at a coupling between the reference light source and the taper, which should be examined. As the reference light source may also be used as an experimental semiconductor structure, specifically adapted tapers can be examined to this structure. The measurement principle of the construction is described relatively easy and with a few words: the optical power of a reference light source is absorbed by the measurement setup and compared with the maximum optical power, which can be coupled from the reference light source using a tapered fiber.

The actual measurement therefore consists of two measuring steps, whereby the individual results of the calculation of the coupling loss are necessary for the complete result. At first, the construction of the measurement setup is described and afterward the specific procedure. Figures 7.22 and 7.23 represent the measurement setup as well as the procedure.

For receiving the optical power measurements, a measuring system is necessary, which is adjusted to the requirements of the taper measurement station. It consists of the individual components of an integrating sphere, detector and an optical power meter. The detector is located on the upper inside of the sphere and allows the inclusion of the radiation flux within the integrating sphere.

The sphere is mounted on a three-axis adjustment table to allow a manual coarse adjustment in the x-, y- and z-direction. It consists of three adjusting modules, whereby a lifting table does the optimization of the z-direction and the optimization of the x- and y-direction was performed using vertical positioners. The drive of the individual modules is done with micrometer screws. The reference light source can be, for example, a DFB laser. It forms with a temperature sensor a Peltier element and the connections for the external control a structural unit, referred hereinafter as the laser unit.

The laser unit is located on a system for fine adjustment of the measuring system. If you control the system by suitable software, an optimization of the x-, y-

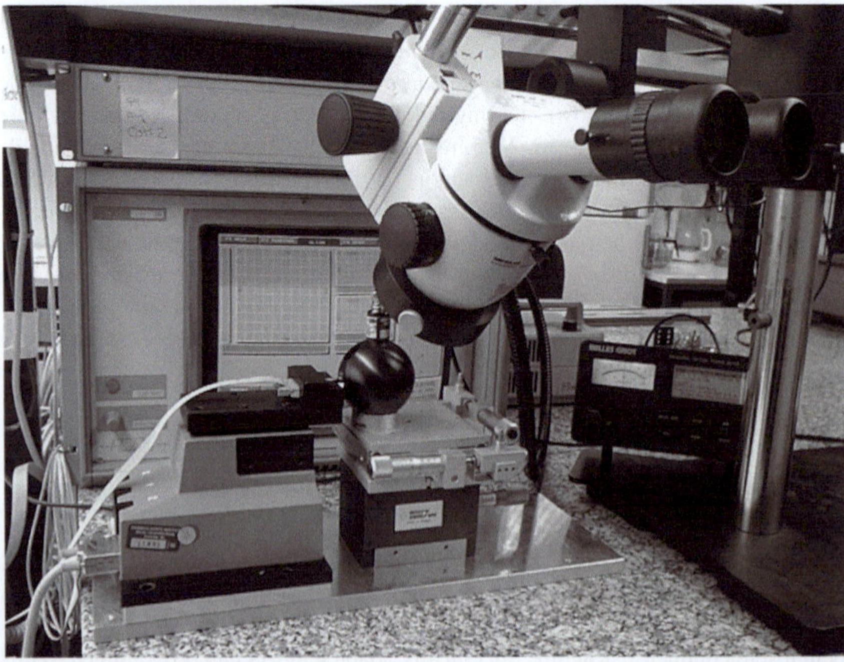

Fig. 7.22 Taper measuring setup overview

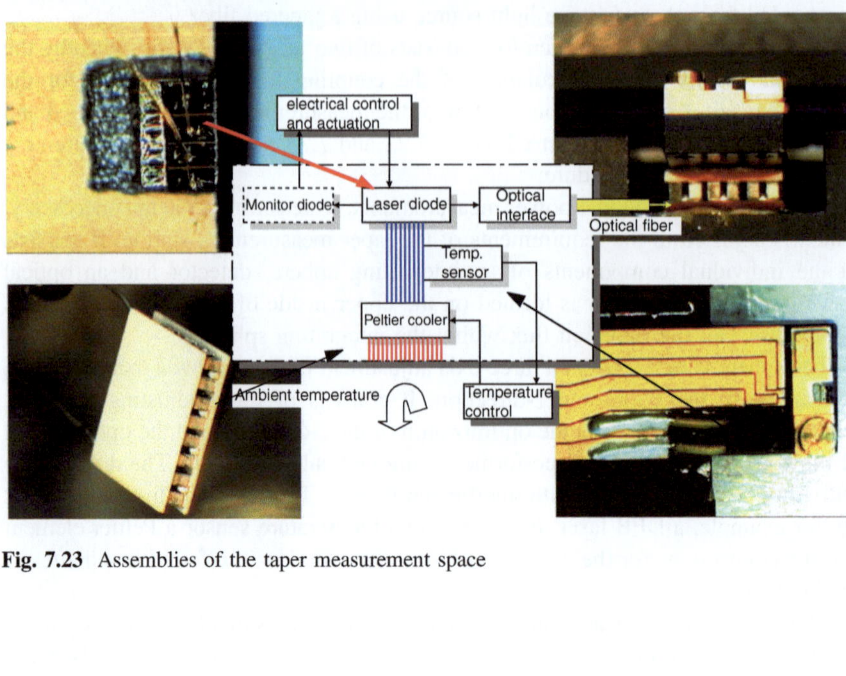

Fig. 7.23 Assemblies of the taper measurement space

and z-axes up to the submicron range is possible. The required step size, which is important to execute the laser unit over the slide system, should be in the range of 0.1–100 nm.

The drive of the carriage or its positioning in automatic mode is performed by a piezo-electric adjuster. They allow, as opposed to the micrometers of the three-axis adjustment, not only the manual adjustment, but also an automatic search of the maximum intensity and a two-dimensional scanning of the light intensity in the x- and z-direction.

7.4.3 Measurement of the Reference Light Source

For coarse adjustment, a three-axis bench is used, as already indicated. For this purpose, the front part of the laser unit is to be aligned directly in front of the opening of the integrating sphere, which is located on the front part of the spherical wall (Fig. 7.24).

Thereafter, the laser unit moves with the aid of the carriage system into this opening. For that purpose, the required step size is entered through the control software and carried out by manual operation of the positioning of the piezo-electric adjuster. The optimal position of the laser within the sphere is determined by the structural design of the laser unit. Using a stereomicroscope, the adjustment procedure can be observed and assessed by visual criteria. Once you have set the parameters for the external control of the laser unit with power supply and temperature control, only the reference value of the optical power on the measurement system must be determined. Obviously, the selected parameters must be stable until the end of the measurement. The recordings of the measured values are saved in the units dBm or mW. For the following second part of the measurement, the measuring system has to be rebuilt. Instead of the integrating sphere, the fiber taper is placed in front of the laser unit. The integrating sphere is now removed. Instead of the sphere, a specially designed fiber holder with coarse positioning stage can now be installed. The fiber holder assumes the responsibilities to include the fiber taper, to permanently fix it, and to save it from tension. For this purpose, a metal spring with a V-groove on the front upper end of the fiber holder is located. The V-groove is dimensioned, that a fiber with a core diameter of 125 μm, without the primary and

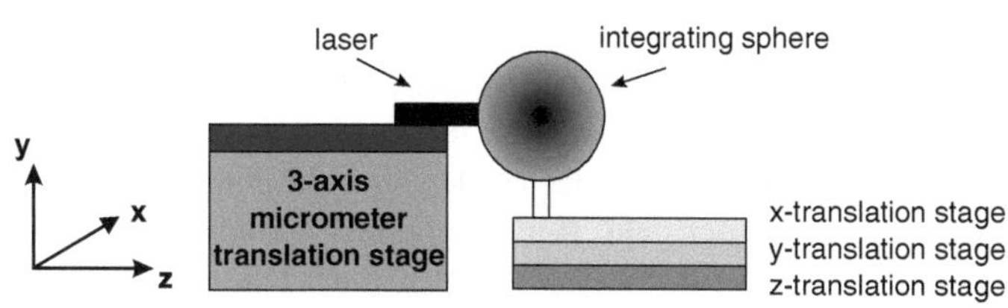

Fig. 7.24 Measurement of the optical performance of the reference light source

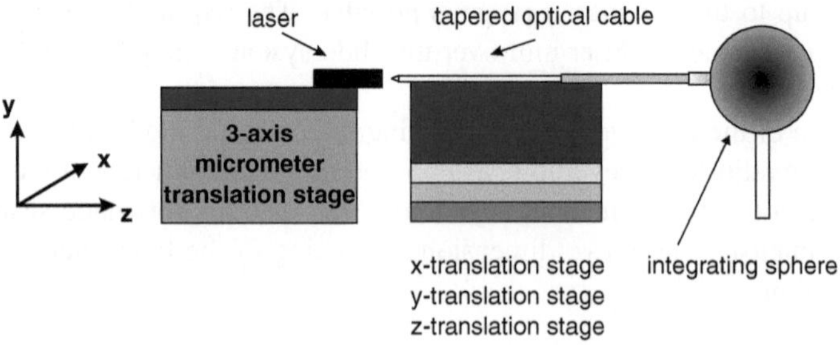

Fig. 7.25 Measurement of the optical performance which is coupled to the fiber

secondary coating, may be loaded. Thereby, the tapered fiber end juts about 5 mm beyond the edge of the fiber holder and is fixed by magnets. The non-tapered fiber end provides, according to the fiber manufacturer, a bare fiber end or end with an optical connector. For receiving the second comparative value, all the light coming from the other end of the taper has to shine in the measurement system (integrating sphere). Here, the optical power is measured (shown in Fig. 7.25).

7.4.4 Measurement of the Reference Light Coupled to the Fiber

After the modifications, the coarse adjustment of the measuring arrangement can be done by the three-axis adjustment. Fiber taper and laser source are converged up to about 2 mm, and the coupling process is visually observed by a stereomicroscope. The tip of the fiber taper should be situated in front of the waveguide structure of the laser source (Fig. 7.25).

The monitoring the coupling process is performed using a stereomicroscope. This device is only used for a control, not to optimize the fine adjustment of the measuring arrangement by the micrometer stages. The evaluation criterion is to couple the maximal power from the laser source to the fiber. Therefore, obviously two crucial conditions are relevant for the implementation of the second measurement. On the one hand, the laser runs during the entire measurement in continuous operation with constant parameters. On the other hand, the power, which is coupled via the taper, is permanently observed. The optimum positioning of fiber taper and laser source in the y-direction is done as already mentioned by the manual actuation of the piezo-electric stage and is assessed in terms of performance. The step size has to be set by software. In the x- and z-direction, an automated adjustment is possible. Within the selected step size, the control software initializes an automatic scan of the laser field. With the help of a continuous feedback of the injected power by a photodetector, a two-dimensional field of the laser power can be recorded. From the measured values, the maximum power can be determined

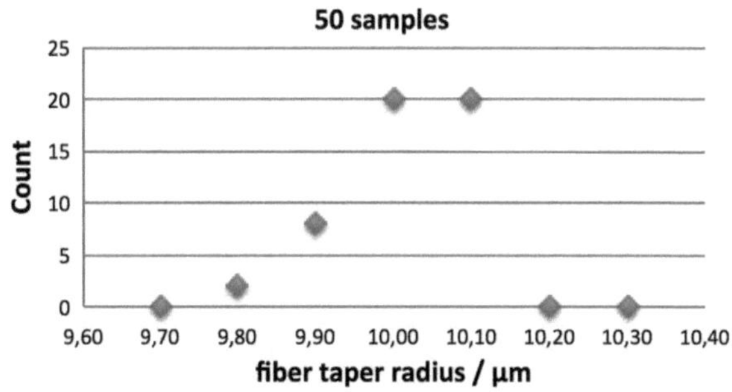

Fig. 7.26 Stable drawing process of 50 samples with 10 µm radii with for 50 tapered fibers; standard deviation $\sigma = 0.16$ µm

and the taper will be positioned into the optimum position. If the adjustment is completed successfully, only a second control measurement must be recorded. Therefore, the non-tapered fiber has to be removed from the detector and fixed through the aforementioned bare fiber and FC/PC adapter into the integrating sphere. The recording of the measurement values is therefore done under the same conditions as in the first step, that is, the value is directly comparable with the first measurement. If both values were recorded in dBm, we get the fiber coupling loss of tapered fibers by the difference of these values:

$$\frac{P_{\text{Loss}}}{\text{dB}} = \frac{P_{\text{Reference}}}{\text{dBm}} - \frac{P_{\text{Coupling}}}{\text{dBm}} \tag{7.1}$$

Figure 7.26 describes typical coupling losses of 50 tapered fibers, fabricated by the former shown hand-made production method. It can be recognized a low variability of the results, which are in this case between 9.9 and 10.2 µm.

Due to the complexity of the relationships, which are needed to understand, the presentation will be described in simplified terms. The integrating sphere, as this spatially integrating photometer is called, is a hollow sphere in which the inner surface is reflecting the light broadband high and diffuse over a wide range of the spectrum. The characteristics of the sphere can be obtained substantially from the coating of the inner surface, which forms a Lambertian reflector, and the spherical shape of itself, which causes the spatial integration. These two design features ensure that every point of the inner surface reveals the same illumination (Windel et al. 2005).

One can measure it at any point by a detector, to get information about the total irradiated light output. Apart from the simple determination of the total capacity at relatively complicated radiation intensity distribution (e.g. fiber or laser), two potential sources of error in measuring performance have contributed to the application of the integrating sphere as measurement instrument for the taper measuring station.

As mentioned before, the emission cone of the semiconductor laser is not approximately rotationally symmetrical. The smaller the radiating surface, the greater

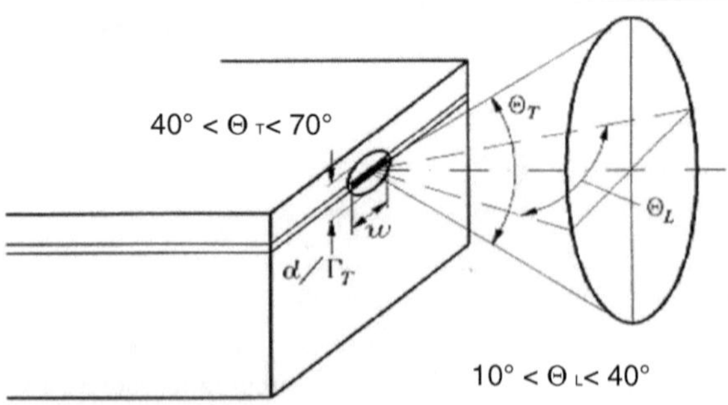

Fig. 7.27 Divergence of the laser emission field

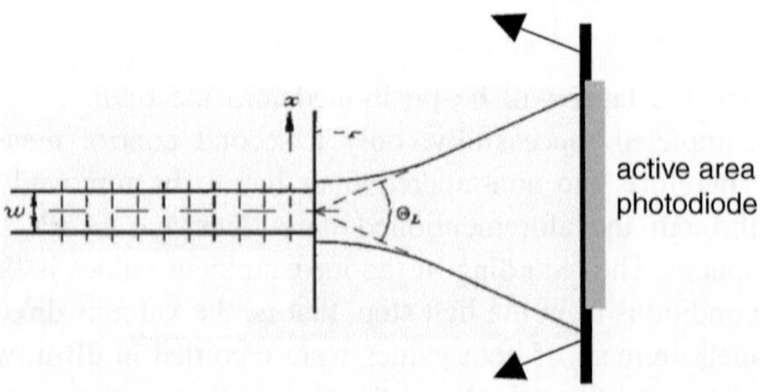

Fig. 7.28 Performance measurement with a photodiode

the divergence of the radiation pattern due to the diffraction increases. Since the area in the lateral direction is more distinct than in the transverse direction, the latter provides the larger beam divergence. Usually is the opening angle of the far field in the ranges of 20°–30°, in contrast to the single-mode fiber of about 5.5° (Fig. 7.27).

The radiated cone with lateral beam width up to 70° may not exceed a maximum distance of about 7 cm to display on the active layer of a photodetector (typical expansion 1 cm) (Fig. 7.28). Thereby, only 50 % of the power density is considered. In order to achieve higher accuracy, the space hast to be drastically reduced.

Therefore, mismatches between the active zone of the photodiode and the field of the laser beam are appropriate. The highly accurate power measurements, as it is required for the taper measuring setup, seem questionable with a photodiode. Another source of error results from the emission cone of the laser diode, which is widened in the far field. The just described wavefronts within the laser change outside of the exit area in curved wave fronts. This means that field components, which are located outside of the core box, come upon the interface of the photodetector at an increasing angle. At the exceedance of the acceptance angle, which provides a statement about the still possible transmission into the medium, it is no

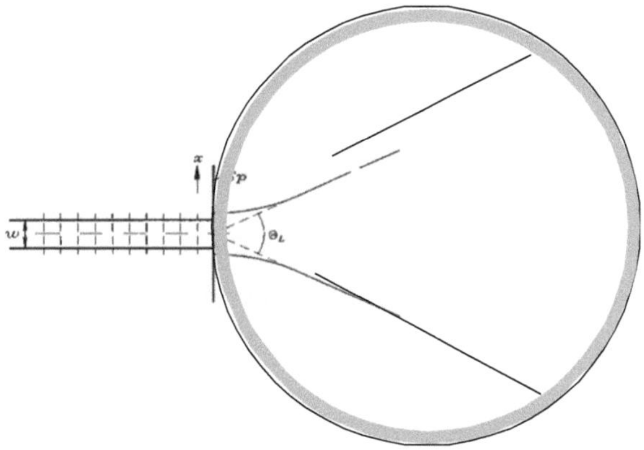

Fig. 7.29 Performance measurement with an integrating sphere

longer possible for the power components to reach the active zone of the photodetector. They are reflected and remain unconsidered in the determination of performance (Fig. 7.28).

If one considers that the integrating sphere excludes these sources of error (Fig. 7.29), it also allows a compensation of different power density profiles. The integrating sphere should indeed the preferred instrument to determine the power characteristics.

References

Eberlein: Lichtwellenleitertechnik, p. 212. Expert Verlag, Esslingen (2000)

Fischer, U.H.P., Windel, T.: Multimode fibers with integrated optical mode field adapters for 40Gbit/s optical ethernet systems. In: Proceedings of SPIE—the International Society for Optical Engineering, vol. 6126. San Jose, CA. Conference code 67347 (2006)

Honecker, J.: Untersuchung der Koppeleffizienz und der Reflexionen zwischen Monomode-Fasern und integrierten Modenfeldtransformern auf InP für LWL-Empfänger im Bereich 40-160 Gbit/s. Elektrotechnik. HTW Berlin, University of Applied Sciences, Heinrich-Hertz-Institut Berlin (FHG) (2000)

Windel, T., Hemrungrote, S., Fischer, U.H.P.: Integrated optical mode field adapters at the end of single/multimode fibers. In: Mouroulis, P.Z., Smith, W.J., Johnson, R.B. (eds.) Proceedings of SPIE—the International Society for Optical Engineering, vol. 5874, pp. 1–8. San Diego, CA (2005)

Chapter 8
Active Adjustment Techniques

Abstract In this chapter, basic components of active adjustment tools will be introduced. Here, the advantages, disadvantages, and applications of mechanical carriage systems are described for micropositioning and afterward deeply analyzed. Furthermore, an overview of existing possibilities of fiber-chip fixation in regard to the long-term stable welding process is shown. At the end of the chapter an example of the application of microwelds is illustrated and described in detail for the use in modules for optical communications systems. A coupling machine in combination with laser welding performs the fixing of the fiber-chip connection. Hereby, an adjusting welding technique, which is called "strain-reducing" welding, is introduced.

8.1 Micrometer Positioners

8.1.1 Mechanical Positioners

In modern optics laboratory, it is, as shown in the previous chapters, necessary to adjust optical fibers and OEIC with submicron accuracy for each other in order to achieve satisfactory coupling efficiencies.

Even with the best conventional adjustment means, such as micrometer screws, this can be achieved in consideration of the whole mechanical structure. Better resolution can be achieved with differential micrometer screws (0.1 μm).

It not only requires adjustment, but also special long-term stable fiber holders, to allow linear and angular movements with high resolution and precision. Three linear movements (translations) and three angle degrees of freedom are needed in order to optimally adjust the fiber prior to the OEIC, as shown in Fig. 8.1.

In these six degrees of freedom, however, the shift must be done uniformly and independently. Advantages and disadvantages of the different mechanical slide modules are summarized in Table 8.1.

© Springer-Verlag Berlin Heidelberg 2015

U.H.P. Fischer-Hirchert, *Photonic Packaging Sourcebook*,
DOI 10.1007/978-3-642-25376-8_8

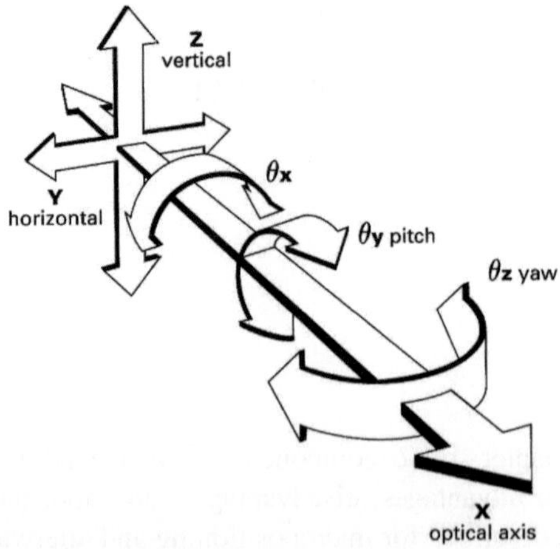

Fig. 8.1 Six degrees of freedom of movement

Table 8.1 Advantages and disadvantages of mechanical slides

	Advantages	Disadvantages
Dovetail slides	Pneumatic suspension possible	High friction
Bearings	Backlash by preloading, self cleaning	Greasing necessary low surface contact
Roller bearings	Backlash by preloading, carry large weights	Pollution

It is the experience of the manufacturer of the linear slide applications that were applied to the submicron applications which are needed for the fiber-chip coupling of optical single mode devices using the following structural principles:

- Dovetail slides (Fig. 8.2)
- Ball bearings (Fig. 8.3)
- Roller bearings (Fig. 8.4)

Important for the movement are the following boundary conditions (Fig. 8.5):

- Lateral roughness (non-rectangular motion)
- Flatness in vertical view.

If the roughness is too large, the guided slide starts to wobble sideways. If the guide is not ground flat enough, the carriage changes its altitude. Both effects reduce the accuracy of the singular translational step movement of each axis and thus adjust two more dimensions—a very undesirable interference effect. To prevent temperature drift, it must be ensured if possible with the choice of materials, that it has a low coefficient of expansion, because otherwise the optical coupling would degrade.

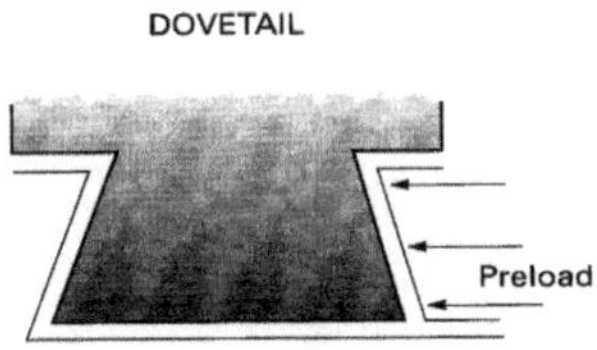

Fig. 8.2 Dovetail slide

Fig. 8.3 Ball bearings

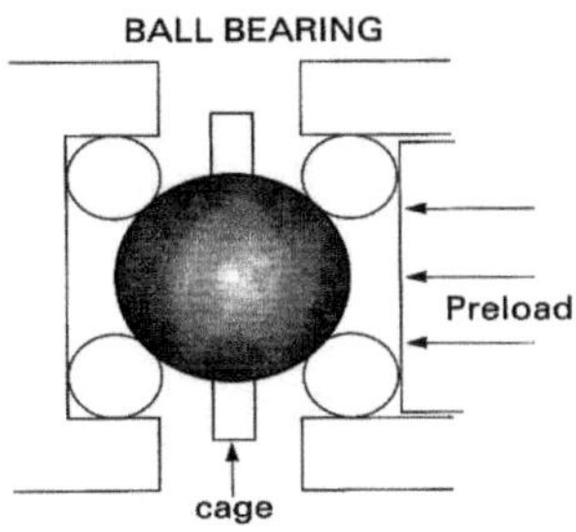

Fig. 8.4 Roller bearings

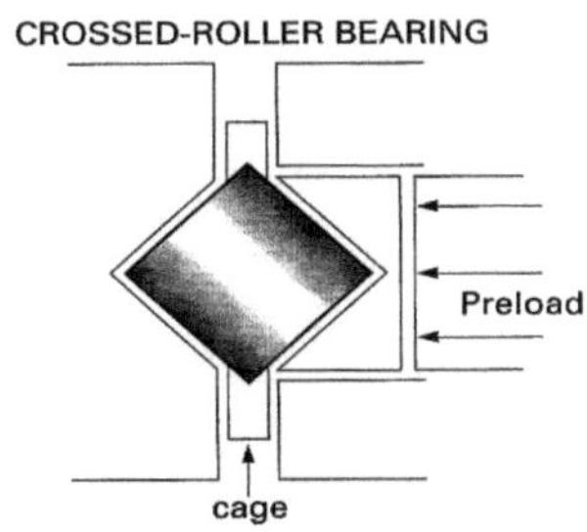

Piezoelectric adjusters are ideal to make precise high resolution adjustments, with accuracy better than 1 μm. Piezoelectric materials are crystals that polarize when they are mechanically compressed. Particularly strong is the piezoelectric effect in tourmaline, quartz, Rochelle salt. Conversely, these actuators can be deformed with a voltage applied to the crystal in a predetermined direction. The effect is used, for example, in piezoelectric speakers, or gas lighters. In Figs. 8.6, 8.7, and 8.8, this special behavior is shown. Path length of 10–50 μm at 1000 V voltage is common. But there are also low voltage piezoelectric drives on the market, which work with lower voltages up to 100 V, but offer the same displacement. Unfortunately, all piezoelectric drives show a hysteresis in the deflection, which is very annoying the precise predictable positioning of the elongation. This effect can only be overcome by active feedback processes that are used in today's commercial systems. Further all insulators show a change in length when a voltage is applied to opposite end faces. However, this electrostrictive effect is very small compared with the

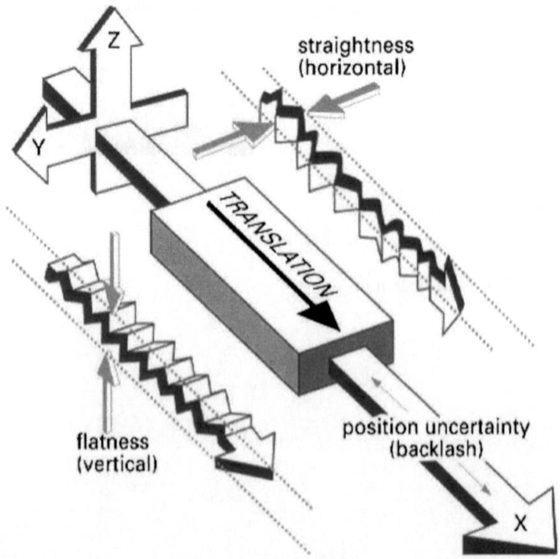

Fig. 8.5 Influence of roughness in case of linear guide

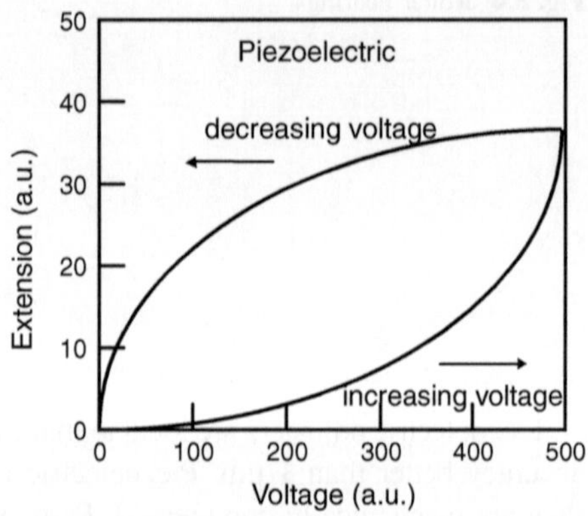

Fig. 8.6 Piezoelectrical effect with hysteresis

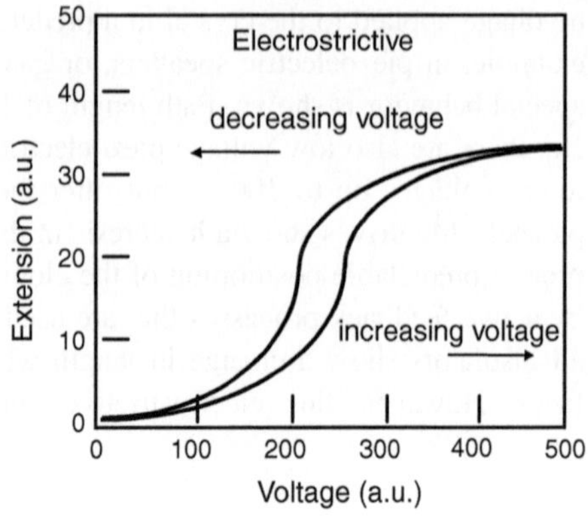

Fig. 8.7 Electrostrictive effect with strong nonlinearity and hysteresis

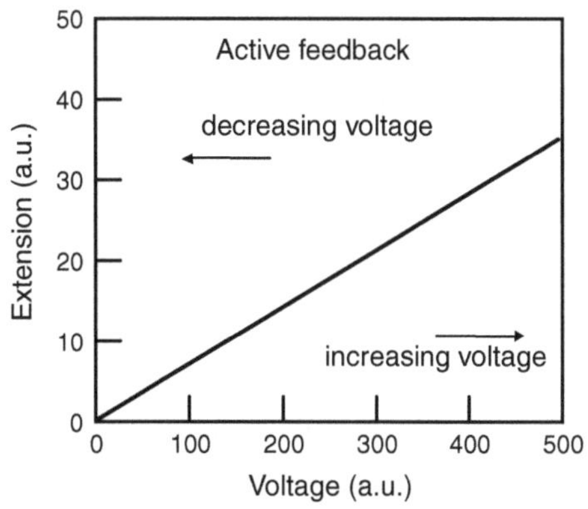

Fig. 8.8 Bypassing the hysteresis by active feedback

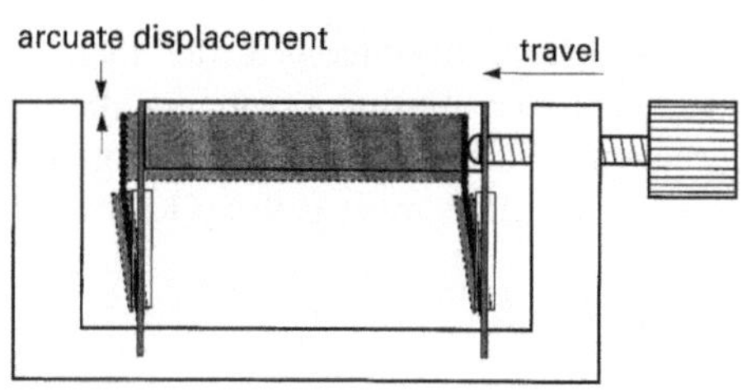

Fig. 8.9 Backlash-free guidance in a plane through solid joints

piezoelectric effect and also has a strong hysteresis. However, these materials are not suitable for high resolution placement machines which can be used for professional fiber-chip-coupling.

Solid joints (Fig. 8.9), which are deflected by piezo-driver, secure the uniform translation in one axis. An overview of the application of the different motion systems is shown in Fig. 8.10. Ball, roller, and plain bearings are used meaningfully only to the movement steps of at least 0.5 μm.

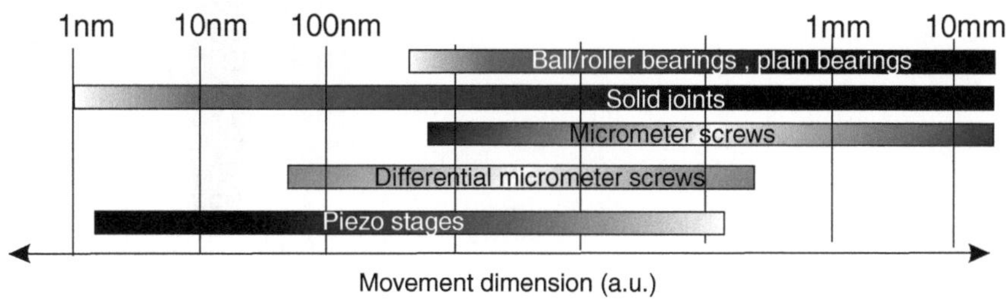

Fig. 8.10 Uses and limitations of various motion systems

The operating range of less than 1 μm expands the solid joints, as these are free from backlash. In the actuator area, micrometer screws up to 0.5 μm resolution and a special variant, the differential micrometer screw, with accuracy up to 0.05 μm are widely used. In this special variant, two threads are moved toward each other, so that only the path difference of the two-threaded movement is received in the external translation. However, the absolute adjustment of these screws is very low. If step sizes are significantly below 0.1 μm, the utilization of piezo-adjusters is useful, as they have a lower limit of about 1 nm.

By combining the piezoelectric effect and a micrometer screw, New Focus (Fig. 8.11) has developed a special adjusting screw whose path accuracy is better than 0.1 μm and the same time has a wide adjustment range of more than 4 mm. Moreover, the user can also change the adjustment by hand.

In the optical design technology, most adjusters for several axes and angles are used simultaneously. The industry has developed appropriate complete adjustment systems, keeping the needs of professional fiber-chip-coupling setups. A six-axis adjustment (three translational axes and three angles) from Melles Griot is shown in Fig. 8.12. These systems achieve resolutions better than 0.02 μm and angular resolutions better than one arc second using a combination of piezo-technology for high-resolution displacement with low total and conventional micrometer screws for the expansion of the overall adjustment path up to several millimeters.

Fig. 8.11 Micrometer screw with the piezo-drive

Fig. 8.12 Six-axis nano-positioning system

8.2 Laser Microwelding

8.2.1 Laser Welding Methods

Laser welding is a preferred method for long-term stable fiber-chip-coupling and can be performed in various methods. Here, the following classifications of welding processes will be presented according to DIN 1910 (Kulina et al. 1993):

- **Fusion welding processes**
  Gas welding, arc welding, laser welding, etc.
- **Welding by the introduction of pressure**
  Resistance spot welding, cold welding, friction welding, etc.
- **Weldability of workpiece**
  Before welding two workpieces it should be checked whether the materials can be combined with each other at all. The weldability is judged on the following criteria:
- **Welding tendency**
  It relates to material properties such as the chemical composition, melting point, etc. For many materials, the welding tendency during laser welding is only partially known, and additionally only for particular workpiece thicknesses. Therefore, it is often required to investigate the weldability of the specific application.

- **Welding certainty**
  The examination of whether mechanical stresses occur or how the seam arrangement must be used and how durable the weld is to be is a second important criteria point that must carefully checked with the material question.
- **Welding ability**
  Whether a weld on the selected mechanical dimensions is possible, is limited by the production characteristics such as joint shape and welding sequence (Kulina et al. 1993).

8.3 Criteria for the Choice of Welding Methods

The selection of the method by which the workpieces to be welded can, among other things, happen according to the following technical parameters, which influence each other (Kulina et al. 1993):

- Required power density
- Required energy per unit length
- Desired seam geometry

Process-related power densities:

- Manual arc welding: about 10^4 W/cm^2
- Plasma Fine Welding: about 10^6 W/cm^2
- Electron beam welding: about 10^7 W/cm^2
- Laser beam welding: greater than 10^8 W/cm^2

The last value of 10^8 W/cm^2 represents the highest power density, which is useful to use in laser welding. However, there are far larger power densities possible for laser material processing. If the power density is lower, the action time must be longer, because heat—by the thermal conductivity of the material—is dissipated into the workpiece and thus the process is no longer available. At higher heat inputs wider welds are produced due to the greater amount of energy and thus greater local extent of the melting zone. The deep welding effect on beam welding methods by focusing the energy to a relatively small diameter allows a good depth-to-width ratio of 10 to 1. By comparison, fusion welding processes allow only a depth-to-width ratio of less than one to one.

Advantages of laser welding methods involve multiple applications. Thus, different materials can be combined with unequal characteristics, such as tantalum, copper, or gold–silicon. The limited hardening of the material occurs only locally, since only a small heat-affected zone is present. There is no or only a small formation of coarse grains. Non-ferrous metals also have excellent weldability. For the industrial exploitation, a high feed rate is possible, which is theoretically up to 25 m/min. In addition, this is a non-contact process that avoids tensions entirely by additional mechanical effects.

However, disadvantages are to be considered. In most cases, there are vibration-sensitive structures with low efficiency in energy conversion, because the highly reflective surfaces obstruct the energy coupling. In addition, an accurate prefabrication of blanks is necessary since the method allows only a small tolerance in joint preparation, which is between 0.01 and about 0.1 mm.

8.4 Laser Material Processing

By focusing the laser radiation to a minimum area, a high power density at the workpiece surface (Fig. 8.13) can be generated. The energy absorption in the material is limited by the exposure time of the laser beam. The essential principle of energy transfer to the material is due to the utilization of the absorption of electromagnetic radiation by the free electrons of the metal compound. Here, a local melting temperature is generated in the vicinity of the impingement surface. The degree of absorption of metals is very low and varies from 1 to 5 %. The surface structure and material selection change the absorption considerably.

The effect of radiation on the material will be examined in more detail in the following paragraphs. First, the radiation of the laser is incident on the workpiece surface. Conversion of the optical energy into heat energy occurs, as already mentioned, by interaction with free electrons. Due to the radial heat conduction into the workpiece forms a heating zone. Since the heat cannot easily flow into the material, heat build provoking occurs, because it is fed more energy into the DUT than can be dissipated by heat conduction. The temperature rises until the melting temperature or vaporization temperature is reached. The absorption coefficient

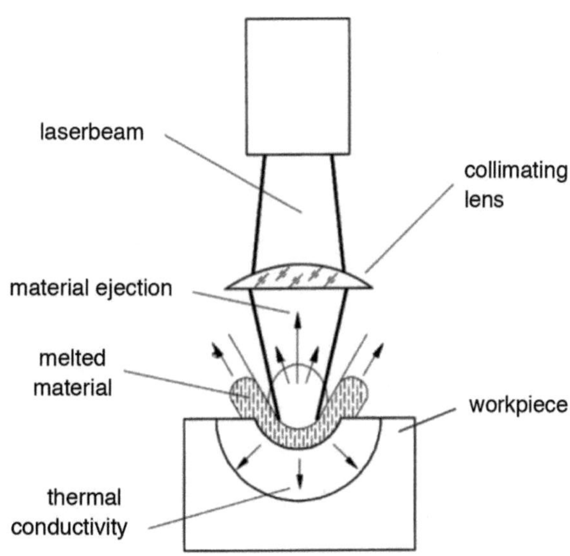

Fig. 8.13 Principle of laser material processing

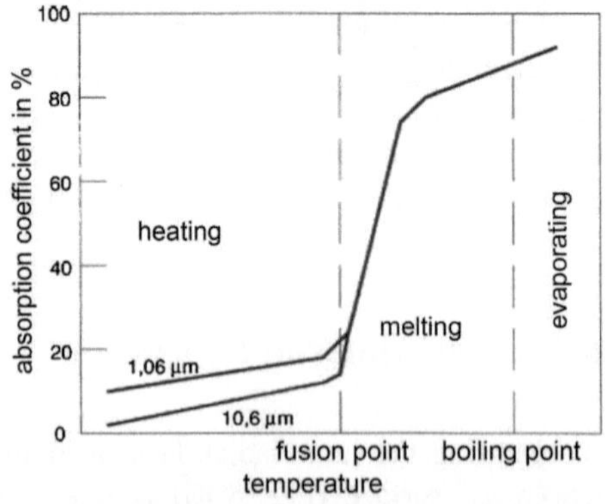

Fig. 8.14 Temperature dependence of the absorption coefficient in metals

increases with the temperature as well (Fig. 8.14). This further increases the local heat. Upon reaching the vaporization temperature, a steam duct is formed, which penetrates the melt. The vapor pressure displaces the melt and conducts heat deeper into the workpiece into it. Then, the formation of the needle hole in dependency of the intensity is reached.

With a further increase of the energy density, a photo-induced plasma in the metal vapor is formed, which is produced by ionization of the gaseous metal. The energy overshoot at the local heating is so great that arise more and more free electrons by further impact ionization of free electrons with high energy. Finally, an increasing number of energy absorbed by the plasma, until the ionized gas shields the surface of the workpiece (Fig. 8.15) and a further increase in temperature is no longer possible.

The Table 8.2 gives an overview of the use of lasers in industrial materials processing. "Insert" is the process that is responsible for the welding of the fiber-chip connection that is most suitable.

There are several types of joint seam have been developed which are suitable for specific applications. The fillet weld and the penetration weld are sketched in

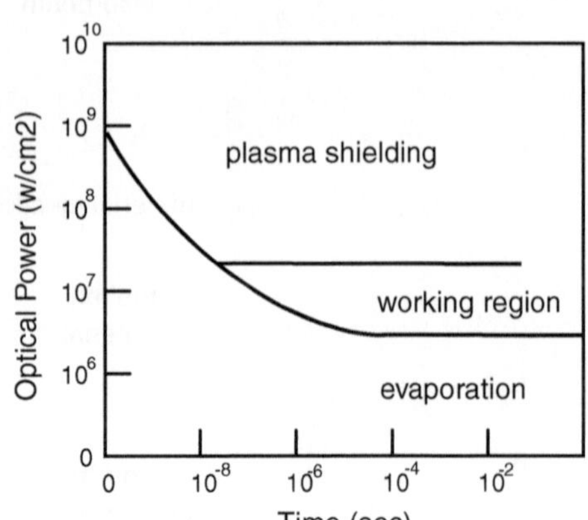

Fig. 8.15 Screening effect at high energy densities

Table 8.2 Laser welding processes at a glance

Forming	Method	Process variants
Surface treatment	… with furnish … without furnish	Alloying, coating, curing
Split up	Laser cutting, laser fusion cutting. cutting, laser sublimation	
Fixing	Welding, soldering	Spot-, seam welding
Eroding	Labeling and marking, trimming, structuring, drilling, and ablation	Of thick film resistors … of circuits … in the micrometer range

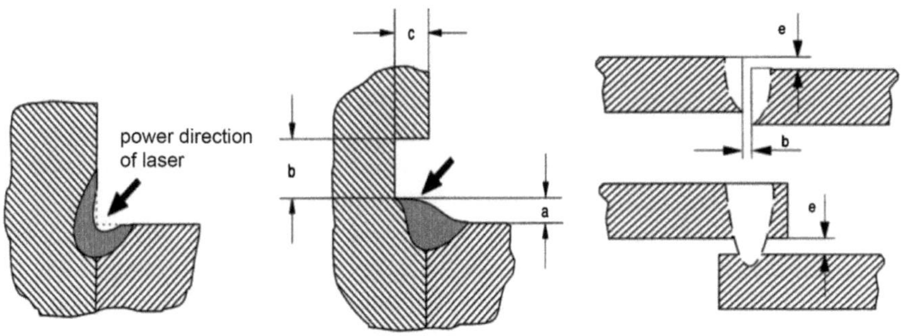

Fig. 8.16 Fillet weld (*left*) and penetration weld (*right*)

Fig. 8.16, which is exemplary for all procedures. The fillet weld is particularly prone to cracking, but is favorable for laser spot welding with very small spot diameters. If you want to create larger welding points, you have to admit that the material expands in the direction of the joint. For this, at the joint a meltdown point is favorable.

Reducing the focus diameter of the laser light on the material to less than 1 mm is called a microlaser welding. Typical weld spot diameter and weld widths are in the range below 800 μm. The reason for the reduction of the light spot is to minimize size, form, and position deviations of the welded parts: A small light spot heated even a small area of material and so brings little additional stresses in the workpiece.

The surface tension in the melt increases during the crystallization due to the vapor pressure of the plasma. This effect can be greatly reduced using a small light spot diameter of about 1 mm. The strength of the joint is hereby of secondary importance. This method of small fused surfaces can realized only a fraction of the holding forces of the conventional welding technology. The usual method for laser microwelding (Becker et al. 2001) is to use a Nd:YAG laser (wavelength: 1.064 nm), which makes a spot weld. For better durability and low distortion, usually two laser pulses are used.

Fig. 8.17 Welds on a laser module

The pre-pulse cleans and heats the surface and causes an increase in the absorption coefficient of the plasma. The main pulse then serves local melting and joining the parts. A big advantage is the very low production of the laser-induced plasma in order to keep the distortion as low as possible.

Figure 8.17 is an example of laser welding and is illustrated in a transmission module of optical communications. Here, the alignment pins of a laser module for fiber-chip coupling of the laser were fixed with (Eckhardt 1999). The achieved accuracy of the fiber before fixing the chip waveguide was better than 1 μm.

8.5 Industrial Multi-point Laser Welding

In the industrial sector, the laser welding is already widely used for realizing long-term stable optical couplings. Since the total cost of such an automated system is in the region of EUR 800.000, such method can be profitable only in the context of mass production. Only micro-welding is applied to reach a low distortion of the welding partners. To realize this setup, several optical laser beams are used at a distance of 90°, 120°, or 180° to be focused on the weld, resulting in a more uniform heating of the components.

The distribution of the radiation energy is realized by the help of a mirror unit (Fig. 8.18, here: 1:3 splitter). It shares the main beam into three equal intensity partial beams, which are guided via optical fibers to the laser heads.

This makes it possible, with butt cut-ended fibers to couple InP and GaAs OEICs with an efficiency of about 10 % in mass production machines. However, to connect OEICs with the tapered glass fibers, a default limit of ±0.5 μm is observed, which is not possible to realize with the systems described so far.

To keep the offset below 0.5 μm, it is necessary to reduce the flange components significantly. Even more, welding points can be added with a defined delay after the first laser weld. This technique is called laser hammering (Anthamatten et al. 1995).

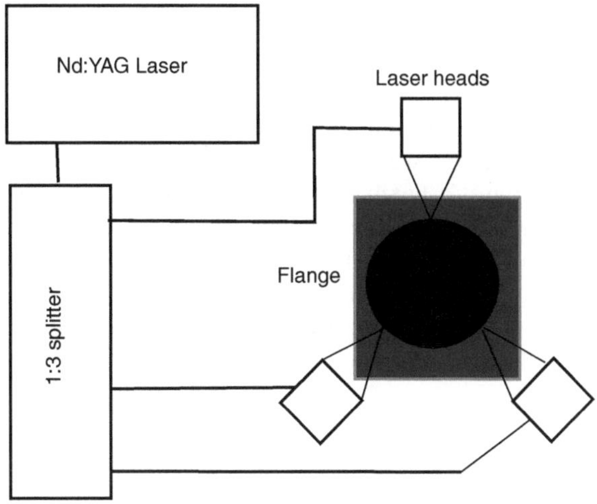

Fig. 8.18 Basic design of three-point welding arrangement

The weld geometry has a strong influence on the size of the expected offset and at the same time on the stability of the welding point. The flanges generally should be connected to each other with sharp edges. In Fig. 8.19, three different weld geometries are shown. The butt weld gives the best adhesion, coupled with very low distortion. However, it has the disadvantage of being poorly applicable within enclosures.

The lap weld produces good adhesion, but the distortion is much greater than with the butt weld. Since the first material must melt the lower lying flange, one also needs (Anthamatten et al. 1995) higher laser energies. The fillet weld is a good choice because it is very flexible to use in geometry management.

Unfortunately, the adhesion and thus the long-term stability are not always sufficient. An alternative is to connect the two latter weld geometries, the lap–fillet weld. Hereby, the welds were carried out successfully by Fischer et al. (2001). Results are consistent with distortion less than 1 μm at 95 % of all welds have been achieved here. For all welding geometries, gaps between the mating parts must be avoided as these may prevent uniform melt both parts.

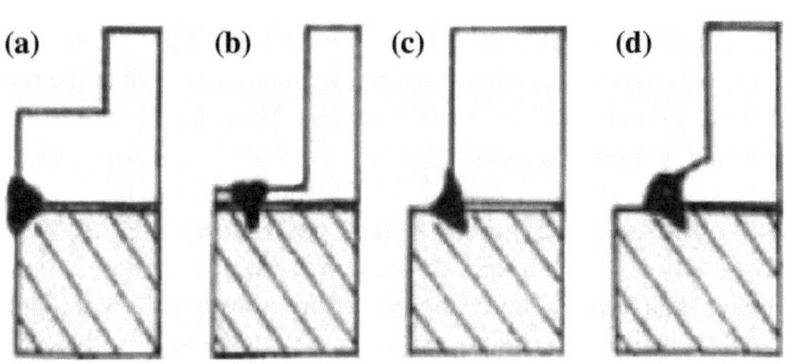

Fig. 8.19 Welding geometries: **a** butt weld, **b** lap weld, **c** fillet weld, **d** lap–fillet weld

8.6 Laser Micro-welding for Modules with Tapered Fibers

However, investment costs for this laser welding equipment are considerably high. Thus, a low-cost coupling machine was designed which is able to work with butt-ended fibers as well as with lensed (tapered) fibers. For laboratory use and rapid prototyping, a flexible design of the modules is needed which is able to adapt different OEICs with changing dimensions to an existing module type and a stable mechanical setup for the fiber-chip coupling. In this chapter, we describe a novel type of laser welding machine, which works with two laser beams to perform semi-automatically the optical-tapered fiber-chip connection of wave-fed OEICs within butterfly (BFY) standard housings.

8.6.1 Coupling Concept

A comparison of the optical fields of a butt-ended standard monomode fiber (SMF) and optical fields of edge the emitting laser diodes shows a great mismatch (see Chap. 4). This mismatch is the reason for the very low coupling efficiency of approximately 10 %. This low efficiency can be overcome by a better adoption of the two optical mode fields with lenses. In our case, we used lenses made at the end of the fiber by melting the glass fiber and pulling it by hand. This kind of fiber end is called fiber taper (Kato 1973) and works like a lens with diameters from 20 to 50 μm (see Chap. 4). With this tapered fibers, a coupling efficiency of more than 50 % can be realized. Unfortunately, a high precision of better than ±0.5 μm is necessary to mount the tapered fiber in front of the OEIC without additional losses. Also the near working distance of the fiber end to the OEIC facet of less than 20 μm makes the adjustment process very harmful. Only a short contact of the fiber end to the facet will destroy the OEIC.

Our method of manufacturing the fiber tapers is a very well-known and low-cost technique, which can be performed by using standard splicing equipment (Kuwahara et al. 1980). The optimal shape of the taper and the radius of curvature of the hemispherical end can be controlled by arc-discharge voltage and drawing speed. In addition to the good lens characteristics, the round end suppresses unwanted optical back reflections below −55 dB. The emitted optical field of the cylindrically formed devices is in average less symmetrical to the longitudinal axis than the melted ones. This causes a more complex fiber-chip adjustment. Here, the fiber must be corrected in three Cartesian and three angular axes.

Using these lensed fibers, a lateral shift of ±1 μm between OEIC waveguide and the fiber core results in an additional optical power loss of 1 dB. Therefore, the mechanical resolution of the coupling mechanisms is a critical value that must be one order better than 1 μm. The fixing procedure after coupling should not introduce additional displacements and must be stable enough to fix the coupling mechanism, which is important for good long-term stability.

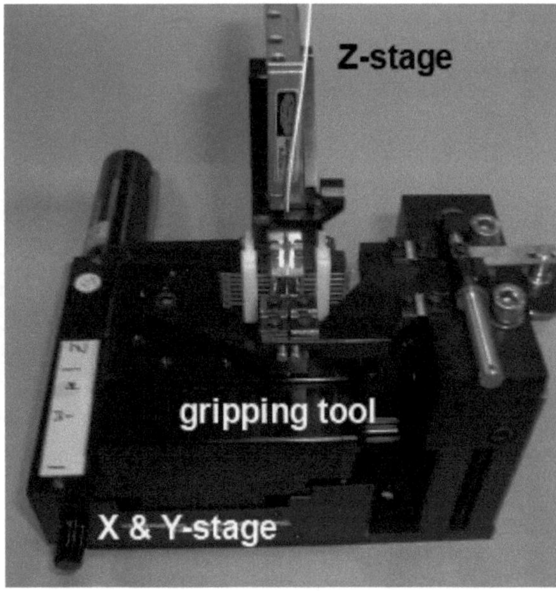

Fig. 8.20 Coupling machine

In Fig. 8.20, the basic elements of the newly developed coupling machine are depicted (Fischer et al. 2002). Two micrometer stages were mounted on the top of an optical bench against mechanical distortions. With the help of the two DC motor-driven stages with 50 nm resolution (Newport), the fiber can be moved precisely in x and y direction. On the top of the stages, a fixation for the BFY-module is situated which carries the housing in a rectangle position. The SMF is fed from the top via a cannula into the inner part of the housing. The fiber is fixed in a clamp. A third micrometer stage adjusts the fiber in perpendicular direction to the OEIC facet. At both sides of the BFY-module, the electrical pins are fixed to connect the bias current of the OEIC and the temperature controller. The adjustment of the fiber end to the chip waveguide is performed by the use of a sliding joint so-called flange (Fig. 8.21). The constructed (Krips 2000) manipulator arm is small enough to grip the flange with the fiber inside the housing. The grip force can be adjusted to that value which is needed to have a strong mechanical contact and which on the other hand is low enough to enable a smooth sliding of the parts. If the frictional resistance is too high, the adjustment will not be possible with sub-μm resolution. After final adjustment, the flange is fixed by two laser welds as indicated in Fig. 8.22. Here, the two laser heads and the focused beams are placed. The focus length is 80 mm and the diameter of the welding spots is 500 μm. The beams are tilted by 30° to the normal axis to reach the middle of the flange without power loss due to the shadowing of the housing. To find the maximum intensity of the fiber-chip coupling, the "autoalign" software of Newport for x- and y-axes was used. In Z-direction, the adjustment is performed manually.

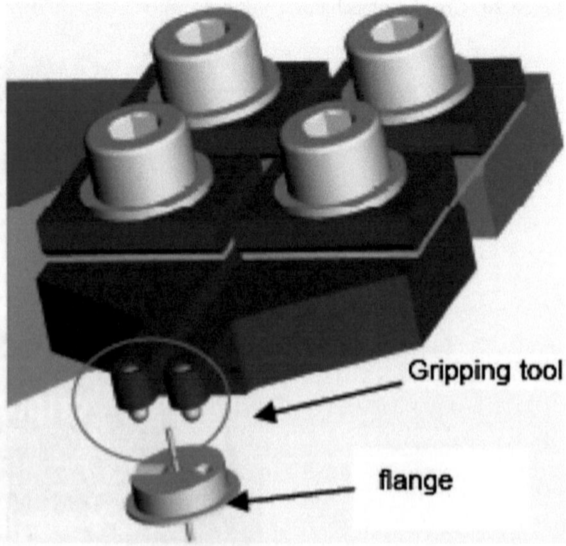

Fig. 8.21 Lateral translation mechanism for the welding flange

Fig. 8.22 Laser focusing head in position to BFY-module

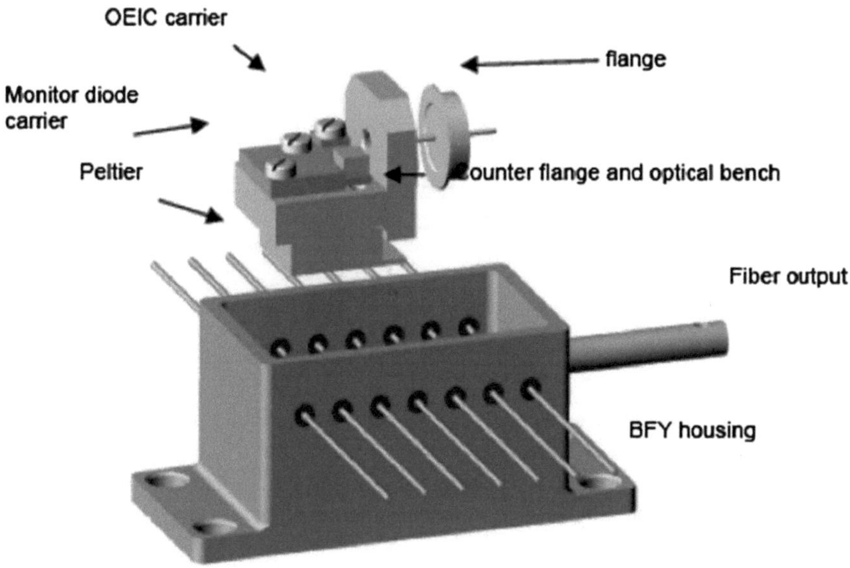

Fig. 8.23 BFY-module configuration

8.6.2 Module Setup

The module consists of a heat sink, which includes the OEIC on a carrier, the tapered fiber, the counter-flange, and the monitor photodiodes as well as the thermistor (Fig. 8.23). The carrier is located on the top of a Peltier cooler for the thermal stabilization of the OEIC. For the operation of waveguide-fed integrated photodiodes instead of laser diodes, there is no need for temperature control and the Peltier cooler is omitted. To minimize the mechanically twist of the optical bench, all used parts are made of stainless steel.

To compensate the difference in the extension coefficients of the OEIC (InP) and the heat sink, the OEIC carrier is made of copper. The used Peltier cooler has a heat pump performance of 2.2 W at 20 °C, which is fixed by glue at the bottom side of the lid of the housing to prevent heat circulation inside the module. Additionally, the BFY-module has to be contacted at the bottom of a heat spreader for use in high-temperature environment (+85 °C).

A screw fixes the OEIC carrier so that it can be removed from the module quickly. The laser bias, the Peltier currents, and the temperature sensor are contacted via the BFY pins. The optical fiber is fed into a metal cannula (ø 350 μm) while the end of the SMF is locked in a clamp for adjusting the longitudinal axis. After longitudinal adjustment, the fiber is fixed by glue or solder to the cannula.

The fixation by glue introduces no additional stress to the coupling region, but restricts the temperature range of the packaged module to +85 °C. This is due to the relative low melting point of the glue. Using solder has the disadvantage of

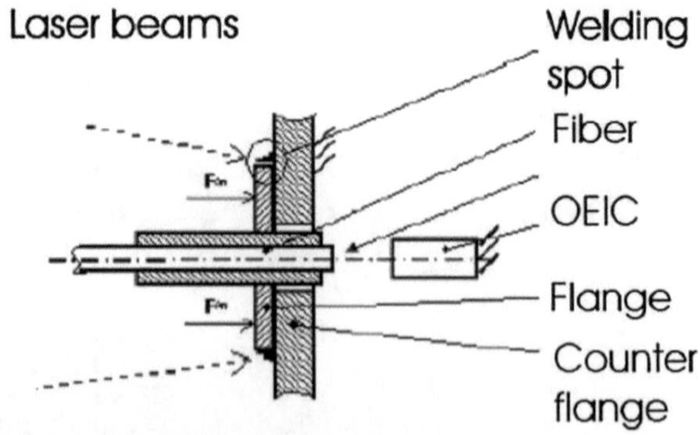

Fig. 8.24 Principle of the arrangement of OEIC and Taper

high-temperature differences in the coupling region. Without effective heat borders, this will damage the OEIC (Fig. 8.24).

To correct the polarization direction of the light in polarization-maintaining fibers or to correct the angle of a spherical-tapered fiber, the fiber can be rotated additionally around the longitudinal axis.

8.6.3 Flange Setup

The parts, flange and its counterpart, are represented in illustration (Fig. 8.24) schematically. The volume shrinking while solidifying and cooling of the welding points is due to a relative movement and consequently a shift between the welding parts.

The illustration in Fig. 8.25 schematically represents the movement of a welding part during cooling, indicated by arrows. Evaporating of superficial strangely materials, fat, corrosion-component, dust, etc. can also lead to an additional shift. Aim of this examination was to optimize the influential factors on welding time so that a minimal displacement will occur.

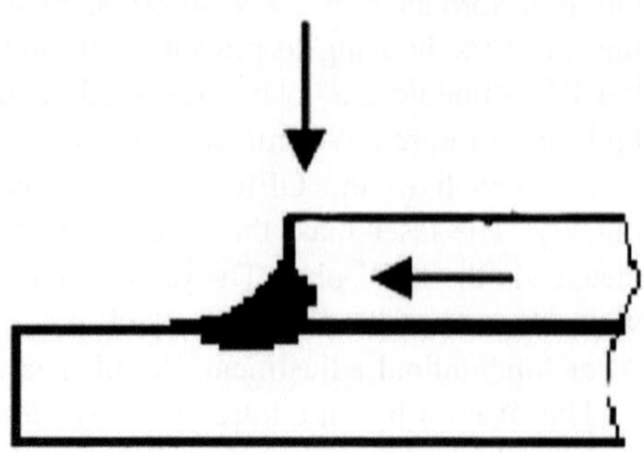

Fig. 8.25 Fillet weld shift

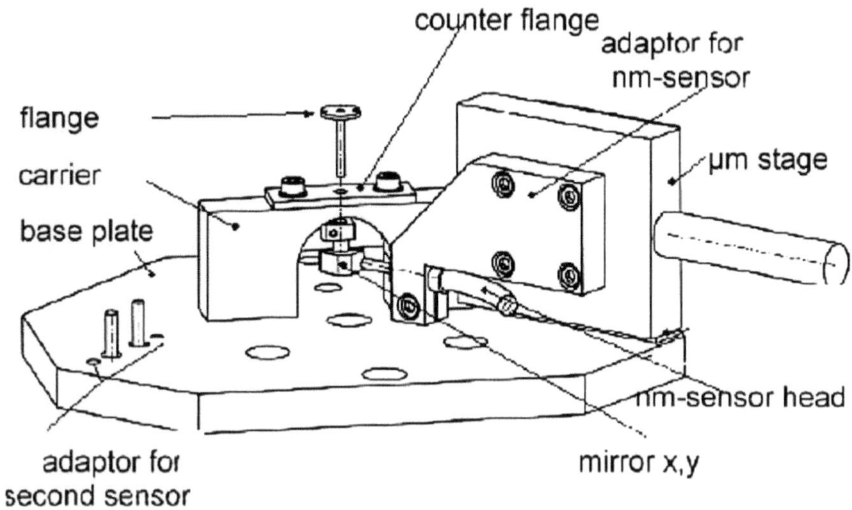

Fig. 8.26 Welding test setup

8.6.4 Welding Results

As shown before, the at most allowable dislocation for the fiber-chip connection of a butt fiber amounts to: ±3.0 µm and of a tapered fiber to ±0.5 µm. To measure the dislocations in dependence of the different welding parameters, we constructed a test setup, which is shown in Fig. 8.26.

The counter-flange with the cylinder-pen end is removed from the base plate. The flange with the cylinder-pen is represented in more detail in Fig. 8.27. The position of the flange in the entire construction is depicted on top of the carrier and counter-flange. The flange disk consists of stainless steel 1.4301 with 0.02 %

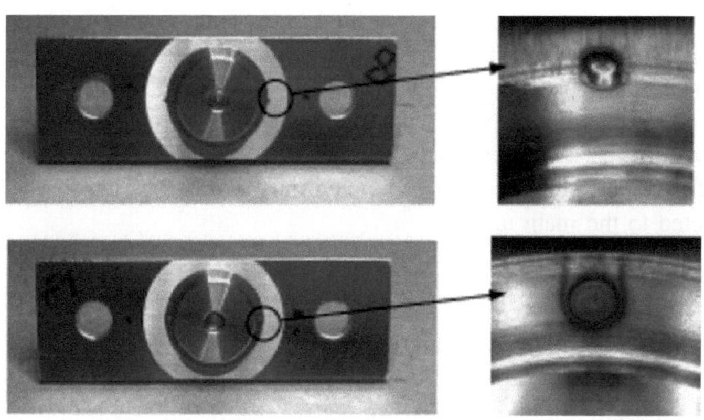

Fig. 8.27 used test flange configurations (*upper*) fillet weld, (*lower*) lap–fillet weld

Part	Number	Produced by / surface	Symbol
Flange	10	straight polished	
Flange	10	diffused polished	
Counter flange	10	diffused polished	
Counter flange	10	simply milled	

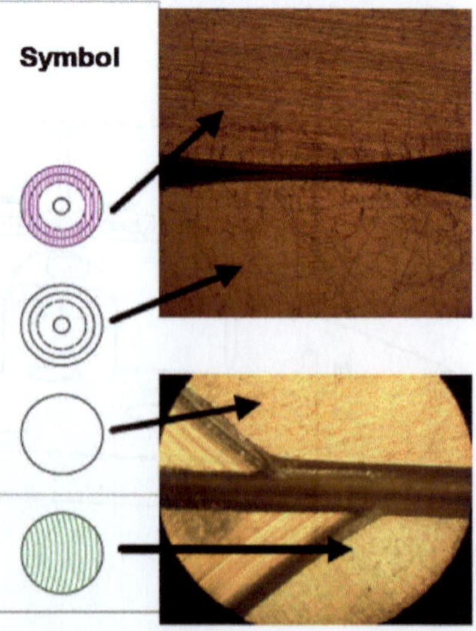

Fig. 8.28 Surface quality of test flanges

carbon. In a second sample, Vacon as basic material instead of stainless steel was used. In Fig. 8.28, the surface qualities of the used flanges and counter-flanges are depicted. Three basic configurations are proved: simply milled, straight polished, and diffused polished, which are used in different combinations.

The thickness of the flange disk is reduced at the welding points to 0.2 mm. Since several tests should be connected, a solvable connection of the counter-flange was intended to the test edition. At the end of the cylinder-pen, a mirror element is placed.

Consequently, the mirroring surfaces follow the movements of the test flange during welding. With the micrometer stage, a fixed distance is put in between nm-sensor top and mirror surface. The second μm stage is not depicted in Fig. 8.26, merely their mounting position, two drilled holes and two pens. The used fiber-optical sensor (Tetra Ltd.) has a measuring range of more than ±42.0 μm with a resolution of 30 nm.

It is used for measuring a static and differential distance optically. The difference originates from the measurement before and after welding. The signal is proportional to the distance between sensor top and mirror surface. Two sensor heads can be connected to the main unit.

8.6.5 Initial Welding Results

With this test setup, we measured the dislocation shifts presented in Fig. 8.29 with several parameter combinations:

Fig. 8.29 Mean values for welding dislocations at four parameters combinations

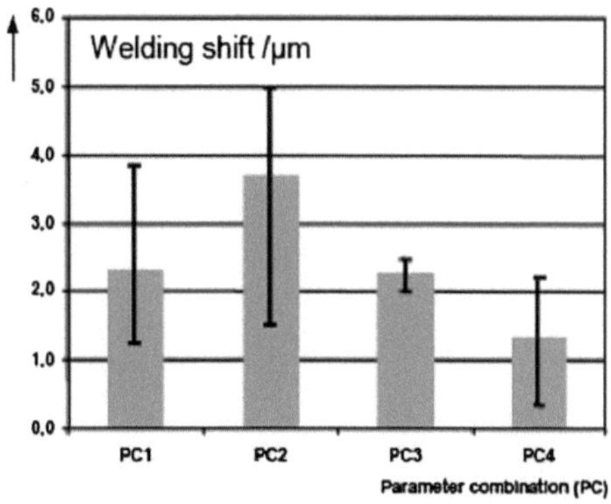

(a) Contact pressure between the two flanges,
(b) Laser energy, and
(c) Different types of welding flanges.

The combinations are listed in Table 8.3. Using the combination PC4, we measured the smallest dislocation shift values by using a lap–fillet weld. The laser power is relatively high in comparison with the other combinations (PC 1–3); on the other hand, the pulse length is very short (2 ms). It is also obvious that the contact pressure is only half of the value of the other combinations. It is surprising that we observed a significant smaller pressure force between the two flanges after welding. It can be interpreted as an indication that the flanges are pressed closer together due to the shrinking process during welding.

Additionally, the direction of the welding shift was measured, which can be seen in Fig. 8.30. Here, two kinds of shifts can be observed: (a) randomly distributed and (b) directional-oriented ones. The size and the situation of the values can be optimized by the welding parameters. It can be seen that the displacements are showing no random distribution around the origin. These asymmetrically distributed shift vectors are due to an asymmetrical energy distribution between the two welding points. If more energy is attached to the welding point L1 in comparison with L2, more material is warmed up there. Consequently, a bigger volume shrinking takes

Table 8.3 Measured welding shifts

PC-no.	Mean value shift (µm)	Parameter
1	2, 3	$F_{An} = 10$ N; $P_{Laser} = 1, 2$ kW; $t_{Laser} = 18$ ms; lap weld (PC 1)
2	3, 7	$F_{An} = 10$ N; $P_{Laser} = 0, 9$ kW; $t_{Laser} = 12$ ms; lap–fillet weld (PC 2)
3	2, 2	$F_{An} = 10$ N; $P_{Laser} = 2, 4$ kW; $t_{Laser} = 2$ ms; lap–fillet weld (PC 3)
4	1, 3	$F_{An} = 5$ N; $P_{Laser} = 2, 4$ kW; $t_{Laser} = 2$ ms; lap–fillet weld (PC 4)

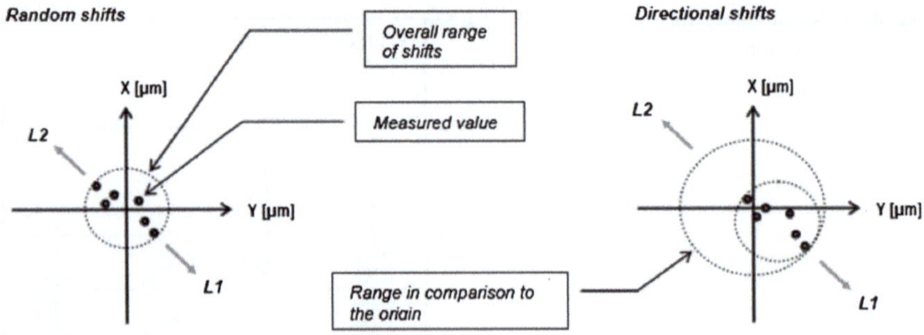

Fig. 8.30 Direction of the welding shifts

place in the welding point L1 while cooling. To improve the results lower than 1 µm, we used Vacon (Ni Co 29 18, 1.3981 after DIN 17745) instead of stainless steel, which has a lower extension coefficient of 4.9×10^{-6} instead of 15.5×10^{-6}.

The yield shown in Fig. 8.31 represents 20 interconnection tests with parameter combination PC4 with Vacon as flange material. Here, the numbers of welds are shown as a function of the dislocation during welding. The distribution shows that most of the welds are within 1 µm shift, but there are also some welds with very high dislocations detected (up to 11 µm). To get a better overview of the welding results, a different type of graph is used. It is shown in the right-hand side of Fig. 8.31. Here, the percentages of welds are drawn in comparison with the dislocation detected to the origin position before welding.

It can be seen that a shift of less than 2.0 µm can be reached with a yield of 80 %, which represents an additional insertion loss of 1 dB at butt-ended fiber coupling. For the coupling with tapered fibers, shifts lower than 1.0 µm are required. More than 50 % of the initial welds are within this region.

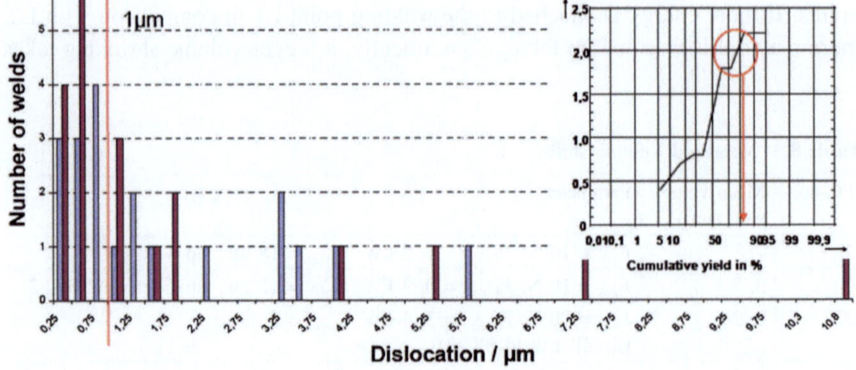

Fig. 8.31 Representation of the distribution of measured shifts with parameters of PC4. The inlet shows the lateral dislocation shift as a function of the percentage of welds performed

The best results have been achieved by using the combination of diffused polished and straight polished flange and counter-flange made from Vacon. Hereby, the straight polishing direction should be applied perpendicular to the axis between the laser spots.

It has also been measured the contact force of the welded flanges. All types of welding combinations showed excellence in mechanical stability (fracture force >60 N). The overall best values are found for the lap weld type where the highest laser power is applied.

8.6.6 *Correction After Initial Welding*

Without further optimizations like "laser hammering," it seems to be not possible to reach a higher yield, which is important for production of laser modules. After initial welding, we firstly measured the direction of the dislocation, which is either to laser L1 or laser L2. In a second step, we tried to redirect the dislocation by setting a second laser spot nearby the first laser weld of that laser, where the shift was in negative direction. The other laser head was faded out. This method was called in literature as "laser hammering." A typical plot of a redirected weld is shown in Fig. 8.32. After an initial dislocation of 0.6 μm, a minimum dislocation of less than 0.13 μm after 6 additional welds was reached. It was possible to reuse all flanges with high dislocations with this technique.

Additionally, we introduced for our knowledge the first time a novel redirecting welding technique, which we called as "strain-reducing" lasering (Müller 2001). Here, we used only the laser when the shift was in positive direction, which is

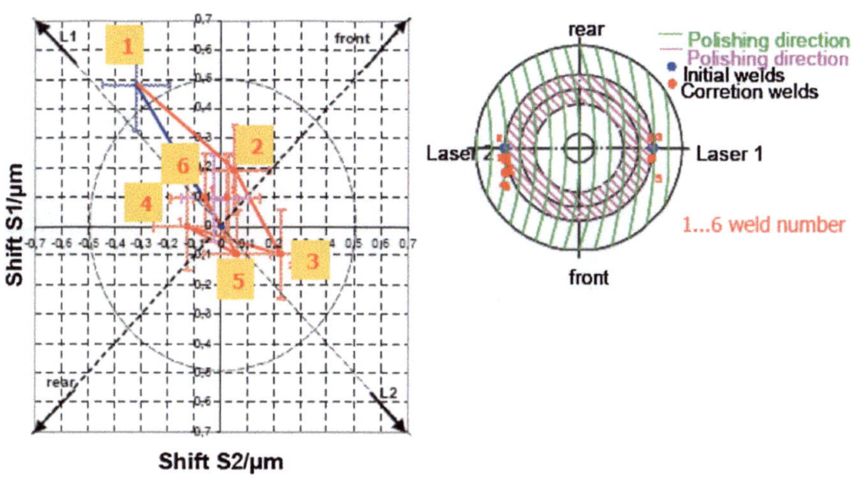

Fig. 8.32 Correction of initial shift by additional laser welds

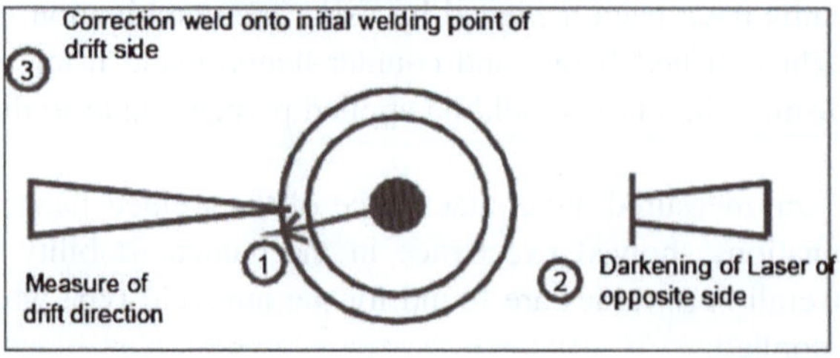

Fig. 8.33 "Strain-reducing" lasering

shown in the sketch of Fig. 8.33. The opposite laser head was darkened to achieve an unsymmetrical heating of the flange connection.

By melting only one welding point a second time, the strain could relief and the shift could be redirected. This technique is very simple because only one laser head must be darkened and no additional translation of the welding laser along the flange must be performed. With this technique, it was possible to lower the initial welding shift with a probability of more than 66 %, which is of the same order as found for the "laser hammering" technique.

In Fig. 8.34, a photograph of the developed laser workstation is shown above. At the top, the two laser heads are situated, while at the bottom the coupling machine with the μm stages can be seen. The butterfly module is placed in the middle of the apparatus, where a microscope camera magnifies the fiber-chip coupling region. For easy loading and unloading, the module from the machine, the stages of coupling machine, and laser heads will drive automatically to defined end positions.

Fig. 8.34 Photograph of the laser workstation

After putting the fiber into the flange tube, the automated search algorithm (Newport) will find within several seconds the point of maximum intensity. Then the fiber will be fixed in *Z*-direction with glue and the welding process will fix the flange in *X*- and *Y*-directions. The whole procedure can be performed within several minutes, which is fast enough for rapid prototyping and low volume production for use in laboratory environment.

8.6.7 Dynamical Shift

To get more information about the dynamics of the shrinking process, we performed additionally time resolved studies during the cooling phase of the welding.

The results are surprising to us, because we noticed a long dynamical phase with very high dislocations: within the first 200 ms after welding the measured shifts are a factor of 5–10 higher than the static measured drifts after thermal equilibrium. We found drifts up to 13 μm with a following static end shift of less than 1 μm.

The power of the thermal expansion of the melted material must be drastically higher in this phase than the frictional forces between the surfaces, because we found that the values of the drift during this dynamical phase are not dependent on the surface quality! All investigations show that the dynamical phase is finished after 200 ms after the initial laser pulse. A typical plot of the dynamical behavior is shown in Fig. 8.35. On the right side, the time resolved curve of the dynamical drift is depicted with a timescale of 10 ms /division. The two curves represent the output of the two sensor heads. A rise time of 6 ms and a decay time of 48 ms are found with a maximum drift of 6 μm.

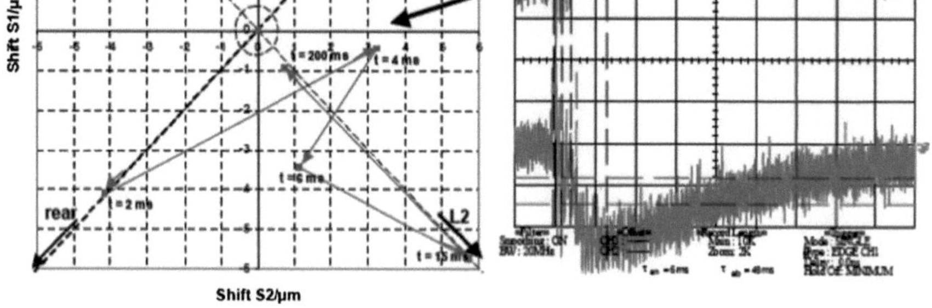

Fig. 8.35 Temporal behavior of the welding shift within the first 100 ms

In the left-hand graph, the drift of the flange is marked in accordance with the right-hand side time resolved curves. It can be seen that in the dynamical phase the flange drifts strong to several directions until the metal solidifies after 200 ms with 0.9 μm shift. By way of summarizing, we consider that it is very important to use stainless steel with low carbon and very low extension coefficient, in combination with moderate high pressure to lock the parts together.

References

Anthamatten, O., et al.: Laser welding for fiber pigtailing with long-term stability and submicron accuracy. Opt. Eng. **34**, 2675–2682 (1995)

Becker, M., Günther, R., Staske, R., Olschewsky, R., Gruhl, H., Richter, H.: Laser micro welding and micro melting for connection of optoelectronic micro-components in laser in Engineering. Springer, Berlin (2001)

Eckhardt, T.: Entwicklung eines DIL-lasermoduls, (1999)

Fischer, U.H.P., Zech, S., Peters, K.: Transmitter modules with reusable fiber-chip coupling method for optical communications systems, http://www.eetimes.com/design/communications-design/4017993/A-Reusable-Fiber-Chip-Coupling-Method-for-Optical-Communication-Transmitter-Modules (2001)

Fischer, U.H.P., Krips, O., Müller, E., Jacob, A.: Laser microwelding for fiber-chip coupling modules with tapered standard monomode fiber ends for optical communication systems. Opt. Eng. **41**, 3221–3229 (2002)

Kato, D.: Light coupling from a stripe-geometry GaAs diode laser into an optical fiber with spherical end. J. Appl. Phys. **44**, 2756–2758 (1973)

Krips, O.: Konstruktion einer Justiereinrichtung zur Ausrichtung von Lichtwellenleitern in einem Gerät zur Chip-Faser-Kopplung (2000)

Kulina, R., Ringelhan, H., Weber, H.: Materialbearbeitung durch Laserstrahl. Deutscher Verlag für Schweißtechnik DVS-verlag, Düsseldorf (1993)

Kuwahara, H., Sasaki, M., Tokoyo, N.: Efficient coupling from semiconductor lasers into single-mode fibers with tapered hemispherical end. Appl. Opt. **19**, 2578–2583 (1980)

Müller, E.: Readjusting Two-Point Laser Welding for Automated Fiber-Chip Coupling (2001)

Chapter 9
Passive Adjustment Techniques

Abstract In the course of the book, methods for optical coupling have been described that realize an active adjustment of the fiber by micromechanical actuator elements. In the present chapter, however, several methods are described that allow a passive fiber–chip connection. These methods include the flip chip (FC) technology and the LIGA technique (lithography, electroplating, molding technique). FC technique has also the great advantage of allowing a batch processing of the optical and electrical connection structure for a mass production of future optoelectronic communication engineering applications.

9.1 Flip-Chip Technique

The flip-chip (FC)-bonding technology or controlled collapse chip connection technology (C4) was developed in 1964 by IBM for the assembly of hybrid modules. The aim was to replace the then unreliable and unprofitable manual wire bonding method (Hauffe et al. 2001). The FC-bonding technology enables of all existing bonding methods the maximum achievable port density and will therefore play in the future an ever greater importance in the microelectronic assembly and connection technology (AVT) (Wale and Edge 1990; Makiuchi et al. 1993).

In the hierarchy of the different levels of electro-optical packaging (Fig. 9.1), the first-level packaging is situated at the top priority. Here, bare chips are electrically connected to a sub-carrier. Therefore, it is of great importance to use a cost-effective, time-saving contacting method. In the FC-bonding technology, the components can be mounted "face down" (i.e., with the top down, see Fig. 9.2). A visual inspection of the solder joints is somehow difficult to realize in this case. On the other hand, this type of contact allows an unpackaged chip assembly, so-called direct chip attach (Katsura et al. 1991; Goodwin et al. 1991; Lau 1995).

In Figs. 9.3 and 9.4, an overview of the FC-bonding method currently used is illustrated. A distinction is generally made between procedures with and without solder. The solder-less applications are working with adhesive compounds that

© Springer-Verlag Berlin Heidelberg 2015
U.H.P. Fischer-Hirchert, *Photonic Packaging Sourcebook*,
DOI 10.1007/978-3-642-25376-8_9

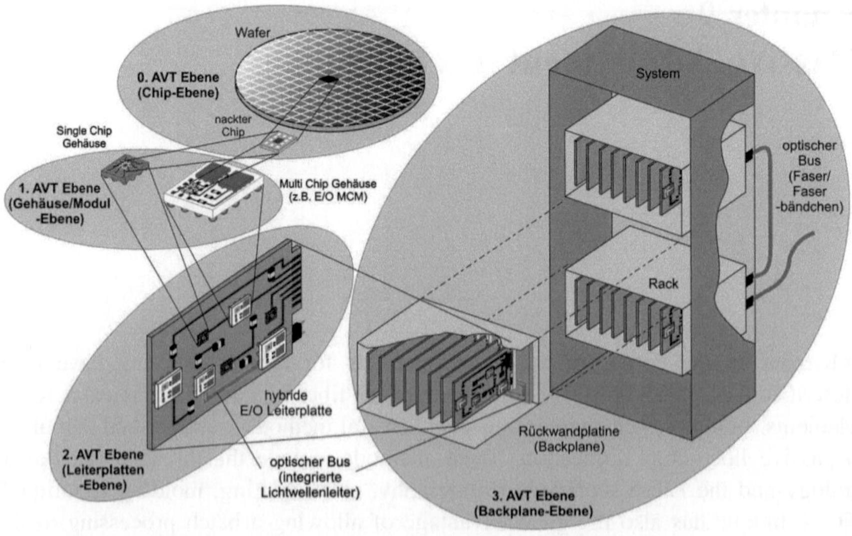

Fig. 9.1 Three basic levels of electro-optical packaging (Nieweglowski 2011)

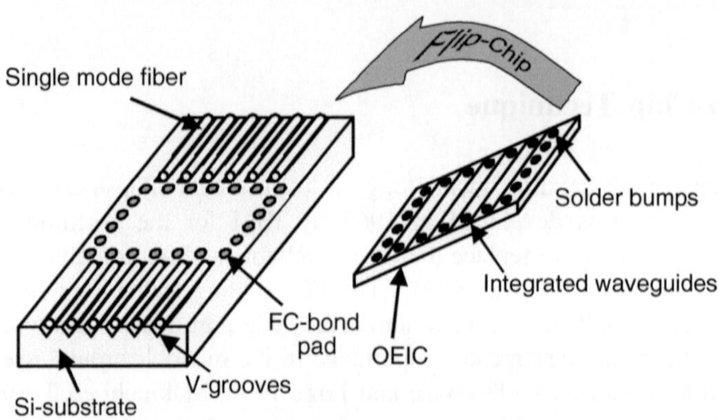

Fig. 9.2 Illustration of the FC-bonding principle

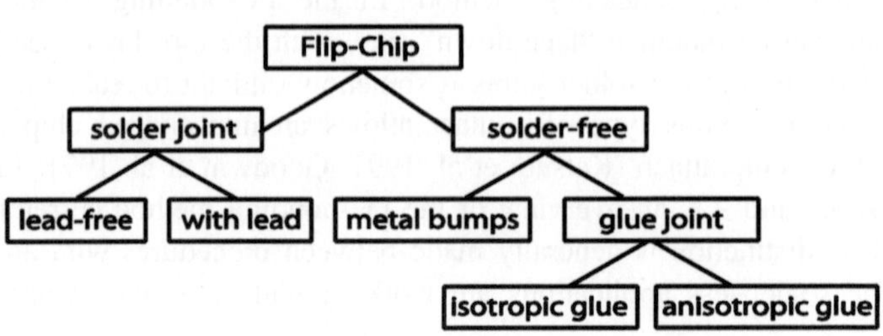

Fig. 9.3 FC-bonding categories by connecting materials

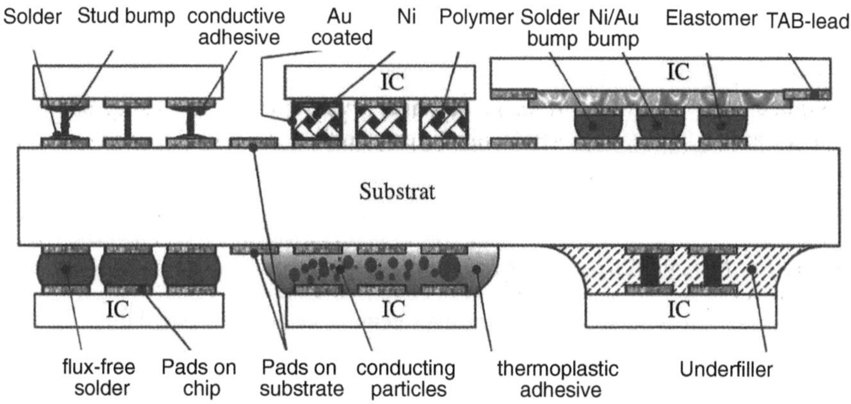

Fig. 9.4 FC connection mechanisms

contact with both, isotropic and anisotropic conductive adhesives. Alternatively, one may operate with metallizations, which are directly pressed together with the opposite side, and thus establish an electrical connection. For experimental setups, the stud bump process (an approximately hemispherical increased solder joint) is very much applicable. The bumps are realized with the help of a commercially available ball bonder (see Chap. 6).

Gold bumps are applied to the substrate, which are bonded to the counter pad (the starting point of wire connection), which is electrically conductive. For large-scale applications on flexfoile, very flexible bumps are necessary using gold-coated polymer bumps that can be produced after the bonding for inexpensive costs. A distinction is made between the choice of the solder in conventional lead–tin compound and those without lead content. In the latter, solder gold–tin alloys have prevailed.

With the help of this ball bonder, bumps of gold (gold bumps) are applied, which are glued electrically conductive to the counter pad (the starting point of the wire connection). For large-scale applications on flexible foils, bumps pads also from gold-coated polymers are on the market. These can be produced particularly inexpensive and also remain very flexible after bonding.

In the soldering process, a distinction is made between the choice of the solder in conventional lead–tin compound and those without lead content. In the latter, gold–tin and silver–tin alloys have prevailed. Lead–tin compounds are internationally forbidden due to physical long-term damage caused by the heavy metal lead, except for special applications, which are documented in the so-called RoHS directive (Commission and Document 2002).

In their dimensions, small and reproducibly manufacturable geometries, FC bond pads with sizes of about 100 μm can minimize RF side effects, which constitute the superiority of FC-bonding technology to conventional wire bonding techniques especially in the range of frequencies above 20 GHz (Sutherland et al. 1993; Dantartas et al. 1997).

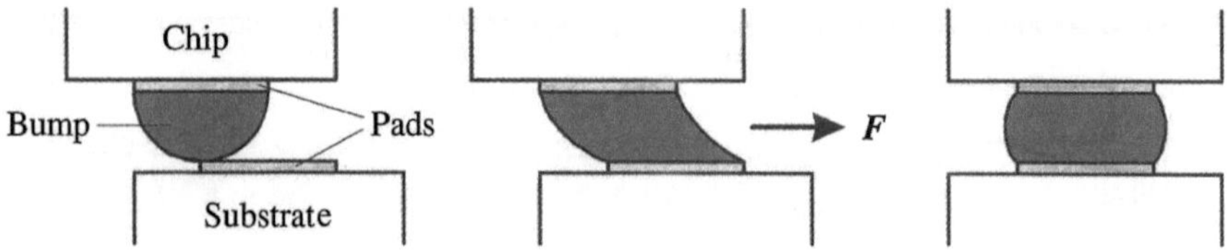

Fig. 9.5 Formation of surface tensions of the liquid bumps with resultant formation force at self-adjustment

A further advantage of FC-bonding technology is the possibility of self-adjustment. It is based on the action of surface tensions of the melted solder, formed during the FC bonding (see Fig. 9.5). In the pre-adjustment of the samples, one needs to ensure that the solder bumps overlap by about 50 % to the opposite, symmetrically arranged metallization pads. This reduces the effort for the pre-adjustment and thus also the costs that arise at the chip mounting by the FC-bonding technology (Nishikawa et al. 1994).

The self-centering of the deflected liquid solder bumps was shown from Whale (Wale and Edge 1990) at the first time for waveguides with lithium niobate devices. Here, it was necessary to realize an offset of less than ±2.5 μm for the single-mode fiber-optic coupling tolerances in lateral directions.

Soldering flux used for the assembly of microelectronic devices can damage the sensitive, optically active surfaces of the photonic devices (eg. lasers, photodiodes). The flux-free process in addition to the required positioning accuracy is the fundamental difference between the microelectronic and the photonic application of this contacting method. The process of the flux-free bonding in vacuum with H_2 reduction has been developed at the Heinrich Hertz Institute and has been confirmed in other institutions (Kuhmann et al. 1996) for lead-free soldering systems.

FC bonding can achieve a simultaneous electrical and mechanical connection between the chip and substrate at the same time. The testing in terms of functionality and reliability before installation is however very difficult. Due to the facedown mounting, the solder from the chip is largely obscured, which has an adverse effect on a visual inspection of the solder joints. The connection between chip and substrate can be made with sealing compounds (underfiller, see Fig. 9.4) that are sealed in order to prevent the unit from environmental influences. The sealing compound provides additional mechanical stability between the chip and substrate.

Without the mechanical stabilization, cracks can occur in the solder joints during temperature fluctuations due to the different coefficients of expansion. If no encapsulants are used, the long-term stability of the connection is very limited. If one uses bonding components, with approximately the same coefficient of expansion, it is possible to achieve stable compounds. A major advantage in the FC technology is the self-adjustment of the liquid bumps during bonding, whereby the adjustment effort significantly reduced. Only slightly more than 50 % of the bond pads must overlap with the bumps in order to achieve a reliable contact.

Once the heated liquid solders wet the bond pad, surface tensions arise in the displaced bump. The resulting forces pull the bond pad to the center of the bump. In the FC bonding, several different techniques for the connection options of the contact surfaces have developed. The most important of these differ in the types of contact between soldering and gluing techniques. To get solderable connections, either the chip or the substrate must be equipped with solder bumps. The generation of these bumps is carried out by different processing techniques. The most important methods are vapor deposition (Beyer et al. 1996), electroplating (Reichel 1998), and wire bonding (Scheel 1999) in a special modified form.

The solder bumps are melted in a reflow process and reach a capped spherical shape on the thin-film metallization. The actual contact is made in a further reflow process. The components to be bonded are successively placed, then aligned, and heated at the end. Thus, the solder melts and connects the two parts, both electrically and mechanically. During evaporation, the desired solder is deposited on the pad metallization. A sufficiently thick resist mask ensures exact position and a sufficient height of the fabricated solder bumps. The mixing ratio of soldering components involved can be set exactly in the evaporation process. The resist mask is removed after the process by a liftoff process, and the solder bumps are reflowed. To increase the height of the bumps, you have to use additional steps in their production.

One of the possibilities is to a higher applied coating layer that increases the thickness of the vapor deposition. Another way to increase the bump height is to choose the structured window areas in the resist mask larger than the metallization areas, creating more solder material. After reflow, the excess solder is withdrawn from the non-wettable surface of the substrate on the pad and shows the typical shape of a bump ball (Fig. 9.13) which has a greater radius due to the increased amount of solder.

The electroplating of solder bumps requires a different approach. The entire substrate with its contact pads is covered with a continuous metal layer. Thereafter, a photoresist is applied and patterned on the corresponding pad areas. The continuous metal layer causes the electroplating of solder material. The layer deposition takes place in the photoresist windows on the exposed and conductive metal surfaces.

The photoresist can be removed easily because it remains freely accessible from above for solvent. Finally, the uniform metal layer is etched on the substrate, while maintaining the solder bumps are used as a mask and will remain. Here, according to the reflow process the bumps are formed to the curled spherical shape. Applying this type of coating layer, a formation of a significantly higher layer thickness than in the process of coating technology to the vapor deposition (Fig. 9.6) can be achieved.

In a third option which is produced with a ball, the bumps wire bonder as shown in Figs. 9.7 and 9.8 can be seen. The bumps are placed individually on the metallization, and the case must withstand thermal loads and mechanical pressure, so that they do not detach from the substrate. Every single bump must be accurately positioned with great expenditure of time. Thus, this method is only suitable for small series. The bump heights also vary according to the possibilities of different ball bonder.

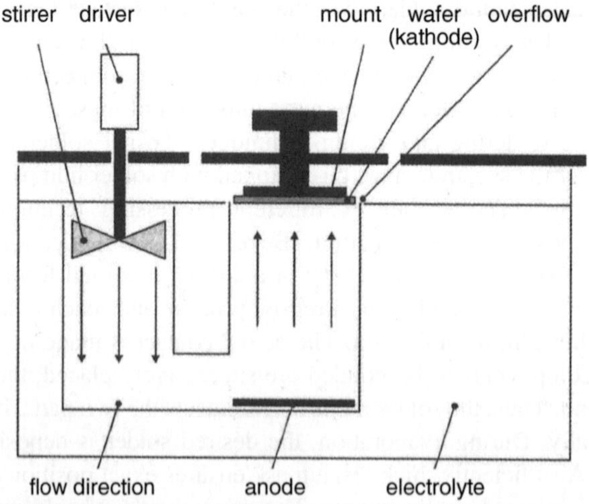

Fig. 9.6 Electroplating plant with electrolyte flow

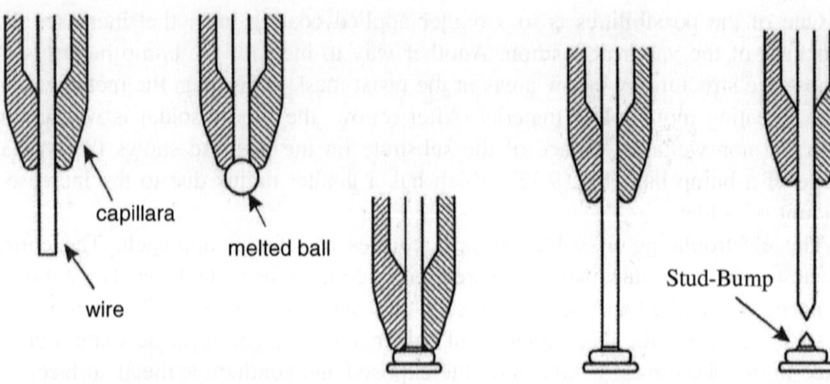

Fig. 9.7 Production of stud bumps

The bonding techniques can be divided into isotropic and anisotropic bonding. At isotropic bonding, the adhesive must be very printed precisely on the substrate (Fig. 9.9). After hardening, an additional underfilling with sealing compound is necessary to obtain a sufficient mechanical stability of the bonds. The contact resistance thus obtained are inferior to those with soldered contacts, but much more better than in the anisotropic bonding. On the other hand, anisotropic bonding is technically the easiest option. Here, no specific pressure is required for the curing, and because of the anisotropy of the system no underfilling is necessary (Fig. 9.9).

Whereas in isotropic adhesives, a high content of equal distributed metal particles ensures that the contacting is carried out in all three spatial directions, the principle of the anisotropic bonding is different (Fig. 9.10).

Fig. 9.8 Stud bump

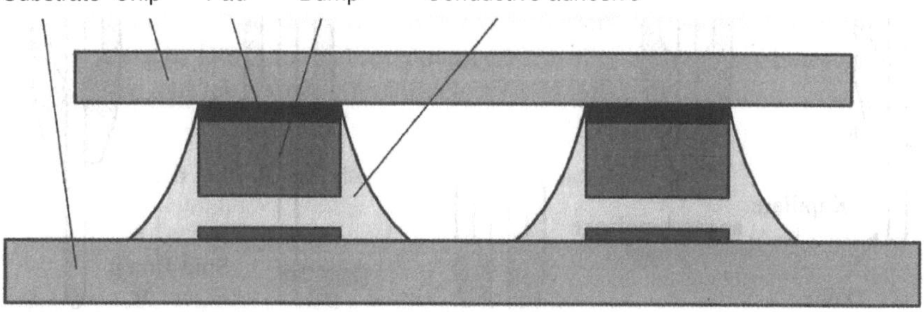

Fig. 9.9 Flip-chip-bonding technique with isotropic adhesive without underfill

Here, the adhesive may be added only with few particles that it is non-conductive in the basic state. Only when the adhesive is compressed under elevated temperature, the conductivity will be stabilized. It is realized only in the pressure direction by direct contact of the metallized particles at the surfaces to be contacted. It must be ensured that a statistically sufficient number of particles per pad area is available. To ensure a flat pressing of the chips on the carrier material, the particles must have a very small diameter distribution. They must also be elastic for optimal contact. Therefore, the particles are fabricated using as small plastic ball form, which is coated by a metallization. As a result that the plastics material is elastic, the particles can be deformed easily.

To avoid short circuits, the degree of filling of the adhesive with conductive particles may be so high that no direct particle contacts occur.

Table 9.1 summarizes the main advantages and disadvantages of the three contact types for the FC technique together.

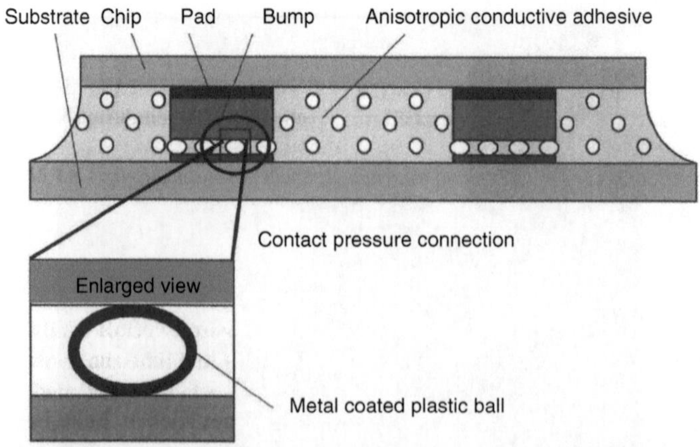

Fig. 9.10 Anisotropic bonding process with image cut from a contact print connection between the path of the metal-coated plastic ball and the bump

Table 9.1 Characteristics of different flip-chip contacts

Contact	Benefits	Disadvantages
Soldering	Lowest contact resistance	Most expensive method
	High-quantity feasibility	Technically difficult
		High investment costs
		Not flexible
Isotropic gluing	High current densities possible	Precise printing technology
	Flexible quantities possible	Underfilling
	Contact resistance good, but slightly worse than when soldering	Adhesive joint not flexible
Anisotropic gluing	Simplest method	Poor contact resistance
	Low investment costs	Low current density
	No precise pressure required no underfilling	Small quantities
	Flexible adhesive joint	
	Ideal for display applications	

9.1.1 Flip-Chip Bonder

For the flux-free FC-bonding technology, a FC bonder was developed at the Heinrich Hertz Institute for Telecommunications, Berlin, in cooperation with the Ferdinand Braun Institute. The bonder is designed that the necessary environmental parameters for bonding can be realized easily. For this purpose, a fine adjusting station for ICs (Rudolph 1999) was modified so that it was possible to perform FC bondings under vacuum conditions. The alignment principle of the adjusting station of the company Finetech (Gruner 2014) is based on the image reproduction of two-layered images in a field of view. The substrate is placed on an adjustable table, and the chip on a movable lever was fixed using vacuum. The chip and the carrier substrate can simultaneously displayed and their position to each other. This is located under a specially adapted microscope using a beam splitter, by where the underside of the chips with the metallization, and the upper surface of the substrate may be viewed simultaneously with the solder bumps (Figs. 9.11 and 9.12).

The chip is fixed using a vacuum suction and placed on a pivot arm on the substrate. Using a positioning stage, the chip can be adjusted to the substrate with an accuracy of up to 2 μm in the *x*- and *y*-directions.

A heater, a halogen lamp from Osram HLX 64635 type, is implemented below a quartz glass plate operating with 150 W at 15 V, realizing temperatures up to 550 °C. The focal point of the lamp can be accurately aligned to the substrate and the chip, which are located directly above the glass plate. The heating of the sample occurs by the absorption of thermal radiation. The heater lamp is operated with a regulated power supply. It is (type CN76060) controlled by a PID controller of the company Omega receiving the necessary temperature values for the regulation as converted analogue values from a non-contact infrared sensor from Raytek (type Thermalert TX).

The required vacuum system for the bonding process consists of an annular vacuum chamber having a bottom opening which can be swung over the sample. A glass plate made of zinc selenide (ZnSe) is the top plate, so that the temperature

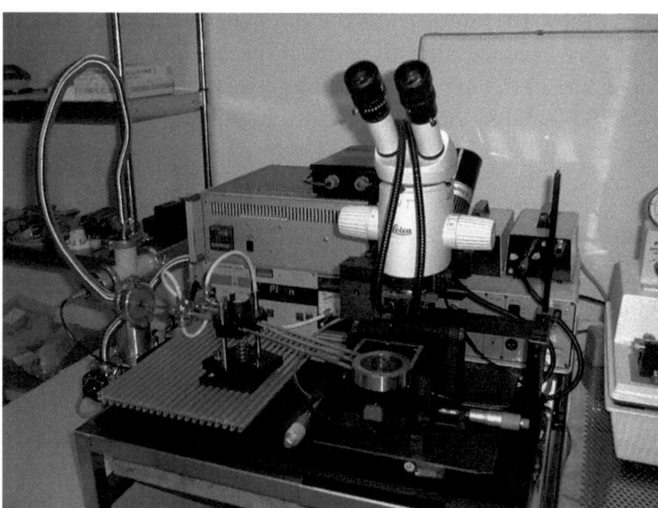

Fig. 9.11 Flip-chip-bonding workplace in HHI, Berlin

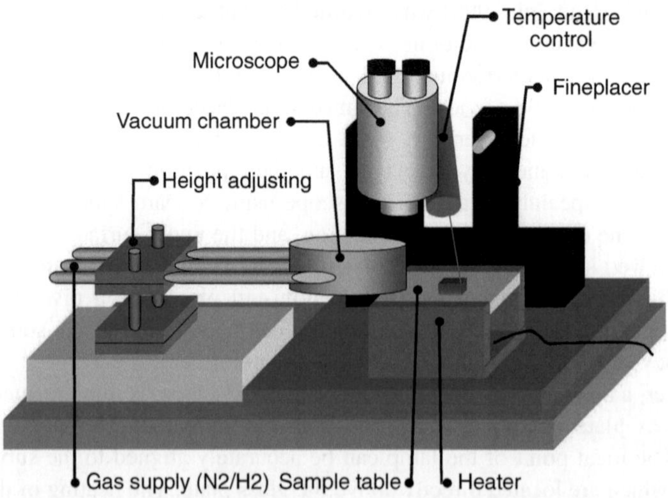

Fig. 9.12 Illustration of the main components of the flip-chip bonder for bonding process

sensor can measure the process temperature during bonding. ZnSe is transparent, in contrast to silica glass for infrared radiation.

An additional connection provides the supply of the inert gas (nitrogen, N_2 or forming gas, a mixture of nitrogen and hydrogen (N_2/H_2) in the ratio of 95:5 percent by volume) for the reduction of the oxide films of the solder bumps.

The vacuum system consists of a rotary vane pump BHV 10 Leybold as a backing pump and a turbo molecular pump TMH 065 with the associated drive electronics Balzers TCP 015 as the main pump. The backing pump generates an initial vacuum of 2×10^{-2} mbar and the turbo pump a vacuum of 2×10^{-5} mbar. The vacuums were measured directly at the respective pump, in which chamber a residual air pressure of 0.3 mbar prevails. More information about the setup of the FC bonder can be found in the thesis of Dirk Rudolph (Rudolph 1999).

9.1.2 Reduction of the Oxide Layers During Bonding

FC-bonding technology, in addition to the favorable electrical properties (low side effects), the ability to center by surface tensions of the liquid solder during the bonding of the solder bumps is deflected. This allows for a highly accurate, self-aligning fiber–chip coupling.

A risk of this process is the formation of solder surface oxide layers, which prevent free flow of the liquid solder and the exact centering. Natural oxide layer with a layer thickness of 3 nm to 5 nm on the solder surface should be reduced during the bonding (Maly 1997). When the melting point is exceeded, the oxide layer is broken by volume expansion and allows the wetting with the metallization.

The remaining oxide layers now prohibit the self-adjustment process and the optimum centering.

There are several approaches for the reduction of the oxide layers. Here, the use of flux in the solder must be omitted in comparison with the usual bonding of microelectronic components. Flux contains chemically active substances such as inorganic or organic acids or salts that can damage the sensitive, optically active surfaces in photonics.

With the use of tin–lead (SnPb in a mixing ratio of 60 to 40) as a solder material, all sorts of oxides such as SnO, SnO_2, PbO, and PbO_2 are present in almost equal amounts. The oxidation kinetics studies of tin and reduction operations of the oxide by the supply of hydrogen were firstly examined by Maly and Kuhmann (Maly 1997; Kuhmann and Pech 1996).

As a result, the bondings are successfully carried out at temperatures of 250 °C with a heating time of less than two minutes. The hydrogen is supplied to the recipient as a forming gas. This resulted in adjustment accuracy of the chip layer, and the better were compared to the substrate than 3 μm. Fulfilling this requirement opens the way for the possible use of the flux-free FC technology for coupling of optical components.

9.1.3 Flip-Chip-Bonding of Optical Components

By the help of self-adjustment during the soldering process further opportunities for using FC technology can be found in photonics. Using this technology, single-mode fibers can be coupled with waveguides with the required alignment tolerances in the submicron range. Studies of Wale (Wale and Edge 1990) have shown for the first time that the required tolerances of ±2.5 μm could be met for a single-mode fiber-optic coupling in the lateral direction. The elaborate and costly active coupling of fibers and waveguides with optical signal processing in high-volume production can be nowadays replaced by the FC technology.

The anisotropic wet chemical etching of silicon is an established process in microsystem technology (Beyer et al. 1996; Menz 1997; Steckenborn et al. 1991) and allows fixation of fibers, fiber arrays, and lenses with very high precision. The FC-bonding technology in combination with high-precision patterned V-grooves represents an inexpensive way to produce self-adjusting electrical and optical connections.

Small bond geometries provide excellent high-frequency characteristics in the frequency range above 20 GHz. Unwanted side effects are significantly reduced and thus the FC-bonding technology is well suited for use in the optical assembly and interconnection technology. In photonics, the optoelectronic ICs (OEIC) usually have a low integration density. The focus here, however, depends on the low-loss transmission of high frequencies caused by an optical–electrical signal conversion in the OEIC. The FC-bonding technology is superior to conventional wire bonding techniques in the RF range.

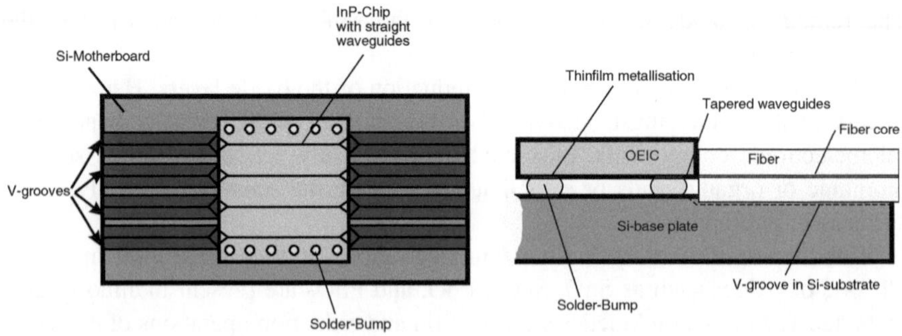

Fig. 9.13 Schematic representation of a fiber–chip coupling using the FC-bonding technology as contacting and self-adjustment technique

The flux is added to prevent oxide layers at the solder bumps to realize good connections between the liquid solder and the connection pads.

However, in photonics, the use of flux must be completely dispensed, for not to destroy the optical properties of the sensitive chip. Flux can penetrate into the surface of the chip and thus change its crystalline properties. Therefore, one reduces the oxide layers instead of flux with the addition of hydrogen.

The flux-free process in addition to the required positioning accuracy is the main difference between microelectronic and photonic application of this contacting method (Fig. 9.13).

9.1.4 Metallization

In the FC technology, chip and substrate are directly connected by solder bumps. In the method used here, the soldering between opposing metallization (bond pads Fig. 9.14), (Reichel 1998; Hügli 1998) takes place. The bonding pad on the substrate and the solder bump is called ground metallization or bump-limiting

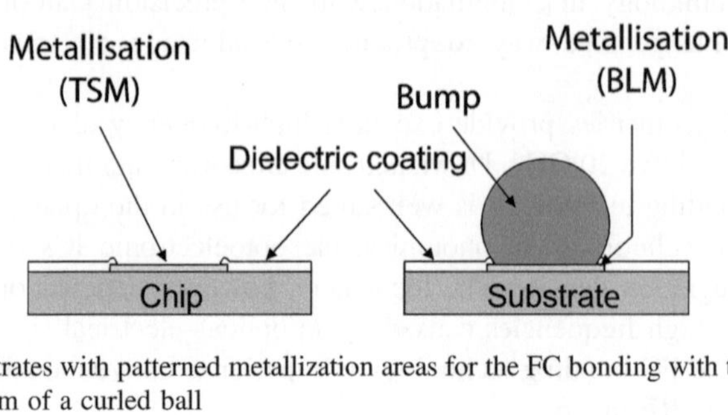

Fig. 9.14 Substrates with patterned metallization areas for the FC bonding with the molten solder bump in the form of a curled ball

metallization (BLM). The pad metallization on the opposite chip is called the top surface metallization (TSM), since applying the chip on top of the solder bumps.

Metallization is used for electrical contacting of the active regions of the integrated circuit and the realization of the electrical connection of the active regions with each other. Additionally, the metallization connects the active parts of the chip (Input/Output, Power, Ground) to the outside world.

The basic layer structure usually consists of multiple film layers with different functions and requirements. Most representatives are generally metallization (several layers possible), dielectric diffusion barrier, adhesion promoters, passivation, and contact pad, among others.

At the metallization, the following requirements are important:

- Low-resistance ohmic contact to silicon (N, P doped)
- High conductivity and high current-carrying capacity
- Good technological processing (structuring, simple processing)
- Good adhesion
- High mechanical and chemical stability
- High temperature resistance

Passivation layers (Menz 1997; Biletzke 2000) have the duty to protect the underlying active and passive layers from external influences. They serve as protection against mechanical damage, corrosion protection of metal deposition, and diffusion barrier against impurities. These layers consist mainly of silicon oxide and silicon nitride or partially polymers. Typical layer thicknesses for silicon oxide/ silicon nitride are up to one micrometer, and polymers are between 2 and 4 μm. In addition to the protection functions of the passivation layers, they have a high ohmic resistance of 1×10^{14} W/cm for $SiO_2/Si_x N_y$ and more than 1×10^{16} W/cm in the polymer.

Adhesion promoters take over the task to connect the conductive material to the carrier substrate, and the thermal and mechanical effects occur in the further production process. Layers as diffusion barriers are used to prevent a possible diffusion of solder components in the metallization pads or reverse.

In coating technology, a basic distinction between two types of coatings is as follows:

- Physical vapor deposition (PVD)
- Chemical vapor deposition (CVD)

Microsystem technology also often uses hybrids of the two methods.

9.1.5 Lithography

Lithography (Beyer et al. 1996; Menz 1997; Schade 1991; Widmann et al. 1996) means structuring by ablation of certain substrate regions. The most important processes are the exposure, the coating with photoresist, developing and etching with the equipment and materials needed for this. To transfer a planar structure by a photolithographic method are used photoresist and mask.

Photomasks are made of a material transparent to radiation and a structured, highly absorbing or radiation opaque layer. The wavelengths of the radiation used in each case comprise the mercury spectrum (200–600 nm), X-rays (0.1–80 nm), electron beams (0.001–0.08 nm), and ion beams (far greater than 10^{-4} nm).

The light-sensitive silver halide emulsions, which are most commonly used for photomasks, are provided for feature sizes of 3 μm. Most usual are emulsions on large glass plates from which the photomasks are made by cutting the plate into smaller parts. Silver as a light-sensitive material is too expensive. Therefore, a transition is made to chromium masks. Here, the contrast is better than at the masks of silver halide. Resist masks are distinguished in that they are resistant to further processing steps such as film growth and etching.

The photosensitive layer (photoresist, photoresist) is deposited uniformly on the wafer to be patterned. One possibility is the separation of the liquid phase and is called the spin coating (Fig. 9.15). In this method, the coating material is applied in liquid form to a rapidly rotating substrate disk so that, after drying, leaving a uniform film. The viscosity and the rotation speed of the substrate can adjust the layer thickness.

The most important substances for adhesion and alternating the masks are as follows:

- Photo- and radiation coatings
- Resists are used as a preservative, particularly when it comes to adhesion masks, and the substrate selectively protects against aggressive attacks.

The coating consists of a polymer and a solvent. Additives are used to modify the properties. In the photolithographic process step, the additives upon irradiation change the molecular weight of the resist. UV light irradiation (l < 300 nm) results to selective solubility differences in the resist. Exposed and unexposed portions have different resistance to selected solvent (developer). Depending on the development option, a distinction in positive and negative resists occurs.

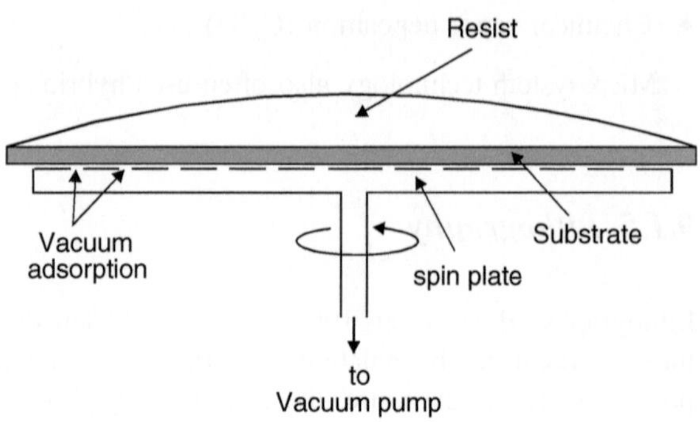

Fig. 9.15 Schematic representation of the spin coating

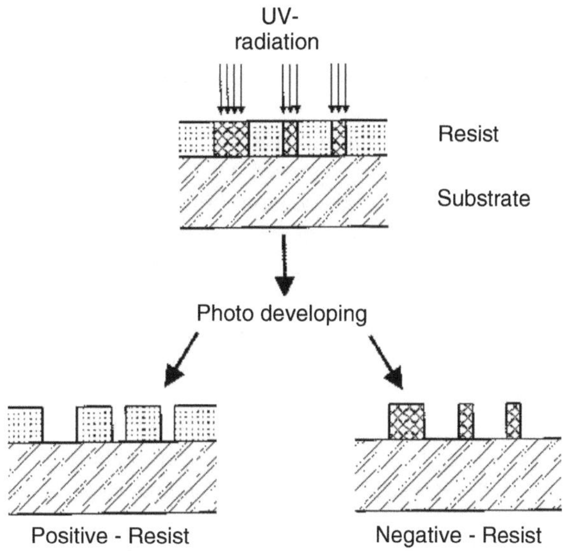

Fig. 9.16 Structural development of positive and negative resist

When positive resist only the irradiated area with the corresponding developer can detach, and the non-irradiated area is retained, i.e., it creates a positive image of the mask. Just the other way around, it behaves as the negative resist (Fig. 9.16).

The positive resist is superior with regard to the negative resist due to the different composition and different photochemical processes in the photoresist reactions. The positve resist has therfore a higher resolving power at the same layer thickness. Thus, the positive photoresist is capable to resolve smaller structures than the negative resist and is therefore widely used in the IC-technology.

9.1.6 Coatings with Increased Layer Thickness

For the deposition of solder on the pads, an additional photoresist layer may be applied again. In contrast to the coating in the thin-film metallization, here a thicker layer of coating must be applied. Contrary to the high bumps at thin-film processing, coatings with increased layer thickness are used due to the required higher frequencies. Liquid coatings are applied by spin techniques in different layer thicknesses on the wafer. The viscosity of the paint and the speed of the spin have direct influence on the film thickness of the paint, and after a certain time, a defined layer thickness is realized.

Thus, certain parameters are defined for different coating materials, which allow a nearly homogeneous layer thickness. For larger differences in height on the substrate surface, special coatings must be used with good edge coverage to ensure a good pattern transfer during exposure. An alternative is the application of solid resists using resist films, which are used among other things in the PCB manufacturing.

A further requirement for the nature of the coatings is due to the liftoff technology. It is an important process in connection with the vapor deposition technology. The formation of sharp edges after exposure and development of the resist in the subsequent evaporation process lead to the fact that all coating is covered by a further coating layer. Thus, the coating cannot be removed. The aim of the liftoff technique is to resist the offer to the solvents after evaporation. The processing criteria of the coatings are crucial for the edge formation. The coating is illuminated with parallel light through the mask openings. Depending on resist layer thickness, the exposure energy is varied in order to penetrate the whole layer. By varying the exposure energy in combination with the development time, you will get differently shaped profiled edges.

The developed areas are dissolved in the developer process first. If you keep the lacquer for a long time in the developer, then the partially exposed edge regions are slowly developed. This represents an undesirable development, which broadens the structure (Biletzke 2000).

As a result of the development, you should get undercut and concavely formed edges that cause shadowing during deposition process, so that the resist remains free on its edges and is accessible by a solvent. This process of removal is called liftoff technique.

9.1.7 Solder Deposition

The solder material consists of two components and is intended to lead to a certain mixing ratio. In the vapor deposition of a particular layer thickness, the two metals with a specific weight ratio be brought into the vapor deposition. Calculating the evaporation surface and the desired layer thickness, the desired amount of solder material for evaporation can be exactly determined (Fig. 9.8).

Usually, the solder material is made of components of tin (Sn) and lead (Pb) for the FC technique. As can be seen in Fig. 9.17, the two metals form at 183 °C an eutectic weight proportion of 62 % tin and 38 % lead. An example of vapor-deposited PbSn solder with diameters of 80 μm for use as a FC substrate is shown in Fig. 9.18.

As solder materials, SnPb or SnAg alloys with about 5 % or less by weight of lead or alloys with compositions in the range of the eutectic for application are used. The formation of intermetallic phases between solder and metallization of the joining partners is often regarded as a prerequisite for a good wettability as well as the expression of a mechanically stable solder joint. Sn forms with all metallizations used for the FC-bonding technology intermetallic phases. The pad metallization platinum (Pt) forms with tin following intermetallic phases: Pt_3Sn, PtSn, Pt_2Sn_3, $PtSn_2$, and $PtSn_4$.

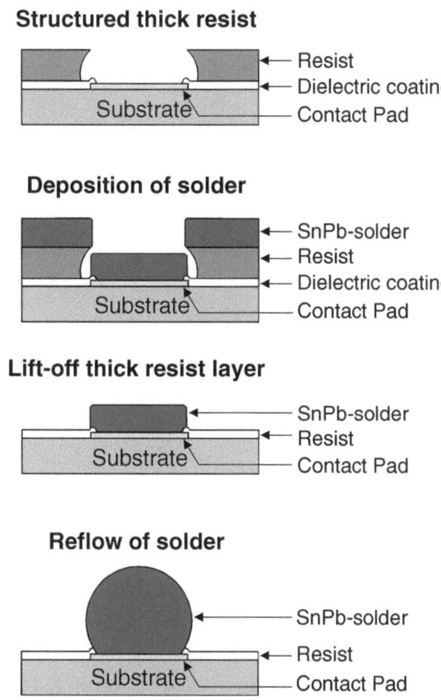

Fig. 9.17 Process sequence for making the solder bumps with liftoff-using thick resist and a vapor deposition process

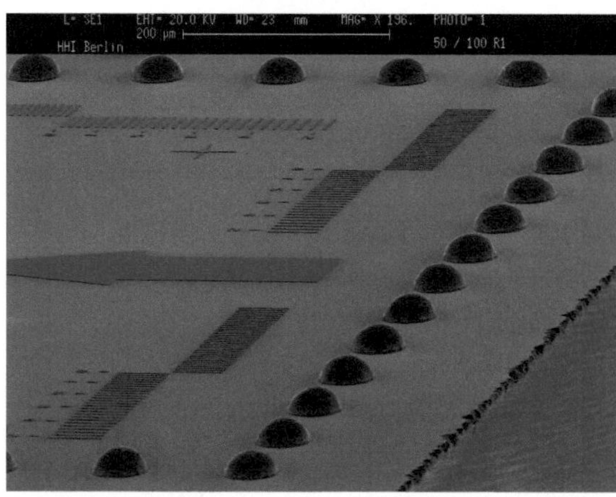

Fig. 9.18 Evaporated bonds after the reflow process

9.1.8 Bonding Process

The bonding or contacting process for the FC-bonding technology can be divided into two parts: the pre-adjustment and the actual bonding operation, in which the components are joint together by melting the solder.

At the beginning of the pre-adjustment, the samples are parallelized. Auto-collimation and movement of the substrate holder by a displacement unit do this.

Fig. 9.19 Heating and positioning

The samples are then adjusted in the *x*- and *y*-directions to each other by moving the lower sample table via adjustment with micrometer screws (as shown in Fig. 9.19).

After the pre-adjustment (±10 μm), the component is placed by a pivoting of the retaining arm carrying the chip to the substrate. Here, the chip still moves in relation to the substrate by a few microns. Then, the vacuum chamber is closed and evacuated. An optimal compromise for a sufficient self-alignment effect is in the range 10^{-2} to 10^{-3} Pa of vacuum pressure. Afterward, the hydrogen is admitted, while the sample is heated until the solder melts.

After bonding, the accuracy of self-adjustment of substrate and chip can be easily checked using the vernier scale, as shown in Fig. 9.20. A micrograph after

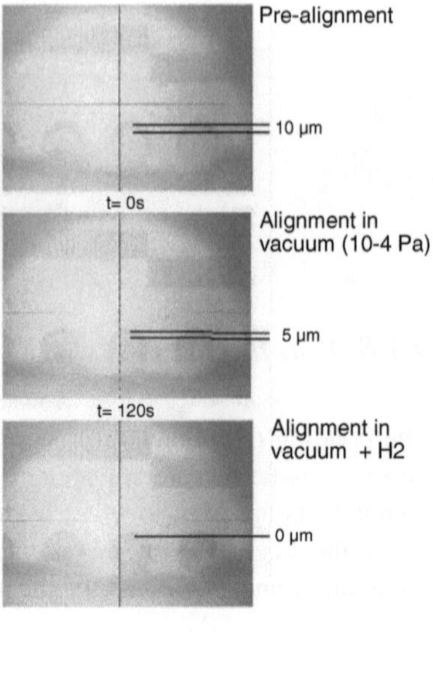

Fig. 9.20 Alignment of the chip and the substrate in bonding under vacuum and after addition of hydrogen

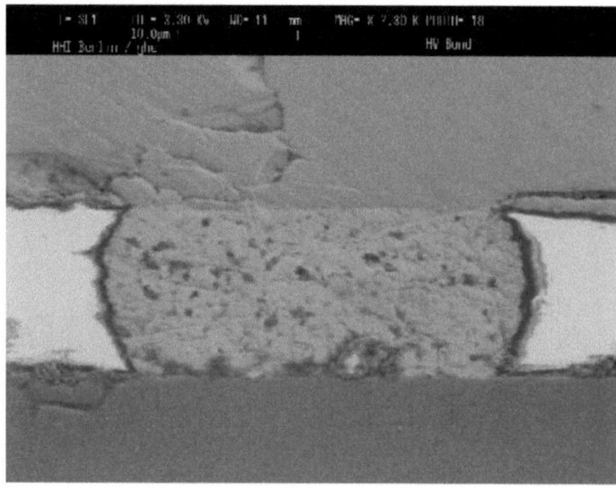

Fig. 9.21 Micrograph of a Pb/Sn bonds with successful self-adjustment

the self-adjustment of a bond is shown in Fig. 9.21. It is easy to recognize the optimal uniform barrel-shaped structure of the compound. Inhomogeneities are also not visible, so you can expect a good mixing of the two metals to an eutectic. It can be also seen that an oxide layer was formed after the bonding.

9.2 LIGA Technique

Another method for the passive fiber–chip coupling can be realized by means of the plastic molding technique LIGA. At the Institute for Microstructure Technology, IMT (Mohr 2014; Saile 2009) in Karlsruhe, the LIGA process was developed in 1985. LIGA is the essential process steps:

1. X-ray lithography,
2. Electroplating, and
3. Plastic molding technique.

These three steps are the way to a potentially low-cost mass production of microcomponents. They are characterized in the following sections. The individual steps for the manufacture of LIGA structures are shown in Fig. 9.22 and are given as follows: Beginning with the production of the mask and the radiation of X-rays on the development of the irradiated mask and the filling of the voids produced by electroplating for preparing the molding tool. After its removal from the mold the last step consist of the molding process using the tool as a master form for the mass production of molded platic devices.

The base of the LIGA process is the plastic molding technique. It leads to a low-cost mass production. The deep lithography in X-ray and electroforming produced metallic microstructures are here used as molding tools in order to produce shape-persistent copies of the primary structure inexpensively in large quantities.

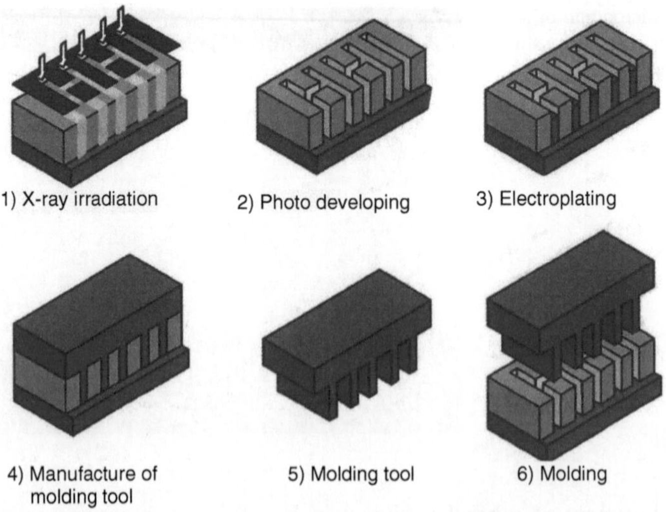

Fig. 9.22 Process steps of the LIGA process

A decisive disadvantage of this method is the high cost for the production of the molding tool, which is due to the high cost of the use of the synchrotron radiation.

On the other hand, a highly precise manufacturing of the tool for the preform is necessary. For this reason, the cost of up to €50,000 for one imprint can occur. This can be profitable only in a mass product, because the number of manufactured components divides the cost per part.

Today, the vacuum embossing of plastics is used for the molding of microstructures. The vacuum embossing has been developed significantly and now represents an interesting alternative to injection molding for special applications, especially in micro-optics and sensors.

The range of materials for the plastic molding ranges from thermoplastics with unique optical properties (acrylic, PMMA, polycarbonate) to materials having particularly high chemical resistance (epoxy–phenol resins, polyvinylidene fluoride (PVDF) and other fluoropolymers) to polymers with high temperature resistance (e.g., polysulfones, polyether ketones).

The embossing technique makes it possible to produce in a gentle way metal or plastic microstructures directly above a corresponding electronic evaluation circuit. This is an almost monolithic structure to be referred to technique, which does not affect the electronic properties of the underlying structure of the microcircuit. The advantage of this integration technique is the combination of the LIGA technique, the silicon microelectronics, and silicon micromachining to produce industrial products.

These devices show good properties to avoid the disadvantages of inflexible monolithic integration, and on the other hand, compared to monolithic integrated circuits, high costs of a hybrid construction.

LIGA properties

The LIGA technique allows the production of components whose properties are to be listed in key words here below:

- Free customizable lateral geometries of the structures
- Structure heights greater than 1 mm
- Smallest lateral dimensions up to 0.2 μm
- Aspect ratios of free-standing individual structures or detailed structures about 1:50 or 1:500
- Surface quality in submicron roughness better than 30 nm
- Different materials: polymers (e.g., PMMA, deuterated PMMA, polycarbonate), metals (e.g., Ni, Cu, Au), and alloys (e.g., permalloy)

The possibility of extremely high aspect ratio to produce a structure of height to width, for example, in a bar structure with a 4-μm-wide side walls, which are 400 μm high, is significant.

In the application of a reflection grating is shown that can be produced small structures whose dimensions are significantly less than 0.5 μm. Here, the steps are only 0.25 μm high and very regular, so that this lattice structure can be used in an optical spectrum analyzer.

9.3 Laser Structuring of Si and PMMA

Another way to structure submicron plastics topologically is the laser patterning. In Fig. 9.23, a description of the method is indicated. Instead of working with an expensive master form, in which preparation for prototype devices is too expensive, one can erode the plastic using short-wavelength laser radiation.

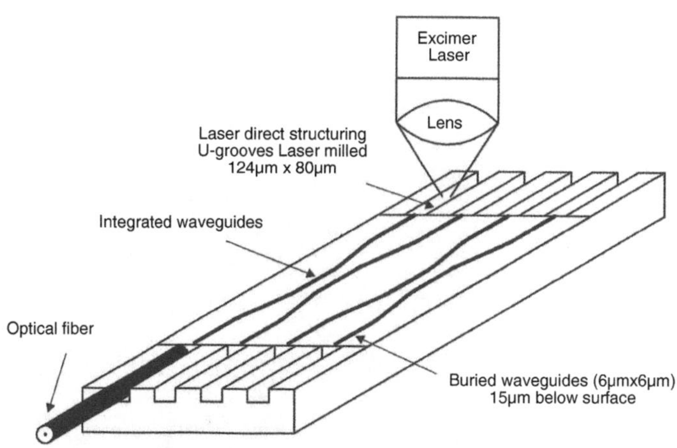

Fig. 9.23 Laser-etched U-grooves in PMMA

Excimer lasers offer a suitable laser sources, whose output wavelengths are between 350 and 190 nm. It is also necessary to apply only very short pulses, so that no melt is formed. The short wavelength smashes with the high photon energy the long hydrocarbon molecules into their individual components. The molecules evaporate directly from the material and are thus removed from the PMMA. The residues have to be removed with a vacuum device.

The thickness of the removal depends on the pulse duration and the photon energy of about 0.1 µm/pulse. With this technique, waveguides in single-mode technology and U-grooves for the reception of glass fibers can be quickly and inexpensively formed for prototypes. A photograph of such a laser machining of PMMA can be seen in Fig. 9.24. The grooves are intended for receiving the fibers

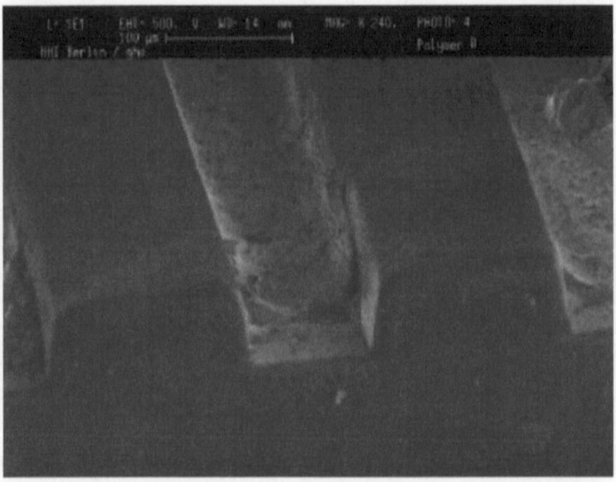

Fig. 9.24 U-grooves made with 125 µm width in PMMA with excimer laser

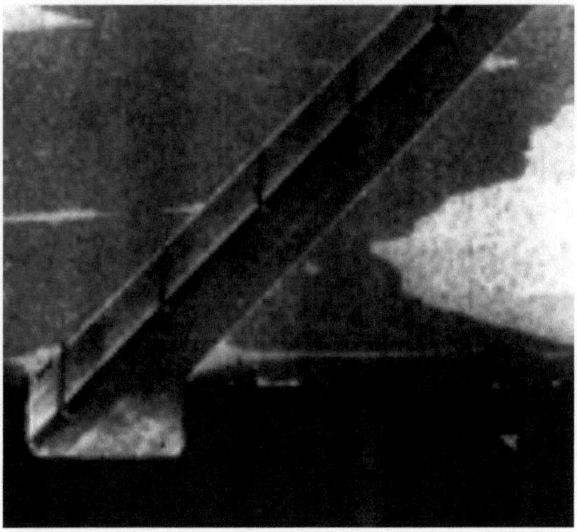

Fig. 9.25 U-grooves 125 µm wide, produced by RIE etching of silicon at the University of Hamburg-Harburg

and having a width of 125 μm. The same technique, but with a reactive ion etching (RIE) as erosion, can be used in the processing of silicon (Fig. 9.25). Here, one has stamped U-grooves for inserting the fiber into the material. This way of structuring is also successfully applied (Keil 2000; Zhang et al. 2004; Vannahme et al. 2010) in PMMA components.

A further possibility for the vertical structure of silicon is irradiation with a high-energy femtosecond laser. It is referred to the method of "laser etching," since the radiation ablates the material as an etching process. Initial experiments showed, however, that silicon is to be processed very badly and you have to use ns pulses, so it cannot come to a melt of the material. The silicon needs to be immediately vaporized into the gaseous state, which is only possible with very high photon energies.

References

Beyer, W., Eigler, H., Eigler, B.: Moderne Produktionsprozesse der Elektrotechnik, Elektronik und Mikrosystemtechnik. Expert-Verlag, Renningen (1996)

Biletzke, M.: Einführung eines Prozesses zur Besichtung von 3-dimensional vorstrukturierten Substraten zur Herstellung von Metallisierungen und Bondpads (2000)

Commission, E., Document, C.: Directive 2002/91/EC of the European Parliament and of the council of 16 December 2002 on the energy performance of buildings. Off. J. Eur. Union 65–71 (2002)

Dantartas, M.F., Blonder, Y.H. Wong, Y.C.C.: A self-aligned optical subassembly for multi mode devices. IEEE CPMT, part B. **18**, 552–557 (1997)

Goodwin, M.J., Meseley, A.J., Kearly, M.Q., Morris, R.C., Kirkby, C.J.G., Thomson, J., Goodfellow, R.C., Bennion, I.: Optoelectronic component arrays for optical interconnection of circuits and subsystems. J. Light. Technol. **9**, 1639–1645 (1991)

Gruner, M.: Finetech Bonder. http://www.finetech.de/products/micro-assembly/fineplacerr-pico-ma.html (2014)

Hauffe, R., Siebel, U., Petermann, K., Moosburger, R., Kropp, J.-R., Arndt, F.: Methods for passive fiber chip coupling of integrated optical devices. IEEE Trans. Adv. Packag. **24**, 450–455 (2001)

Hügli, K.: Semiconductor Industry Sourcebook. Semiconductor Industry Sourcebook. Balzers Process Systems, Liechtenstein (1998)

Katsura, K., Hayashi, T., Ohira, F., Hata, S., Iwashita, K.: A novel flip-chip interconnection technique using solder bumps for high-speed photoreceivers. J. Light. Technol. **8**, 1323–1327 (1991)

Keil, N, et al.: Jahresbericht Heinricht-Hertz-Institut., Berlin (2000)

Kuhmann, J.F., Pech, D.: In situ observation of the self-alignment during FC-bonding under vacuum with and without H2. IEEE Photonics Technol. Lett. **8**, 1665–1667 (1996)

Kuhmann, J.F., Harde, P., Pech, D., Poittroff, W., Preuß, A., Adolphi, B., Wirth, T., Oesterle, W.: Fluxless Flip-Chip Bonding for the Photonic Assembly: Comparison between Evaporated SnPb(60/40) an AuSn(80/20) Solder. Micro System Technologies, Potsdam (1996)

Lau, J.H.: Flip Chip Technologies. McGraw-Hill, New York (1995)

Makiuchi, M., Norimatsu, M., Sakurai, T., Kondo, K., Yano, M.: Flip-chip planar Ga InAs/InP p-i-n photodiode array for parallel optical transmission. IEEE Photonics Technol. Lett. **5**, 518–520 (1993)

Maly, K.: Untersuchungen zur Oxidationskinetik von flüssigen SnPb(60/40)-Lot in Abhängigkeit des O_2-Partialdruckes (1997)

Menz, M.: Mikrosystemtechnik für Ingenieure. VCH Verlagsgesellschaft mbH, Weinheim (1997)

Mohr (2014) LIGA IMT. http://www.fzk.de/imt/liga/d_index.html

Nieweglowski, K.: Beiträge zur Aufbau- und Verbindungstechnik für optische Kurzstreckenverbindungen. Verlag Dr. Markus A. Detert, Templin (2011)

Nishikawa, T., Ijuin, M., Satoh, R., et al.: Fluxless soldering process technology. In: 44th Electronic Components and Technology Conference, pp. 286–292, Washington, DC (1994)

Reichel, H.: Direktmontage: Handbuch für die Verarbeitung ungehäuster ICs. Springer, Berlin (1998)

Rudolph, D.: Umbau eines Fineplacers zum Flip-Chip-Bondplatz zum Test von flussmittelfreien Lot- und Klebebondungen (1999)

Saile, V.: Introduction: LIGA and its Applications. LIGA and its Applications, pp. 1–10 (2009)

Schade, K.: Mikroelektrotechniktechnologie. Verlag Technik GmbH, Berlin (1991)

Scheel, W.: Baugruppentechnologie der Elektronik: Montage. Verlag Technik, Eugen G. Leutze Verlag Saulgau, Berlin (1999)

Steckenborn, A., Winkler, T., Jantke, G., Arndt, F., Schlaak, H.F.: High precision wafer orientation for micromachining. Microsyst. Technol. **91**, 467–471 (1991)

Sutherland, J. et al.: Optical coupling and alignment tolerances in optoelectronic array interface assemblies. In: 45th Electronic Components and Technology Conference, pp. 577–583, Las Vegas, NV (1993)

Vannahme, C., Klinkhammer, S., Kolew, A., Jakobs, P.J., Guttmann, M., Dehm, S., Lemmer, U., Mappes, T.: Integration of organic semiconductor lasers and single-mode passive waveguides into a PMMA substrate. Microelectron. Eng. **87**, 693–695 (2010)

Wale, M.J., Edge, C.: Self-aligned flip-chip assembly of photonic devices with electrical and optical connections. IEEE Trans. Comp. Hyb. Manufac. Techn. **13**, 780–7886 (1990)

Widmann, D., Mader, H., Friedrich, H.: Technologie hochintergrierter Schaltungen. Springer Verlag, Berlin, Heidelberg (1996)

Zhang, C., Yang, C., Ding, D.: Deep reactive ion etching of PMMA. Appl. Surf. Sci. **227**, 139–143 (2004)

Chapter 10
Optical Motherboard

Abstract In this chapter, the hybrid integration of optoelectronic components on a suitable substrate such as silicon, ceramic, glass, or PCB is set out in detail. So far, this technology is succeeded only in approaches to produce a wide range of applications with this technology. The large number of additional technologies such as flip chip technology, vapor deposition, and silica etching makes the production of complex component groups very expensive and complex. A potential approach to lower the complexity and hence the fabrication costs is the integration of optics into printed circuit boards. The key for wide adoption of optics on board-level is the development of compatible processes for integration and assembly. The chapter addresses these developments and shows the supremacy of PCB-integrated optics for applications where high energy efficiency and bandwidth density are particularly in demand.

10.1 Flip-Chip Technique

The production of optoelectronic components for telecommunications is not only limited to the design of individual components. Analogous to the electronic circuit technology, it evolves toward ever more complex structural and functional groups that can be distinguished in terms of their construction manner into two groups (Fig. 10.1):

- Monolithic integration
- Hybrid integration

Monolithic integrated assemblies are made of the same material "of a piece." The advantages of this process are the mechanical robustness, since no additional optical interfaces are available on the board, and the number of construction technical additional technologies is low. Disadvantage is the low overall yield of functional components, and the high cost of epitaxy can be devised due to large OEICs and the fact that the components are realized with compromises due to

© Springer-Verlag Berlin Heidelberg 2015

U.H.P. Fischer-Hirchert, *Photonic Packaging Sourcebook*,
DOI 10.1007/978-3-642-25376-8_10

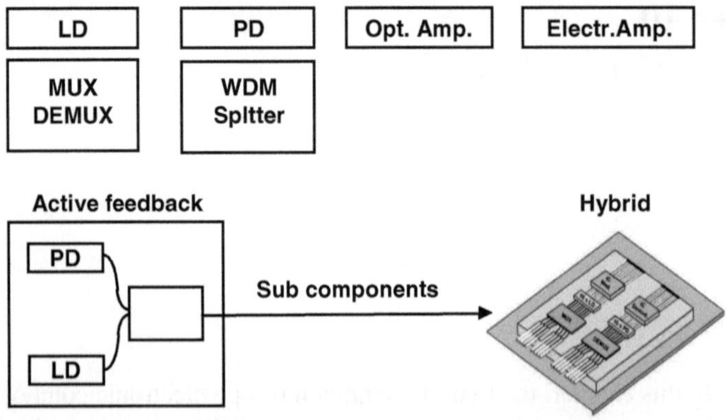

Fig. 10.1 Comparison between the hybrid and monolithic integration

processes. The components cannot be optimized for their basic functions. The large number of individual components in the overall function causes the poor yield. If only a single component is defective in the circuit, the whole OEIC is inoperative.

In the hybrid technique, one needs for the function not make any compromises, since the components are pre-manufactured individually. Thus, a very high yield is inherent in the system. However, the mechanical reliability suffers because of the many electrical and optical interfaces on the board. In industry, one therefore relies heavily on the hybrid technology.

Here in this chapter, the hybrid integration of optoelectronic components on a suitable substrate such as silicon, ceramic, glass, or TMM is set out in detail. It will start with the basic structure of such a hybrid system, which is sketched in Fig. 10.2 with several components:

- Support plate (motherboard)
- V-grooves for glass fibers or mechanical guides for optical connector
- Optical waveguides on silica based
- Electrical connections for bias feeds

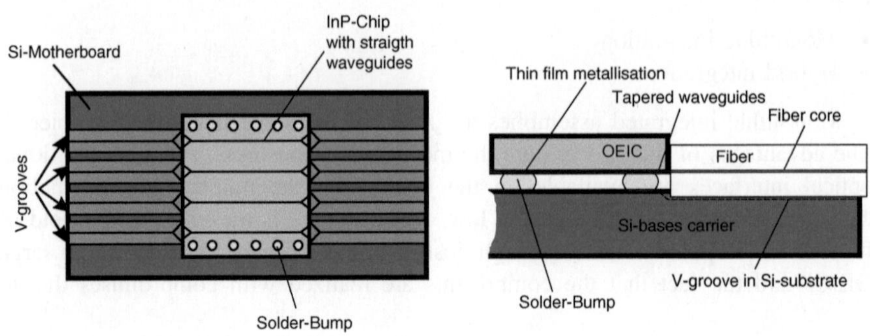

Fig. 10.2 Motherboard assembly

- RF connectors
- Electrical ICs
- Optoelectronic components with spot-size expanders

The passive adjustment replaces in the future active fiber–chip coupling using micromanipulators by using the FC-bond technology for complex and conventional applications. If several lensed fibers must be positioned in front of an OEIC by fiber–chip coupling, the successive adjustment must be performed manually with elaborate devices in three axes and three angles.

10.2 Three-Level Concept

10.2.1 First Level: FC-Bonding and Tapered Waveguide

In the first step of the motherboard approach, an InP chip is applied to a silicon substrate with V-grooves by FC bonding (see Figs. 10.2 and 10.3). InP chip in this example has four optical inputs and outputs, which are optically connected to the tapered straight waveguide. The Si substrate material has V-grooves for receiving the fibers and for the solder bumps and metal coatings for the electrical connections. After FC-bonding process, the glass fibers are placed in the grooves, pushed up, and fixed up to about 10 μm on the facets of the InP chip. In this configuration, the attenuation of the optical coupling is measured. Examples of communications engineering components are the laser array of GTE Laboratories Incorporated (Fig. 10.3) and a transceiver module (receiver and transmitter module) from Bookham (Armiento et al. 1991; Morris 1997).

A batch operation for high volume is only possible with extremely high costs, and the component must be operated electrically and optically to measure the coupling efficiency. The proportion of working time increases linearly with the number of fibers. An additional complicating factor is that the adjusting space must be at least 500 μm between the fibers. Besides the advantage of the passive adjustment, FC bonding offers the possibility that the fiber distances reduce to 250 μm and the distances of the electrical connections reduce significantly below 100 μm (see Fig. 10.4).

The different field distributions of single-mode optical fibers ($n = 1.5$, diameter 9 μm) and InP waveguides ($n = 3.2$, Sect. 0.2 '2 μm) are the cause of the low coupling efficiencies. Measured coupling losses are of about 85 % (see Chap. 3)—for the single-mode fiber-optic coupling with truncated separated fiber (butt coupling)—including the Fresnel losses, but neglecting the propagation losses.

Because of these low coupling efficiencies, the industry uses frequently lensed fibers that reach much higher efficiencies of about 50–70 %. On the other hand, this increased coupling efficiencies lead to increased sensitivity of the lensed fibers against maladjustment (3 dB additional coupling loss of about 1 μm offset).

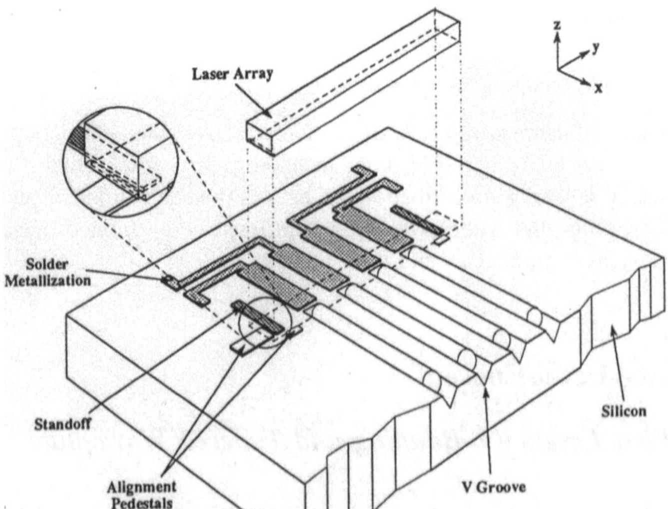

Fig. 10.3 Quad laser array on Si motherboard

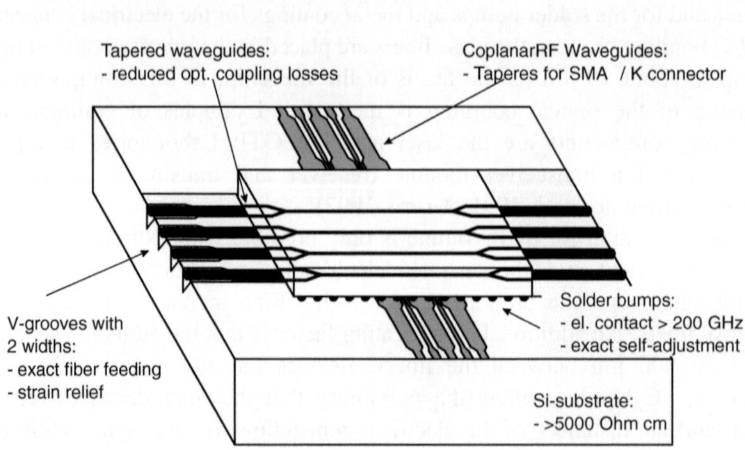

Fig. 10.4 FC-bonded photonic chip with optical inputs at a distance of 250 μm

Due to production-related fluctuations, lensed fiber ends have often curvatures, which then lead to the "squint" of the fiber (see Sect. 7.2). This "squint" must be compensated by active adjustment. In the case of a coupling of fiber arrays to waveguides in the narrow 250-μm grid, this is very difficult to realize.

A more effective instrument for the efficient coupling of several fibers to photonic devices is waveguide tapers in InP (see Figs. 10.5 and 10.7), to ensure a good field adaptation between the waveguide and fiber, and thus lead to increased coupling efficiency at the same time relaxed alignment tolerances (Trommer et al.

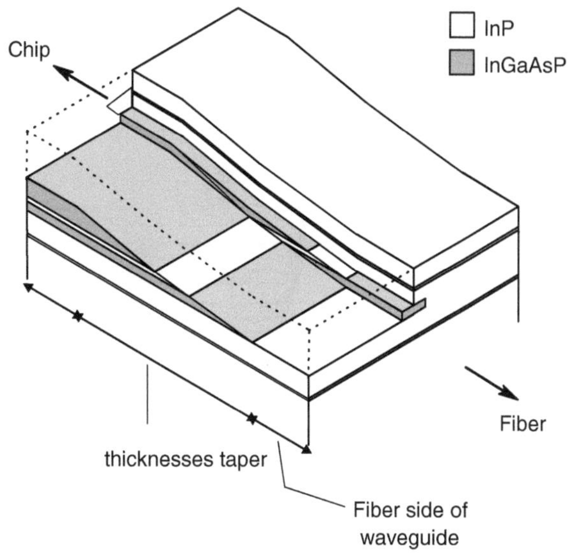

Fig. 10.5 Buried waveguide taper

1999). A novel photolithographic manufacturing process reduces the coupling loss down to 0.5 dB. An overview of the losses and the alignment tolerance for 1 dB excess loss for different configurations of SMF fibers and tapered waveguides is summarized in Table 10.1.

The combination of the coupling and insertion losses shows that alignment tolerances of about 2 μm are easy to achieve with the use of InP waveguide tapers with overall coupling efficiencies greater than 50 %. Other configurations such as the waveguide–fiber coupling or the waveguide taper–fiber coupling are due to the high losses less useful in conjunction with the FC-bonding technology because of the high demands on the alignment accuracies.

10.2.2 Second Level: Silica Waveguides

In the second stage, the existing optical ICs on the motherboard (here silicon) form a passive optical network of silica waveguides. In Fig. 10.8, an arrangement of two

Table 10.1 Coupling losses and adjustment tolerances in comparison with different fiber–chip configurations (Honnecker 2000)

	Fiber–Fiber	Waveguide–Fiber
Coupling losses	0.1 dB	0.5 dB
Alignment tolerances x-direction (±1 dB)	2.7 μm ± 0.1 μm	2.0 μm ± 0.3 μm
Alignment tolerances y-direction (±1 dB)	2.7 μm ± 0.1 μm	2.0 μm ± 0.3 μm
Alignment tolerances z-direction (±1 dB)	28 μm ± 1 μm	30 μm ± 0.3 μm
Alignment tolerances angle (±1 dB)	1.05° ± 0.1°	1.4° ± 0.1°

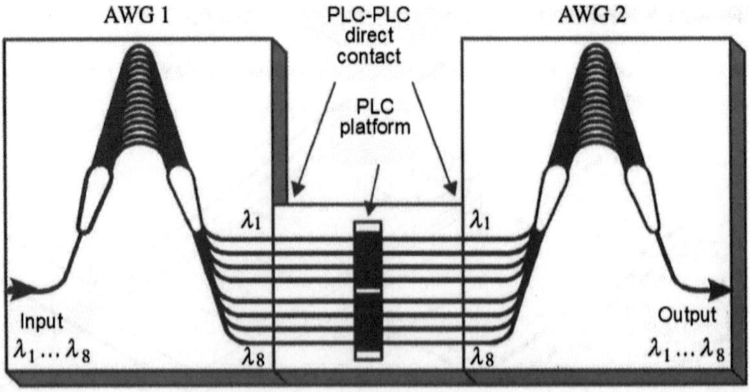

Fig. 10.6 Hybrid integrated semiconductor optical amplifiers on a motherboard with two AWGs

AWG components can be seen, between which optical semiconductor laser amplifier has been included in order to minimize the intrinsic losses (Glebov et al. 2005). The structure is not made in this example comprising a base plate, but rather of three parts. It facilitates thereby the difficulties in structuring. After the SiO_2 components, since the amplifier base plate with their metallization and solder vapor deposits can be made externally for flip-chip-bonding pads (Fig. 10.6).

10.2.3 Third Level: Addition of RF Lines and Electrical ICs

In the third stage, additional electrical leads are deposited on the motherboard, for operating the OEIC and to modulate. Now, it is possible to position electrical driver ICs adjacent to the transmitter laser and the receiver diode, in order to obtain a short transmission path for the RF. A sketch of this structure is shown in Fig. 10.7. Here,

Fig. 10.7 Principle design of an hybrid optical motherboard

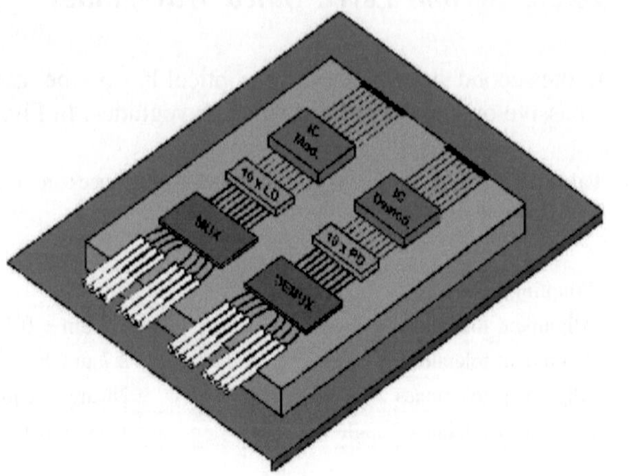

on the left side, the fiber inputs are located, which are guided in V-grooves. The waveguides on the motherboard are made in silica technology and fit with their optical field close to the SMF. The optical add/drop multiplexer is also prepared in silica technique. It splits the signals, which are then guided to the receiver by waveguide diodes and transmitter diodes.

Subsequently, the signals are processed in several electrical amplifier stages, modulators, and multiplexers on the board. The path to the plug is also very short to prepare for low-loss operation. In addition, the data rates are so far down-divided that the losses in the high-frequency parts are only weak.

10.3 PCB-Level Photonic Integration

Primarily, the growing bandwidth demand in the future board-level interconnects was the motivation for the development of competitive solutions that combine high performance and low-packaging complexity at low cost. However, conventional electrical solutions constantly evolved and prevented the adoption of the optical transmission despite their indisputable advantages with regard to the achievable data rates and EMI insensitivity. Further development of copper-based signal transmission shifts the implementation requirements toward higher data rates. Commercially viable technologies already enable electrical transmission with a bandwidth–length product of more than 10Gbit/s × m (Berger et al. 2003). These are, however, much more space and power consuming at given data rates than the optical technologies. Recent advances though in optical datacom for PC peripherals prove the emergence of this technology.

The evolution of high-performance computing (HPC) is in the last years the main driver for the integration of optics in electronic systems. Historically, computational power has grown approximately 10-fold every 4 years, and this growth is expected to continue (Schow et al. 2010). The count of processor cores on a server platform for HPC applications has been steadily increased and this trend will continue in future. Therefore, the amount of processed data increases as well. This is accompanied by the requirement for a fast and high capacity interface to exchange the processed data between the multiple cores and the peripheral devices. The challenge is to transmit this huge amount of data between the chips. Thus, current HPC and data centers mainly suffer from two interconnect bottlenecks: the board-to-board and the on-board connection between processors/memory. As a result, the chip-to-chip bandwidth limits the overall system performance of HPC systems (Nieweglowski et al. 2014).

Photonic interconnects are in use since 2005 for fast local distances such as rack-to-rack interconnect (Kash et al. 2010). Following the trend of HPC performance, in 2020, servers and supercomputer systems with exaflop peak performance and with a huge link number of more than 10^8 are expected. These systems require massive parallelization and integration of optical interconnects into the system on different levels (from board to chip level). Despite the persisting growth in the systems' bandwidth demand, the motivation for the introduction changed slightly. Recently, new criteria as energy efficiency and bandwidth density evolved and illustrate the supremacy of optics. The power consumption of such systems has to be drastically reduced in order to decrease cooling requirements, operational costs, and environmental impact.

The future goal power consumption of optical interconnects for exascale system is assumed to be 1 pJ/bit (Kash et al. 2010). The current state-of-art laboratory fiber-based and integrated optical board-level interconnects show power consumption of 7.3 pJ/bit/link (at 20 Gbit/s link data rate) (Doany et al. 2012a) and 9.7 pJ/bit/link (at 15 Gbit/s link data rate), respectively. In order to further minimize the power consumption with increasing performance novel, packaging concepts for integration of optical transceivers with computing elements (processor, memory) into the computing node are needed.

On the other hand, the advantages of optical data transmission are used on chip level in the emerging silicon photonics, were already first commercial solutions are available (Narasimha et al. 2010). This approach requires single-mode technology with submicron alignment requirements. Therefore, the connection of silicon photonics and on-board waveguide technology is challenging.

Optical interconnections, currently used on a rack and backplane level inside large electronic system, are essentially made through fiber ribbons of dozen parallel fibers, each carrying one data stream. The challenge for optical interconnections at board and backplane level is the cost competitive implementation in standard electrical substrates. Integrated multi-mode waveguides are the most promising approach to realize optically functionalized hybrid (electro-optical) printed circuit boards (EO-PCB), which combine fabrication compatibility and high cost-efficiency. Another challenge remains in the effective light coupling from the optoelectronic devices (lasers or photodiodes) into the integrated waveguide. Here, the key development is demanded for the mirror fabrication process, low-loss light redirection, alignment of the active components, and coupling optics.

The compatibility of implemented fabrication processes with existing well-established technologies for the fabrication of substrates (PCBs) and assembly of components seems to be the answer to the wide application of optics at board level.

10.3.1 PCB-Level Integrated Waveguides

In comparison with fiber-based optical interconnects, the integrated planar waveguides enable achieving high channel density and routing of optical path by integration of passive structures (couplers). The integration of waveguides on the board level is desired because elaborate and costly handling is avoided.

For the integration, two principle assemblies are possible: the overlay and inlay integration. The overlay scheme uses separate optical functionalized substrate, which is assembled on the top of the electrical printed circuit board after completed fabrication of both substrates (optical and electrical) including the assembly and soldering of components (Chan et al. 2010). The main advantage of the technology is simplified condition for test and rework of both boards previous to assembly, which increases the yield of the system.

For the polymer-based optical waveguides, low thermal stress is relevant in order to increase the performance and reliability on the optical layer. On the other hand, the overlay technique requires additional process step, which is undesired for SMT (surface mount technology) assemblies. In contrast to the technology, the inlay scheme can reduce the fabrication costs because of fully integration of optical layer into the printed circuit board and feasibility for volume production. On the other hand, the optical layer integration requires the development of novel optical coupling schemes for efficient interconnection into the optoelectronic devices, which will be discussed in Sect. 10.3.3.

10.3.1.1 Optical Materials

A number of materials that are available on the market have been developed, which are suitable for the fabrication of planar optical waveguides. These materials have to meet several requirements related to the conditions during waveguide processing, as well as during the integration in printed circuit boards or backplanes. They have to withstand a long-term exposure of 1 h at 180 °C during the PCB production, and they must endure reflow conditions with temperatures up to 230 °C (Franke and Schiefelbein 2004). The most important demands for waveguide materials are as follows:

- Low optical attenuation
- High thermal stability
- Low birefringence
- Refractive index controllability
- Good adhesion
- Good dimensional stability
- Stable optical properties
- Long durability with stable optical properties
- Low cost
- Environmental compatibility

Optical loss is the key selection factor. Contrary to the optical fibers, where the fabrication technology is already well engineered, the main impact to the optical attenuation has the manufacturing process (intrinsic attenuation is of less importance). The roughness on the interface between the core and cladding material has a dominant influence on the performance of waveguides. The roughness should be $\ll 0.1\ \lambda$ and amounts to less than 20 nm, in order to minimize scattering losses, which are proportional to roughness depth/λ^2.

The target attenuation is should be less than 0,05 dB/cm, to ensure the transmission distances up to approximately 1 m on the board level and backplane level. Up to now, no material could entirely fulfill all these requirements. Both glass and polymers have been investigated as suitable materials for this application. Table 10.2 shows an overview of materials for PCB-integrated waveguide fabrication. Essentially polymer materials have proved oneself, because of theirs advantages in compatibility with current printed circuit board fabrication. The main disadvantage of polymers is low thermal stability and high optical losses for telecom wavelength windows in near infrared range (1.3 μm and 1.55 μm), because of the absorption of CH-group.

On the other hand, a number of materials with low attenuation at the wavelength of 850 nm have been developed, where the preferred well-engineered vertical-cavity surface-emitting lasers (VCSELs) can be used as a laser source for parallel optical transmission.

However, as long-wavelength VCSEL and silicon photonic devices become more mature, materials optimized for 1.3 and 1.55 μm may be required. The embedding of thin glass sheets into the multilayer PCBs is an alternative solution for integration of optical waveguides on board level, which has large potential because of enhanced thermal and dimensional stability.

The glass withstands temperatures >400 °C. The integrated glass waveguides promise improved thermo-mechanical behavior due to better matching of their coefficient of thermal expansion (CTE) to that of FR4 laminate, which determines the reliability of the system. High transparency of the silica glass in the 1.3 μm/1.55 μm region will presumably favor the glass layer concept, when the low-cost long-wavelength laser sources will be available.

The main disadvantage of this material is complicated glass processing, which have to be solved for an efficient coupling concept. The waveguide core can be realized either by glass etching or laser ablation with subsequent polymer filling or by local refractive index modification using silver ion exchange (Schroeder et al. 2003). In particular, two materials are remarkable: Ormocere® (Himmelhuber et al. 2007) and LightLink™ (Moynihan et al. 2003). These are hybrid inorganic–organic materials that combine very high thermal stability (>300 °C) and processing compatibility with standard PCB fabrication.

10.3.1.2 Manufacturing Techniques for Optical Waveguides on PCB-Level

In consequence of limited tolerances of PCB fabrication and limited assembly accuracy of pick-and-place machines, the PCB-level optical waveguides have typically rectangular cross section with dimensions (waveguide width and height) in the 30–100 μm range. The fabrication techniques for the structuring of such multi-mode planar strip waveguides can be classified as in the Table 10.3.

The photochemical patterning is the most widespread technique for polymer waveguide structuring. The definition of core region with increased refractive index can be performed by photopolymerization, photolocking, photobleaching, or photooxidation (Müller 2000). The strip waveguides are typically defined with photolithographic methods by using UV mask exposure (UV-photolithography),

Table 10.2 Overview of optical materials for PCB-integrated waveguide fabrication

References	Material	Refractive index n	Attenuation (dB/cm)	T_G (°C)
Booth et al. (1997), BPA (2000)	Acrylate, Polyguide™	1.46–1.52	0.08–0.12 @ 800 nm	<150
Müller (2000)	BCB	1.56	0.04 @ 830 nm	>350
Müller (2000)	Cytop™	1.34	<0.1 @ 830 nm	108
Schroeder et al. (2002)	Cycloolefincopolymer–Topas®	1.5–1.56	–	140–180
Schroeder et al. (2008)	Glass, Ag-diffused	1.52–1.54	<0.1 @ 850 nm	>400
mrt (2008), Himmelhuber et al. (2007)	Ormocere®	1.53–1.55	<0.1 @ 633 nm	>270
Müller (2000)	Perfluordioxolpolymer	1.33	0.0033 @ 830 nm	–
Müller (2000), Booth et al. (1997)	Polyimide, Kapton™	1.60	0.5 @ 630 nm	ca. 300
BPA (2000)	Polyimide, fluoriert, NTT-AT	–	<0.4	<380
Fujiwara et al. (2007), Ito et al. (2012)	Polynorbornene, Sumitomo Bakelite Co.	1.53–1.58	0.029 @ 830 nm	270
Neyer et al. (2005)	Polycarbonat	1.56–1.60	–	180–200
Müller (2000)	PMMA VQ101	1.49	<0.1 @ 630 nm	102
Yoshimura et al. (1997)	PMMA, deutiert	1.49	0.018 @ 830 nm	100
Müller (2000)	PS, Polystyrol	–	0.16 @ 630 nm	–
Neyer et al. (2005)	Polysiloxane	1.43	0.03 @ 850 nm	>350
Müller (2000)	PVC, Polyvinylchloride	1.60	0.5 @ 630 nm	80
Moynihan et al. (2003)	Siloxane, LightLink™	1.45–1.55	0,21 @ 850 nm	>400
Karppinen et al. (2004), Chen (2005), Schmieder et al. (2003)	Epoxy—SU-8®	1.60	0.6 @ 850 nm	200
Himmelhuber et al. (2007), Schroeder et al. (2006)	Epoxy—EpoCore/EpoClad	1.57–1.58	0.2 @ 850 nm	>180
Müller (2000)	Teflon AF	1.8	<0.1 @ 630 nm	160

UV laser beam (laser direct writing), or focused IR laser beam (two-photon absorption TPA). Regarding the compatibility with conventional PCB fabrication is UV-photolithography the structuring method of the most interest. The fabrication process flow for this method is depicted in Fig. 10.8a.

Table 10.3 Fabrication methods/techniques for optical waveguide structuring

Photochemical patterning	Replication	Trench and Fill	Other techniques
UV-photolithography	Hot embossing	Laser ablation	Ink jet printing
Laser direct writing	UV-molding	Wafer sawing	Aerosol jet printing
Two-photon absorption	Injection molding	Reactive ion etching (RIE)	Ion exchange
Photo-addressing	Nanoimprint	Waveguide-in-copper technique	

UV-photolithography is matured and well-controlled technology with good dimensional accuracy and resolution at relatively low cost. The laser direct writing and TPA are sequential processes, where the photosensitive polymers are selectively modified creating a waveguide channel. These technologies are independent of the mask's size and therefore can be used for large substrate dimensions. In (Lunitz et al. 2001) laser direct written 1-m-long waveguides with low loss of 0.04 dB/cm have been reported.

The TPA process enables 3-dimensional definition of waveguide core between optoelectronic devices (laser or photodiodes) and thereby complex alignment of elements can be avoided (Lunitz et al. 2001; Schmid et al. 2009). The photo-addressing is remarkable technique, where waveguide core is directly UV patterned in polymer film without development step. The waveguides exhibit a high thermal resistance >270 °C and low optical loss of 0.029 dB/cm (Fujiwara et al. 2007; Ito et al. 2012).

For photolithographic structuring of integrated waveguides such polymers as Ormocer® by microresist technology, (mrt 2008; Himmelhuber et al. 2007), acrylates [Polyguide™ by DuPont (Booth et al. 1997)], siloxanes [LightLink™ by Rohm and Haas (Moynihan et al. 2003)], polynorbornene by Sumitomo Bakelite Co. (Fujiwara et al. 2007; Ito et al. 2012), as well as epoxy resins [SU-8® (Karppinen et al. 2004; Shen et al. 2005; Schmieder 2003), EpoCore and EpoClad by microresist technology (Schroeder et al. 2006; Himmelhuber et al. 2007)] have been applied.

The replication technologies (example process flow—see Fig. 10.8b) use either thermoplastic polymer foils (cycloolefincopolymer—Topas® by Hoechst AG (Schroeder et al. 2002), polycarbonat (Neyer et al. 2005) or PMMA) structured with hot-embossing technique or molded UV/thermo-curable polymers (polysiloxane (Neyer et al. 2005), epoxy resin (Seemann et al. 2004)). These techniques enable the integration of coupling (micromirrors, lenses) and alignment elements into the replication tool. These masters are fabricated in LIGA (**LI**thographie, **G**alvanoformung, **A**bformung—Lithography,Electroplating,Molding) or SIGA (**Si**lizium, **M**ikrostrukturierung, **G**alvanik **A**bformung) process (see Chap. 9) and achieve excellent surface quality (roughness <5 nm). This tool can be directly used as a stamp for replication or can be used as a master for the replication of PDMS stamp,

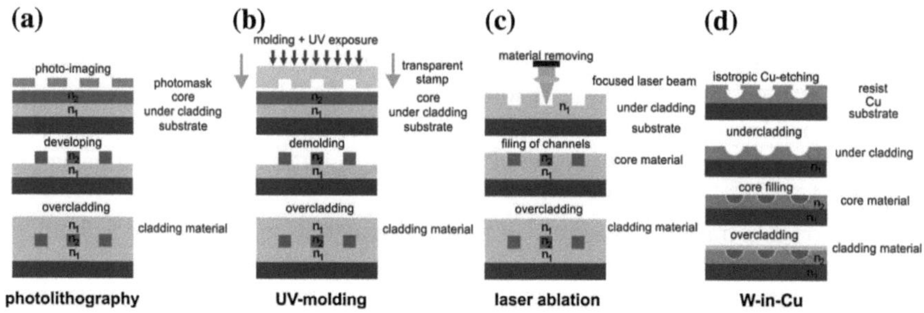

Fig. 10.8 Exemplary process flows for typical representative of: **a** photochemical patterning, **b** replication, **c** trench and fill, and **d** other waveguide fabrication methods

which reduces the abrasion of expensive tool. The replication technologies are only feasible for small substrate sizes, because of strong increase of technology costs with large tool sizes.

The further waveguide structuring possibility—trench and fill method—use selective substrate material removal with subsequent filling of the remaining cross section with waveguide materials (see Fig. 10.8c). The material removal can be performed with laser ablation (Daele et al. 2003), mechanical removal [e.g., wafer sawing (Shen et al. 2005)], or etching (RIE (Yoshimura et al. 1997) or wet chemical etching). The first two methods are limited by processing speed because of the sequential operation. With laser ablation, 45°-deflection surface can be simultaneously fabricated in order to realize waveguide end face for indirect coupling scheme (see Chap. 10.3.3). The integration of polymer waveguides into copper layer with so-called waveguide-in-copper technology is an alternative structuring method, which offers easy implementation and cost-effectiveness (see Fig. 10.8d). The technology is suitable for processing on large substrates, but limited by optical performance (high attenuation—>0.22 dB/cm) and downscaling possibilities (channel width and pitch) (Nieweglowski et al. 2009; Nieweglowski et al. 2010).

The disadvantages of polymer materials over the glass are the higher attenuation and lower thermal stability, although the cost and compatibility issues force the development in investigations on polymers for optical waveguides.

The waveguide core could be implemented by ion exchange. This contactless technique allows increased transparency because of no insertion of extrinsic materials. Also, other methods such as a combination of lithographic printing and etching (wet chemical, RIE or laser) are possible for waveguide definition.

In order to lower technology costs due to selective material deposition, printing techniques have been proposed. These methods are seen as an attractive route to large-scale fabrication of optical waveguides due to the high flexibility. The suitability of this deposition method for waveguide fabrication has been demonstrated as ink jet (Chappell et al. 2008) aerosol jet printing as well as offset printing (Dumke et al. 2011). The challenge here is achieving of waveguide aspect ratio (height to width) of 1, because of trade off between material spreading and good wetting/adhesion.

Alternative to aforementioned technologies, the glass doping use the excellent material properties of glass as a material for optical PCB layer. Selective ion exchange using ion implantation or ion diffusion from salt melt (e.g., $Na^+ \leftrightarrow Ag^+$) can be used. Integration of multi-mode (Schroeder et al. 2008), (Brusberg et al. 2013) and single-mode (Brusberg et al. 2012) waveguides on board and backplane level has been demonstrated.

A reasonable trade off between all physical material properties and technology compatibility will show what material with a proper technology will prevail.

10.3.2 Photonic Packaging for Optoelectronic Devices

In contrast to long-haul optical interconnects, the very short reach interconnects on PCBon PCB level comprise typically transmitter, receiver, optical link, and coupling optics between these components (Fig. 10.9).

The transmitters for parallel optical interconnections include laser diode arrays with drive electronics. In the opposition to the long-haul systems, lasers are direct modulated. Receiver key components are pin photodiodes, avalanches photodiodes, and MSM (metal-semiconductor-metal) photodetectors. The transmitters and receivers for short connections typically do not contain expensive external modulators, multiplexers, or switches, which are typical for high-speed (D)WDM systems.

Lasers

Telecom

For datacom optical transmission using multimode fibers, vertical-cavity surface-emitting lasers (VCSELs) are used as an cost-effective alternative to edge-emitting lasers used in telecom long-reach applications. VCSELs feature simplified fabrication because expansive and complex step of cleaving and die packaging can be avoided. In addition, VCSELs can be tested and packaged on wafer scale without singularizing of single device and hence significantly lowering the final device costs. Circular output beam vertical lasers enhance the optical coupling efficiency, and short laser cavity enables inherent longitudinal single-mode

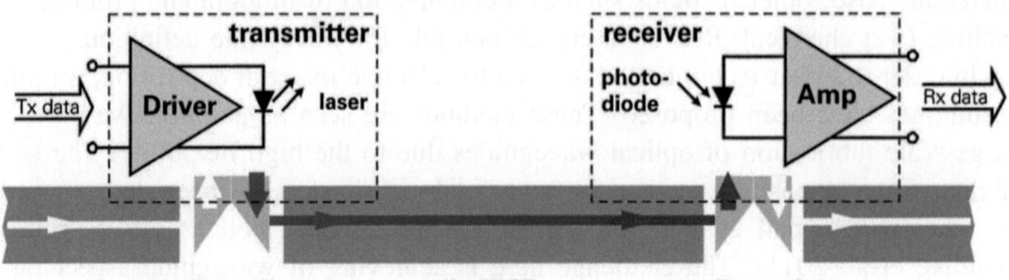

Fig. 10.9 Schematic of board-level short reach interconnect

operation of VCSELs (Lauer et al. 2005). The optical link with direct modulated VCSEL with maximum data rate of 64 Gb/s has been demonstrated (Kuchta et al. 2014).

Gallium arsenide-based VCSELs for the wavelength around the 850 nm are mature and reliable devices, but VCSELs have also a great potential to replace Fabry–Perot lasers and some DFB lasers at 1.3 and 1.55 μm. High data rate modulation with VCSELs in development for both wavelengths has been already (Ortsiefer et al. 2008), respectively, (Hofmann et al. 2009, 2011) shown. However, despite strong demand for VCSELs at long wavelengths, the ability to grow an optimal combination of a good quality active region and high-reflectivity Bragg mirrors is technologically challenging.

The aforementioned advantages of VCSELs predestined these emitters as laser source for board-level links. The vertical emitters can be easily arranged in 2-dimensional arrays allowing building multi-channel transceivers with high aggregated bandwidth capacity for parallel optical transmission with integrated waveguides.

Photodetectors

The photodetectors at the receiver side, typically surface detecting devices, are employed, which need small size of active area for low capacitance and hence high bandwidth requirement. For the bandwidth above 5 Gbps, pin photodiodes active area diameters do not exceed approximately 50 μm. The over-illuminating of the small active area of photodiode leads to coupling loss and bandwidth degradation, because of generation of slow charge carriers in the boundary area. Using focusing micro-optics, which however causes additional costs, can solve this problem. It would be more convenient to use high-speed large area photodetectors. Large area MSM photodetectors are the good candidate, but their speed must be increased in the future.

Optical communication long-haul link systems use InGaAs pin photodiodes or avalanche photodiodes (APD). The short distance datacom systems, working at the 850 nm wavelength, use multi-mode guiding media-integrated waveguides, MM-fibers, POFs, or PCS. These systems require large diameter photodetectors to capture all the light from the multi-mode waveguide or fiber and are based on GaAs or InGaAs absorption layers. It would be most attractive to use a silicon photodetector monolithically integrated a silicon-based CMOS or BiCMOS amplifier in order to achieve cost-effectiveness, high reliability, high performance, and volume manufacturability. Due to poor absorption of silicon at 850 nm in comparison with GaAs, Ge, or InGaAs, it is difficult to design high-speed silicon photodetectors with high efficiency in CMOS process (Schaub 2004). High bandwidth (>34 GHz) good quantum efficiency (42 %) could be achieved with silicon resonant-cavity-enhanced (RCE) pin photodiodes (Schaub et al. 1999), which however uses fabrication technologies deviating from standard silicon process flow. Alternative two CMOS compatible photodetectors has been demonstrated, which use lateral pin and lateral trench structure. They feature low capacitance per unit area in order to achieve multi-Gbit/s performance, while the LTD has improved quantum efficiency (Schaub et al. 2002).

Consequently, optical transmitters and receivers incorporate heterogeneous chips. They conventionally use separately developed III-V lasers or photodetectors and laser drivers or amplifiers fabricated in silicon CMOS technology. This hybrid approach is time and cost-inefficient because of complex packaging processes. Second drawback is that the high-speed performance declines because of interconnect parasitics such as wirebond inductance and bonding-pads capacitance. Flip-chip assembly of chips is one alternative to enhance the performance of transceivers. But significant fabrication challenges in terms of alignment and reliability because of thermal mismatch have to be solved for this approach (Schaub 2004).

Transceiver packaging

Commercially fiber-based 12-channel transmitters and receivers, called Micro-POD [Avago, (Fields et al. 2010)], have been developed, which feature small form factor (7.8 mm × 8.2 mm × 3.9 mm), low power consumption (125 mW/channel), and high aggregated bandwidth of 120 Gbit/s, resulting in high bandwidth efficiency of approx. 13 pJ/bit. In R&D fiber-based transceivers with flip-chip mounted VCSEL and photodiode with aggregate bandwidth in Tbit/s-range have been shown. In Benjamin et al. (2013), 1.3 Tbit/s transceiver module with 12 × 14 array of 8 Gbit/s backside emitting GaAs VCSELs and backside detecting InP phodiodes working at wavelength of 1000 nm is demonstrated. Using optical vias in CMOS driver/amplifier ICs, 850 nm direct flip-chip-mounted opto-electronic (OE) devices (VCSELs and photodiodes) can be implemented and enable optical coupling to fiber arrays. A so-called Holey transceiver with 48 channels (24 Tx + 24 Rx at 20 Gbit/s/ch) fiber-based optical link features high efficiency of 7.3 pJ/bit and high bandwidth density of 31.8 Gbit/s/mm^2 (Doany et al. 2012b), (Doany et al. 2013). In Doany et al. (2011) a 48-channel (24 Tx + 24 Rx), parallel optical modules for board-level integrated optical links with full link power consumption of 9.7 pJ/bit have been demonstrated. Here, optical vias are realized in Si-interposer, where the OE devices and CMOS ICs are assembled using flip chip technology. The coupling into waveguides mounted on PCB is performed with using two microlens arrays, which lower the misalignment requirements for assembly of OE module on PCB.

10.3.3 Optical Coupling for Board-Level Interconnects

Board-level optical interconnects suffer from the lack of robust low-loss coupling schemes, which can support low-cost assembly routines. Compatibility with surface mount technology (SMT) seems to be the answer to this optical coupling challenge. The passive alignment scheme can lead to a volume capable assembly process with an acceptable cost regime. In order to ensure the assembly tolerances typical for SMT processes, a dedicated micro-optics is required for both low-loss light deflection and coupling of the optical signals into active and passive optical components (Nieweglowski et al. 2010)

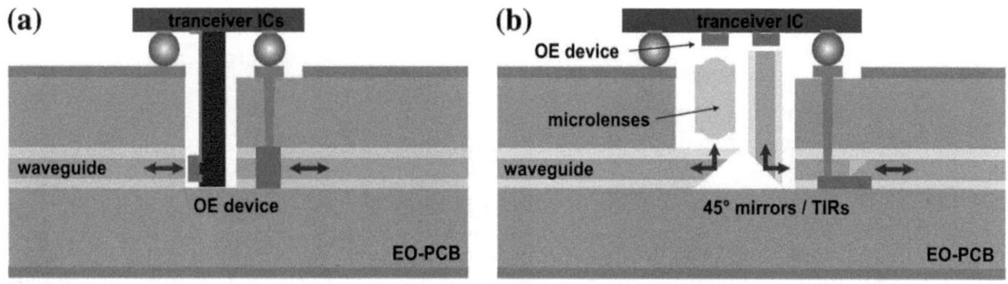

Fig. 10.10 On-board coupling schemes: **a** direct and **b** indirect coupling between OE device and integrated waveguides

Generally, the interface between the integrated waveguides and optoelectronic devices can be performed with direct or indirect coupling. Schematic comparison of these two approaches is shown in Fig. 10.10.

The direct butt coupling features overlapping of optical axes of the waveguide and optoelectronic component. Optoelectronic devices (laser or photodetector) have to be placed directly in front of end facet of waveguide. Here, the effect of beam widening is reduced because of minimizing of axial offset. The main advantage of this technique is no need of use of additional microoptical elements such as lenses or mirrors. On the other hand, the assembly process requires additional positioning in z-direction that poses incompatibility with standard pick-and-place SMT technology. Two approaches for assembly of OE device are possible. The OE device can be embedded into the optical layer of electro/optical printedcircuit board (EO-PCB) or bonded on a sub-mount (Dellmann et al. 2007; Schroeder et al. 2002) which is put into the PCB cavity (Fig. 10.4). In , Schmid et al. (2009) the embedded device was mounted on PCB prior to direct writing of waveguides between active areas of OE components, what eliminates the complex alignment process (Fig. 10.11a). The disadvantage of this approach is complicated thermal management.

The approach depicted in Fig. 10.11b uses additional mechanical elements (metal pins) for passive alignment of the optoelectronic sub-mount with integrated waveguides.

In contrast to the indirect coupling, SMT-compatible transceiver packages can be used, which have a potential to lower the complexity of assembly process and hence alignment/assembly costs. Self-centering forces of melted solder during the soldering can be used for alignment of OE devices, but need additional mechanical components (e.g., spacers, bedstops) for very precise alignment.

The indirect coupling is based on reflecting facets in front of the waveguide for 90° beam redirection. In this case, the optical axis of the optoelectronic device is perpendicular to the waveguide axis. There are several methods applied for realizing of out of plane light deflection. One of these is creation of 45° total internal reflector (TIR) on the end of the waveguide. Such mirrors can be performed by micromachining by wafer sawing (Ishii et al. 2003), (Glebov et al. 2005), grinding/polishing, or laser ablation (Van Steenberge et al. 2004; Chan et al. 2010).

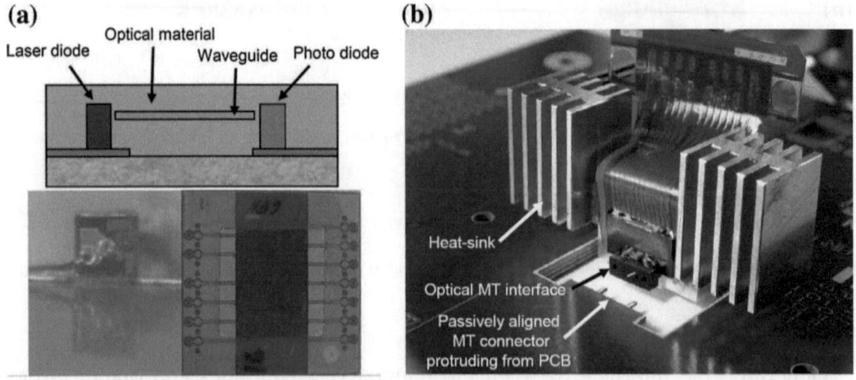

Fig. 10.11 Direct optical coupling with **a** embedded OE device (Langer 2010) and **b** optoelectronic sub-mount (Dellmann et al. 2007)

Another possibility is to fabricate mirrors simultaneously with core definition in the UV-lithography (Immonen et al. 2004) or molding process (Neyer et al. 2005). An additional optical component, for example a prism, can also change beam direction of 90° (Hendrickx et al. 2008; Betschon et al. 2010; Karppinen 2008). These micro-optical elements are placed in the cavity in the front of waveguide end face and hence increase the alignment complexity. The large axial displacement between waveguide and OE device requires collimation and focusing with additional micro-optics (Doany et al. 2012c; Ishii et al. 2003; Karppinen et al. 2006) (Fig. 10.12a) or realization of parabolic mirrors (Betschon et al. 2010), which poses the main disadvantage of the coupling scheme. In order to bridge the free space gap, waveguiding elements (fibers or waveguides) as an optical pillar (Bakir et al. 2008) or waveguides with integrated mirrors/TIRs can be used (Nieweglowski et al. 2010; Bauer et al. 2005; Rho et al. 2004) (Fig. 10.12b). The latter concepts combine, beside beam deflection, the function of passive alignment using mechanical pins and waveguiding from the optoelectronic component to the integrated waveguide. For higher reflection efficiency, the mirror surface can be metalized, e.g., with

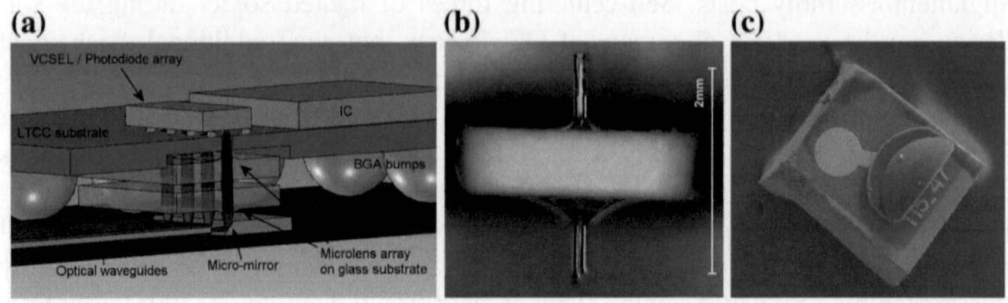

Fig. 10.12 Indirect optical coupling with **a** micromirror deflection and microlens array (Karppinen et al. 2006) and **b** fiber coupling element with (Nieweglowski et al. 2009) and **c** embedded OE device with integrated parabolic micromirror (Langer 2010)

Au-coating. Mirrors have the advantage of being wavelength independent, but can cause high losses due to surface roughness of the mirror surface.

The indirect coupling with embedded active components has been also demonstrated in order to minimize the alignment complexity (Langer2010 ; Chang et al. 2004). In Langer (2010) on a top of VCSEL/photodiode, a molded and metallized parabolic micromirror has been integrated in order to deflect and focus the light into the direct written waveguides with two-photon absorption (Fig. 10.12c).

A diffractive optics (e.g., gratings) is alternative approach for light deflection. Gratings can have high coupling very efficiency (can theoretically diffract up to 90 % of the incident energy in and out of the waveguiding layer (Franc et al. 2006), but only within a small wavelength range and small number of modes. The grating coupling has been already commercially implemented for fiber–chip coupling into silicon photonic integrating circuits (e.g., in active optical cables AOC) (Mekis et al. 2011) with silicon-on-isolator (SOI) single-mode waveguides with submicron dimensions. Furthermore, resonant gratings have been demonstrated for in/out coupling of the light into/from multi-mode polymer waveguides (Franc et al. 2006; Hendrickx et al. 2007).

References

Armiento, C.A., et al.: Hybrid optoelectronic integration of transmitter arrays on silicon waferboard. In: Proceedings of the SPIE, Intergrated Optoelectron is for Communication and Processing, pp. 112–120 (1991)

BPA: Optical backplanes, a global market and technology review 2000–2005 (2000)

Bakir, M.S., Glebov, A.L., Lee, M.G., Kohl, P.A., Meindl, J.D.: Mechanically flexible chip-to-substrate optical interconnections using optical pillars. Adv. Packag. IEEE Trans. **31**, 143–153 (2008)

Bauer, J., Ebling, F., Schroeder, H., Beier, A., Beil, P., Demmer, P., Franke, M., Griese, E., Reuber, M., Kostelnik, J., Park, H., Mödinger, R., Pfeiffer, K., Ostrzinski, U.: Leiterplatten mit innenliegender Optolage - Wellenleitertechnologie und Koppelkonzept. Tagungsband des IV. ITG-Workshop Photonische Aufbau- und Verbindungstechnik (2005)

Benjamin, Y., Hasharoni, K., Mesh, M.: Assembly development of 1.3 Tb/s full duplex optical module. In: Proceedings of the IEEE 63rd Electronic Components and Technology Conference (ECTC), pp. 292–296 (2013)

Berger, C., Beyeler, R., Bona, G.-L., Dangel, R., Dellmann, L., Dill, P., Horst, F., Kossel, M.A., Menolfi, C., Morf, T., Offrein, B., Schmatz, M.L., Toifl, T., Weiss, J.: Optical links for printed circuit boards. In: 16th Annual Meeting of the IEEE Lasers and Electro-Optics Society (LEOS 2003), pp. 61–62 (2003)

Betschon, F., Michler, M., Craiovan, D., Halter, M., Dietrich, K., Kremmel, J., Franke, J., Gmür, M., Paredes, S.: Mass production of planar polymer waveguides and their applications. In: Proceeding of SPIE, p. 76070M. SPIE (2010)

Booth, B.L., Marchegiano, J.E., Chang, C.T., Furmanak, R.J., Graham, D.M., Wagner, R.G.: Polyguide polymeric technology for optical interconnect circuits and components. In: Proceedings of the SPIE Photonics West Conference, pp. 238–251 (1997)

Brusberg, L., Schroeder, H., Pitwon, R., Whalley, S., Herbst, C., Miller, A., Neitz, M., Röder, J., Lang, K.-D.: Optical backplane for board-to-board interconnection based on a glass panel gradient-index multimode waveguide technology. In: Proceedings of IEEE 63rd Electronic Components and Technology Conference (ECTC), pp. 260–267. IEEE (2013)

Brusberg, L., Schroeder, H., Queisser, M., Lang, K.-D.: Single-mode glass waveguide platform for DWDM chip-to-chip interconnects. In: Proceedings of the IEEE 62nd Electronic Components and Technology Conference (ECTC), pp. 1532–1539. IEEE (2012)

Chan, B., Lin, H., Carver, C., Huang, J., Berry, J.: Organic optical waveguide fabrication in a manufacturing environment. In: Proceedings of the IEEE 60th Electronic Components and Technology Conference (ECTC), pp. 2012–2018. IEEE (2010)

Chang, G.-K., Guidotti, D., Liu, F., Chang, Y.-J., Huang, Z., Sundaram, V., Balaraman, D., Hegde, S., Tummala, R.R.: Chip-to-chip optoelectronics SOP on organic boards or packages. IEEE Trans. Adv. Packag. **27**, 386–397 (2004)

Chappell, J., Hutt, D.A., Conway, P.P.: Variation in the line stability of an inkjet printed optical waveguide-applicable material. In: 2008 2nd Electronics System integration Technology Conference, pp. 1267–1272. IEEE (2008)

Chen, Y.-M., Yang, C.-L., Cheng, Y.-L., Chen, H.-H., Chen, Y.-C., Chu, Y., Hsieh, T.-E.: 10Gbps multi-mode waveguide for optical interconnect. In: Proceedings of IEEE Electronic Components and Technology Conference (ECTC), pp. 1739–1743 (2005)

Daele, P. Van, Geerinck, P., Steenberge, G. Van, Put, S. Van, Cauwe, M.: Laser ablation as an enabling technology for opto-boards. In: Proceedings of 53rd Electronic Components and Technology Conference, pp. 1140–1146 (2003)

Dellmann, L., Berger, C., Beyeler, R., Dangel, R., Gmur, M., Hamelin, R., Horst, F., Lamprecht, T., Meier, N., Morf, T., et al.: 120 Gb/s optical card-to-card interconnect link demonstrator with embedded waveguides. In: Proceedings of the IEEE 57th Electronic Components and Technology Conference (ECTC), pp. 1288–1293 (2007)

Doany, F.E., Lee, B.G., Rylyakov, A. V, Kuchta, D.M., Baks, C., Jahnes, C., Libsch, F., Schow, C.L.: Terabit/sec VCSEL-based parallel optical module based on holey CMOS transceiver IC. In: Optical Fiber Communication Conference and the National Fiber Optic Engineers Conference (OFC/NFOEC), pp. 1–3 (2012a)

Doany, F.E., Lee, B.G., Rylyakov, A. V, Kuchta, D.M., Jahnes, C., Baks, C., Libsch, F., Schow, C.L.: Terabit/sec 48-channel fiber-coupled optical module based on holey CMOS transceiver IC. In: Proceedings of the IEEE 62nd Electronic Components and Technology Conference (ECTC), pp. 1499–1504. IEEE (2012b)

Doany, F.E., Schow, C.L., Lee, B.G., Budd, R., Baks, C., Dangel, R., John, R., Libsch, F., Kash, J. A., Chan, B., Lin, H., Carver, C., Huang, J., Berry, J., Bajkowski, D.: Terabit/sec-class board-level optical interconnects through polymer waveguides using 24-channel bidirectional transceiver modules. In: Proceedings of the IEEE 61st Electronic Components and Technology Conference (ECTC), pp. 790–797 (2011)

Doany, F.E., Schow, C.L., Lee, B.G., Budd, R.A., Baks, C.W., Tsang, C.K., Knickerbocker, J.U., Dangel, R., Chan, B., Lin, H., Carver, C., Huang, J., Berry, J., Bajkowski, D., Libsch, F., Kash, J.A.: Terabit/s-class optical links incorporating 360-Gb/s bidirectional 850 nm parallel optical transceivers. J. Light. Technol. **30**, 560–571 (2012c)

Doany, F.E., Lee, B.G., Kuchta, D.M., Rylyakov, A.V., Baks, C., Jahnes, C., Libsch, F., Schow, C.L.: Terabit/Sec VCSEL-based 48-channel optical module based on holey CMOS transceiver IC. J. Lightwave Technol. **31**, 672–680 (2013)

Dumke, M., Craiovan, D., Rieske, R., Franke, J., Overmeyer, L., Wolter, K.-J.: Innovative Produktionsverfahren zur drucktechnischen Herstellung 3D-geführter optischer Wellenleiter. In: Proceedings—112. Jahrestagung der Deutschen Gesellschaft für angewandte Optik (DGaO 2011) (2011)

Fields, M.H., Foley, J., Kaneshiro, R., McColloch, L., Meadowcroft, D., Miller, F.W., Nassar, S., Robinson, M., Xu, H.: Transceivers and optical engines for computer and datacenter interconnects. In: Optical Fiber Communication (OFC), Collocated National Fiber Optic Engineers Conference, 2010 Conference on (OFC/NFOEC), pp. 1–2 (2010)

Franc, J., Destouches, N., Blanc, D., Pommier, J.-C., Tonchev, S., Steenberge, G. Van, Hendrickx, N., Last, A., Parriaux, O.: High-efficiency diffraction grating coupler for multimode optical interconnect. In: Proceedings of SPIE—Photonics Europe. p. 61851F. SPIE (2006)

Franke, M., Schiefelbein, F.-P.: Optical interconnects on level—results based on the German funded project OptiCon. In: Proceedings of the IEEE 54th Electronic Components and Technology Conference (ECTC), pp. 1542–1546 (2004)

Fujiwara, M., Shirato, Y., Owari, H., Watanabe, K., Matsuyama, M., Takahama, K., Mori, T., Miyao, K., Choki, K., Fukushima, T., Tanaka, T., Koyanagi, M.: Novel optical/electrical printed circuit board with polynorbornene optical waveguide. Jpn. J. Appl. Phys. **46**, 2395–2400 (2007)

Glebov, A.L., Roman, J., Lee, M.G., Yokouchi, K.: Optical interconnect modules with fully integrated reflector mirrors. Photonics Technol. Lett. IEEE. **17**, 1540–1542 (2005)

Hendrickx, N., Erps, J. Van, Alajoki, T., Destouches, N., Blanc, D., Franc, J., Karioja, P.: Towards Low Cost Coupling Structures for Short-Distance Optical Interconnections. In: Proceedings of the 16th European Microelectronics and Packaging Conference (EMPC), pp. 247–252 (2007)

Hendrickx, N., Van Erps, J., Thienpont, H., Van Daele, P.: Out-of-plane coupling structures for optical printed circuit boards. In: Optical Fiber communication/National Fiber Optic Engineers Conference, Conference on OFC/NFOEC, pp. 1–3 (2008)

Himmelhuber, R., Fink, M., Pfeiffer, K., Ostrzinski, U., Klukowska, A., Gruetzner, G., Houbertz, R., Wolter, H.: Innovative materials tailored for advanced micro-optic applications. In: Proceedings of the SPIE, Photonics Package Integration Interconnects VII. pp. 6478, 64780E–1–12 (2007)

Hofmann, W., Müller, M., Nadtochiy, A., Meltzer, C., Mutig, A., Böhm, G., Rosskopf, J., Bimberg, D., Amann, M.-C., Chang-Hasnain, C.: 22-Gb/s long VCSELs. Opt. Express **17**, 17547–17554 (2009)

Hofmann, W., Müller, M., Wolf, P., Mutig, A., Gründl, T., Böhm, G., Bimberg, D., Amann, M.-C.: 40 Gbit/s modulation of 1550 nm VCSEL. Electron. Lett. **47**, 270–271 (2011)

Honnecker, J.: Untersuchung der Koppeleffizienz und der Reflexion zwischen Monomodefasern und integrierten Modenfeldtransformern auf InP auf LWL-Emüfängern im Bereich zwischen 40-160 Gbit/s (2000)

Immonen, M., Wu, J., Kivilahti, J.: Fabrication of polymer with integrated micromirrors for out-of-plane surface normal optical interconnects. In: 4th IEEE International Conference on POLYTRONIC 2004, Polymers and Adhesives in Microelectronics and Photonics, pp. 206–210 (2004)

Ishii, Y., Koike, S., Arai, Y., Ando, Y.: Compatible large-tolerance "OptoBump" interface for interchip optical interconnections. IEEE Trans. Adv. Packag. **26**, 122–127 (2003)

Ito, Y., Terada, S., Singh, M.K., Arai, S., Choki, K.: Demonstration of high-bandwidth data transmission above 240 Gbps for optoelectronic module with low-loss and low-crosstalk polynorbornene waveguides. In: Proceedings IEEE 62nd Electronic Components and Technology Conference (ECTC), pp. 1526–1531. IEEE (2012)

Karppinen, M., Alajoki, T., Tanskanen, A., Kataja, K., Makinen, J.-T., Kautio, K., Karioja, P., Immonen, M., Kivilahti, J.: Parallel optical interconnect between ceramic BGA packages on FR4 board using embedded waveguides and passive optical alignments. In: Proceedings of IEEE 56th Electronic Components and Technology Conference (ECTC) (2006)

Karppinen, M., Makinen, J.-T., Kataja, K., Tanskanen, A., Alajoki, T., Karioja, P., Immonen, M., Kivilahti, J.: Embedded optical interconnect on printed wiring board. In: Proceedings of the SPIE Vol. 5453 Micro-Optics, VCSELs, and Photonic Interconnects, pp. 150–164 (2004)

Karppinen, M.: High bit-rate optical interconnects on printed wiring board—micro-optics and hybrid integration (2008)

Kash, J.A., Benner, A.F., Doany, F.E., Kuchta, D.M., Lee, B.G., Pepeljugoski, P.K., Schares, L., Schow, C.L., Taubenblatt, M.: Optical interconnects in exascale supercomputers. In: 23rd Annual Meeting of the IEEE Photonics Society, pp. 483–484 (2010)

Kuchta, D.M., Rylyakov, A.V., Schow, C.L., Proesel, J.E., Baks, C.W., Westbergh, P.; Gustavsson, J.S., Larsson, A.: 64 Gb/s transmission over 57 m MMF using an NRZ modulated 850 nm VCSEL. In: Proceedings of the Optical Fiber Communications Conference (OFC) (2014)

Langer, G.: Integrated Optical Interconnections - R&D project IOI. Austria Technologie & Systemtechnik (AT&S) (2010)

Lauer, C., Maute, M., Hofmann, W., Amann, M-C.: VCSELs emerge as low-cost alternatives to edge-emitters. In: Laser Focus World, June, pp. 127–131 (2005)

Lunitz, B., Guttmann, J., Huber, H.-P., Moisel, J., Rode, M.: Experimental demonstration of 2.5 Gbit/s transmission with 1 m polymer optical backplane. Electron. Lett. **37**, 1079 (2001)

Mekis, A., Gloeckner, S., Masini, G., Narasimha, A., Pinguet, T., Sahni, S., De Dobbelaere, P.: A grating-coupler-enabled CMOS photonics platform. Sel. Top. Quantum Electron. IEEE J. **17**, 597–608 (2011)

Morris, R.C.: Silicon on insulator integrated optic transceivers. In: Proceedings of the SPIE. pp. 56–67 (1997)

Moynihan, M.L., Allen, C., Ho, T., Little, L., Pawlowski, N., Pugliano, N., Shelnut, J.G., Sicard, B., Zheng, H. Bin, Khanarian, G.: Hybrid inorganic-organic aqueous base compatible waveguide materials for optical interconnect applications. In: Proceedings of the SPIE, pp. 50–60 (2003)

mrt: Dataheet: ORMOCER Materialsystem für planare optische Wellenleitung, http://www.microresist.de/ (2008)

Müller, L.: Polymere für in Abformtechnik hergestellte Integriert-Optische Lichtleiter (2000)

Narasimha, A., Abdalla, S., Bradbury, C., Clark, A., Clymore, J., Coyne, J., Dahl, A., Gloeckner, S., Gruenberg, A., Guckenberger, D., Gutierrez, S., Harrison, M., Kucharski, D., Leap, K., LeBlanc, R., Liang, Y., Mack, M., Martinez, D., Masini, G., Mekis, A., Menigoz, R., Ogden, C., Peterson, M., Pinguet, T., Redman, J., Rodriguez, J., Sahni, S., Sharp, M., Sleboda, T.J., Song, D., Wang, Y., Welch, B., Witzens, J., Xu, W., Yokoyama, K., De Dobbelaere, P.: An ultra low power CMOS photonics technology platform for H/S optoelectronic transceivers at less than 1 per Gbps. In: Optical Fiber Communication (OFC), Collocated National Fiber Optic Engineers Conference, 2010 Conference on (OFC/NFOEC), pp. 1–3 (2010)

Neyer, A., Kopetz, S., Rabe, E., Kang, W.J., Tombrink, S.: Electrical-optical circuit board using polysiloxane optical waveguide layer. In: Proceedings of 55th Electronic Components and Technology Conference, pp. 246–250 (2005)

Nieweglowski, K., Henker, R., Ellinger, F., Wolter, K.-J.: Performance of step index multimode waveguides with tuned numerical aperture for on-board optical links. In: Proceedings of SPIE, Optical Interconnects XIV, ser., vol. 8991, pp. 05:1–05:10 (2014)

Nieweglowski, K., Rieske, R., Wolter, K.-J.: Assembly requirements for multi-channel coupling micro-optics in board-level optical interconnects. In: Proceedings of 3rd Electronics System-Integration Technology Conference (ESTC), pp. 1–6 (2010)

Nieweglowski, K., Rieske, R., Wolter, K.-J.: Demonstration of board-level optical link with ceramic optoelectronic multi-chip module. In: Proceedings IEEE 59th Electronic Components and Technology Conference (ECTC), pp. 1879–1886 (2009)

Ortsiefer, M., Hofmann, W., Roenneberg, E., Boletti, A., Gatto, A., Boffi, P., Rosskopf, J., Shau, R., Neumeyr, C., Boehm, G., et al.: High speed 1.3 μm VCSELs for 12.5 Gbit/s optical interconnects. Electron. Lett. **44**, 974–975 (2008)

Rho, B.S., Kang, S., Cho, H.S., Park, H.-H., Ha, S.-W., Rhee, B.-H.: Compatible optical interconnection using 45°-ended connection rods and via-holed waveguides. J. Light. Technol. **22**, 2128–2134 (2004)

Schaub, J.D., Csutak, S.M., Yang, B., Campbell, J.C., Rogers, D.L., Yang, M., Kuchta, D.M., Zier, S.J., Sorna, M.: High-speed optical receivers in advanced silicon technologies. In: The 15th Annual Meeting of the IEEE, Lasers and Electro-Optics Society, LEOS 2002, pp. 772–773 (2002)

Schaub, J.D., Li, R., Schow, C.L., Campbell, J.C., Neudeck, G.W., Denton, J.: Resonant-cavity-enhanced high-speed Si photodiode grown by epitaxial lateral overgrowth. IEEE Photonics Technol. Lett. **11**, 1647–1649 (1999)

Schaub, J.D.: CMOS detectors enable optical interconnects. In: Laser Focus World, pp. 109–112 (2004)

Schmid, G., Leeb, W.R., Langer, G., Schmidt, V., Houbertz, R.: Gbit/s transmission via-inscribed on printed circuit boards. Electron. Lett. **45**, 219–221 (2009)

Schmieder, K.: Aspekte der Aufbau- und Verbindungstechnik Elektro-Optischer Verdrahtungsträger. Verlag Dr. Markus A. Detert, Templin (2003)

Schow, C., Doany, F., Kash, J.: Get on the Optical Bus. IEEE Spectr., **47**(9), pp. 32–35, 52–56 (2010)

Schroeder, H., Amdt-Staufenbiel, N., Cygon, M., Scheel, W.: Planar glass waveguides for high performance electrical-optical-circuit-boards (ECTC)-the glass-layer-concept. In: Proceedings of IEEE 53rd Electronic Components and Technology Conference (ECTC), pp. 1053–1059 (2003)

Schroeder, H., Arndt-Staufenbiel, N., Beier, A., Ebling, F., Franke, M., Griese, E., Intemann, S., Kostelnik, J., Kuhler, T., Mödinger, R., Others: Thin glass based electrical-optical circuit boards (EOCB) using ion-exchange technology for graded-index multimode waveguides. In: Proceedings of the IEEE 58th Electronic Components and Technology Conference (ECTC), pp. 268–275 (2008)

Schroeder, H., Ebling, F., Strake, E., Himmler, A.: Heißgeprägte Polymerwellenleiter für elektrisch-optische Schaltungsträger (EOCB) - Technologie und Charakterisierung. DVS/GMM-Tagung Elektronische Baugruppen - Aufbau- und Fertigungstechnik, Fellbach. pp. 51–59 (2002)

Schroeder, H., Arndt-Staufenbiel, N., Ebling, F., Franke, M., Beier, A., Demmer, P., Söllau, W., Kostelnik, J., Mödinger, R., Pfeiffer, K., Ostrzinski, U., Griese, E.: Temperaturstabile Wellenleiter und optische Kopplung für elektro-optische Leiterplatten. DVS/GMM-Tagung "Elektronische Baugruppen und Aufbau- und Fertigungstechnik," pp. 93–98 (2006)

Seemann, R., Kramer, E.J., Lange, F.F.: Patterning of polymers: precise channel stamping by optimizing wetting properties. New J. Phys. **6**, 111 (2004)

Shen, L.-C., Lo, W.-C., Chang, H.-H., Fu, H.-C., Lee, Y.-C., Chang, S.-M., Chen, Y.-C., Chen, W.-Y.: Flexible electronic-optical local bus modules to the board-to-board, board-to-chip, and chip-to-chip optical interconnection. In: Proceedings of the IEEE 55th Electronic Components and Technology Conf. (ECTC). pp. 1039–1043 (2005)

Trommer, D., et al: A novel, flexible and easy to use technique for fabrication of optical spot size converters for InP based PICs. In: 11th International Conference on Indium Phosphide and Related Materials: Conference Proceedings : May 16–20, 1999, Congress Center Davos. p. Postdeadline Papers, pp. 12–14. Davos, Switzerland (1999)

Van Steenberge, G., Geerinck, P., Van Put, S., Van Koetsem, J., Ottevaere, H., Morlion, D., Thienpont, H., Van Daele, P.: MT-compatible laser-ablated interconnections for optical printed circuit boards. Lightwave Technol. J. **22**, 2083–2090 (2004)

Yoshimura, R., Hikita, M., Tomaru, S., Imamura, S.: Very low loss multimode polymeric. Electron. Lett. **33**, 1240–1242 (1997)

Chapter 11
Fiber Optic Modules

Abstract In this chapter, different module structures are presented which are applied in commercial modules. Usually, module assemblies are classified into the following categories: (1) transmitter modules (laser) with and without cooling; (2) receiver module (photodiode); (3) mixed modules (transmitter or receiver); (4) multi-fiber modules (arrays). For each category, an example is shown in more detail in the following. Previously, however, it is necessary to provide some explanation of the used coupling method.

11.1 Fiber–Chip Coupling Mechanisms in Module Construction

The quality of a coupling and the necessary effort depend strongly on the used type of the coupling element and the optical fiber. It depends primarily on the coupling efficiency between the component and the fiber. In industrial applications, you always have to make a compromise between cost and the achievable coupling efficiency.

If one wants to achieve a high coupling efficiency, it is always associated with high costs in the sale of the entire module. These high costs must be amortized in one or two years. For this reason, the market value of the module will directly determine the cost of the coupling. A mass market allows only very low module costs. Here, only solutions with very poor efficiency can be realized. The expense of the coupling grows inversely proportionally to the fiber cross section: Multi-mode fiber coupling is cheaper than single-mode fiber coupling, and each of these are lower in cost than fiber taper coupling. These are generally cheaper than lens couplings, since additional components (lenses) increase the amount of time for the whole coupling procedure.

11.1.1 Butt Fiber Couplings

The multi-mode fiber (see Chap. 2) is a fiber having a core diameter 50 μm and a numerical aperture of 0.25 (15° aperture angle). Placing this fiber directly in front of

© Springer-Verlag Berlin Heidelberg 2015
U.H.P. Fischer-Hirchert, *Photonic Packaging Sourcebook*,
DOI 10.1007/978-3-642-25376-8_11

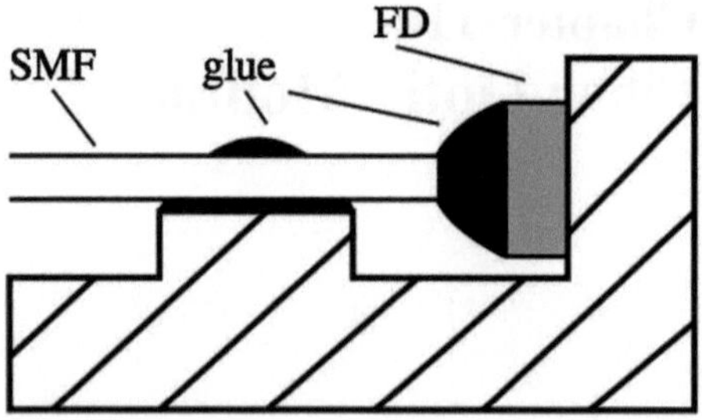

Fig. 11.1 Gluing a glass fiber in front of a photodiode

a photodiode with 50 μm of light-sensitive surface, one can measure an alignment tolerance of ±10 μm for 1 dB excess loss in the lateral direction. (Fig. 11.1). Here, Butt-coupling and gluing typically performs the fixation of the fiber. A second fixation of the fiber must be implemented for strain relief.

To obtain a higher efficiency in laser diodes couplings, than for Butt-coupling, it is affordable to implement optical mode adapter elements like a lens on the fiber end surface. You may also represent the end of an optical connector, which is used in different modules.

11.1.2 Single-mode Fiber Coupling

Transmitter modules with single-mode fiber coupling are typical basic elements used for transmitters in medium- and long-haul optical networks. These modules are typically fabricated by laser welding to realize a stable long-term fiber-chip coupling, which is illustrated in Fig. 11.2. More details of this connection technology have already been dealt with in detail in Chap. 8.

There, it was shown that a butt fiber coupling, however, can achieve only 10 % coupling efficiency. Using a lens arrangement quite a high coupling efficiency of 50–90 % can be achieved. A distinction is made between simple coupling without imaging optical elements and assemblies with field adjustment by lenses or lens systems between OEICs and single-mode fibers. A possibility of coupling with the help of two lenses is shown in Fig. 11.3. In Table 11.1, coupling efficiencies are listed for various lens systems compared to Butt fiber coupling.

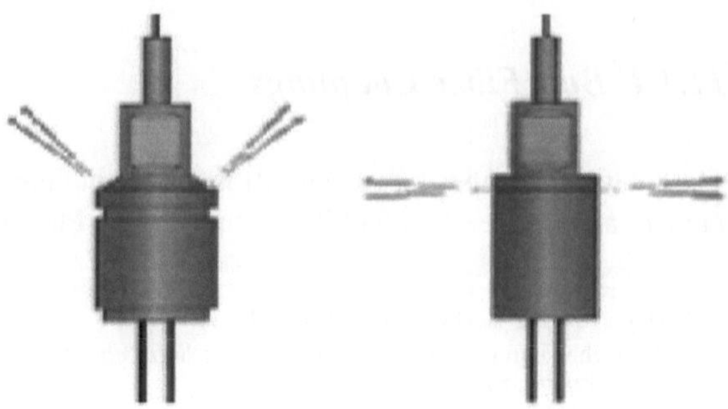

Fig. 11.2 Principle of laser-welded lens coupling

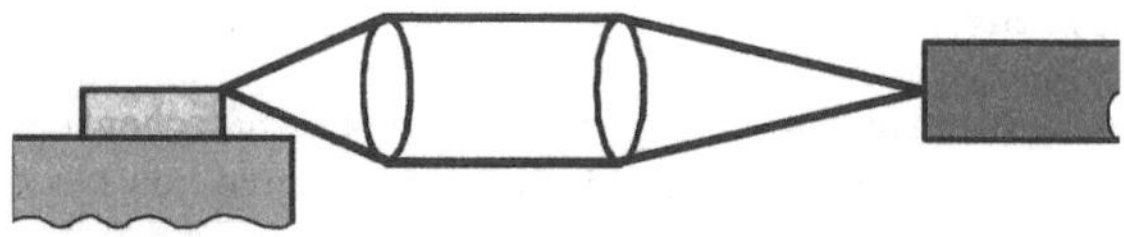

Fig. 11.3 Fiber–chip coupling by lens system

Table 11.1 Properties of different laser fiber coupling systems

Type of coupling (laser fiber)	Alignment accuracy μm for 1 dB additional loss in the lateral/longitudinal displacement	Coupling loss (dB)
Butt fiber (50 μm)	15/50	7–10
Butt fiber (9 μm)	2/20	7–10
One lens system	0.5/5	3
Double-lense system	0.5/5	1–3
Fiber taper	0.3/3	3–5

11.1.3 Fiber Taper Coupling

The fiber taper coupling has the advantage of a mechanically simple design with only one element to be moved. The high efficiency up to 50 % and the low price make the fiber taper very attractive for industrial module structures. The disadvantage is the effect of the very short distance to the component, so that the danger of a destruction of the OEIC is very high. In addition, a high-precision mechanics is required, which forces high manufacturing cost.

Specifically, I would like to focus to a construction of a miniaturized optical bench that was developed and patented in the Heinrich Hertz Institute (Fig. 11.4). In its center, the OEIC is placed on a heat sink that is attached by a screw on the bench. This has the advantage that the costly mechanics is reusable if the chip becomes defective. In addition, the carrier can support OEICs with different dimensions. This is a smart way for quick prototyping in research and development. The optical bench and thus the chip can additionally be temperature-controlled by a Peltier element.

A preloaded brass ferrule in which the tapered fiber is adhesively bonded carries out the optical coupling of the chip and the fiber. The fiber tip is moved by means of two pushers in the *x*- and *y*-directions. Pulling the fiber tube carries out the longitudinal movement. The module is clamped for optical coupling in an adjustment machine (Fig. 11.5).

The brass tube can be adjusted in the *x*-, *y*-, and *z*-directions using piezo-actuators with sub-micrometer resolution. Afterward, the tube which carriers the tapered fiber is fixed with 1-mm screws. After the adjustment, the module can be removed from the adjustment machine; afterwards, the machine can be used again. The adjustment principle and the matching adjustment machine (Fischer et al. 2001) were patented from the Heinrich Hertz Institute (Peters 1994a, b).

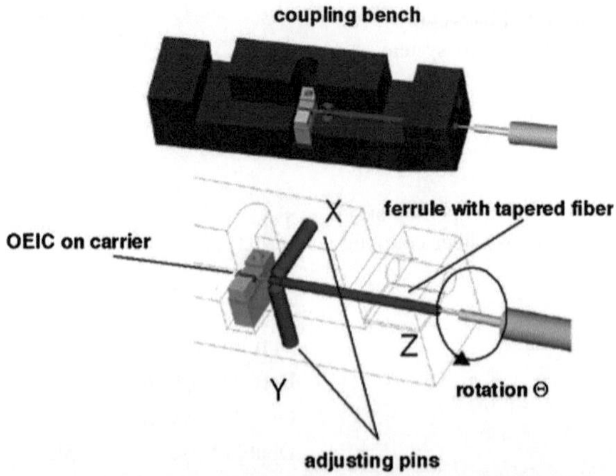

Fig. 11.4 Micro-optical bench

Fig. 11.5 Adjusting bench

11.2 Transmitter Modules with Cooling

Transmitter modules and the laser chip used in the modules exhibit a strong temperature dependence of both the power output and the output wavelength (see Chap. 2) and must therefore be stabilized in its temperature.

At the same time, the output power of the laser can be monitored and controlled by the operating current by the monitor photodiode. In Fig. 11.6, such a scheme is outlined. A control unit measures the current temperature of the laser and regulates the current of a Peltier element, which in turn stabilizes the chip to its working temperature. With commercially available devices, the control accuracy is generally ±0.05 °C.

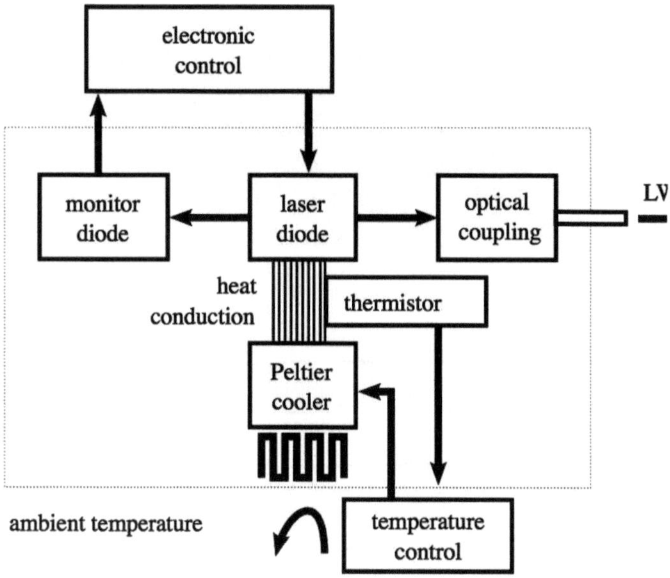

Fig. 11.6 Control of a laser module

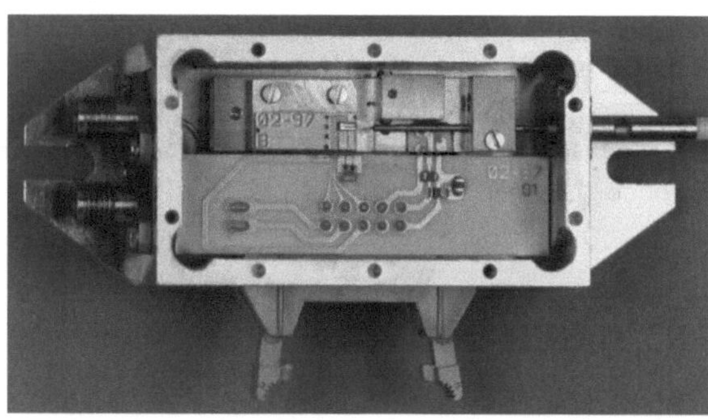

Fig. 11.7 Transmitter module with fiber taper coupling with up to 50-GHz bandwidth

In the next photograph (Fig. 11.7), a transmission module (Fischer 1999) for frequencies up to 50 GHz is shown. Its chip consists of a three-section DFB laser and a monolithically integrated butt-coupled modulator. As an integration concept, the butt coupling was chosen because herewith a separate optimization of laser and modulator is possible.

The drive electrode of the modulator is designed as an electric traveling waveguide. With suitable dimensioning of this line, the optical signal and the electrical modulation signal have the same propagation velocity. Then, the cutoff frequency is essentially determined by the electrical transmission losses. The supply of the electric control pulses is performed via low loss semi-rigid coaxial cables in the module (see Chap. 5). With this component, a transmission of 20 Gbit/s has

Fig. 11.8 Transmitter module with isolator and coupling lenses

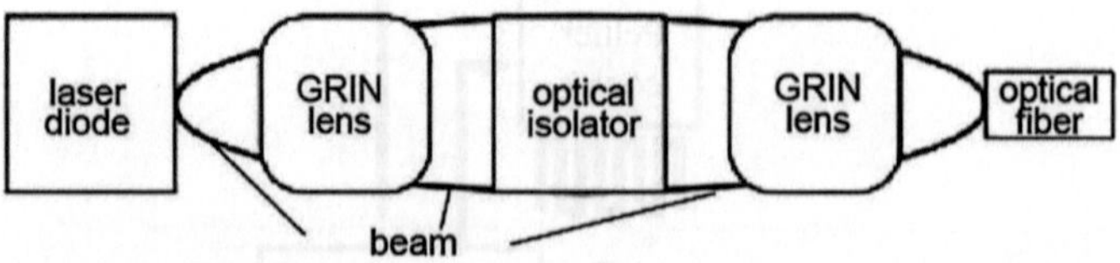

Fig. 11.9 Lens system of the transmitter module of Fig. 11.10

been successfully demonstrated. In addition, the traveling wave concept includes a high potential for bit rates up to 200 Gbit/s (Rosin et al. 1998).

In Fig. 11.8, an example of transmitter modules with a lens system for coupling to single-mode fibers is shown. The construction has the advantage that optical isolators suppress the interfering optical back-reflections may be incorporated into the beam path. The basic structure of the lens system is shown in Fig. 11.9. In the area between the two lenses (here Selfoc/gradient index), the optical isolator is introduced, which guarantees up to 55 dB suppression of the optical reflection.

11.3 Dual Inline Package Modules

Here, a description of the development of the DIL module into a miniaturized package (DIP24M6, 10 × 12 × 30 mm) is explained, which is an offspring of the industrial DIL-housing family. The module includes one optical input and one RF input, several DC-inputs and is temperature controllable. The OEIC is placed at one end of the miniature optical coupling bench on a sub-carrier, which is fixed by a screw bolt shown in Fig. 11.10. The glass fiber is drawn by a thin cannula of 175 μm radius and secured by conductive adhesive or solder. The back end of the cannula is clamped at the end of the optical bench. To realize a mostly parallel adjustment of the fiber tip to the waveguide of the DUT, the cannula is fixed with an angle of nearly 15° to the *x*- and *y*-directions depicted in Fig. 11.11. The hollow needle acts as a spring in *x*- and *y*-directions with lateral feed to of ±300 μm.

Using two adjusting pins, the fiber end can be moved precisely in *x*- and *y*-directions by Piezo-actuators or micrometer screws. The longitudinal elongation

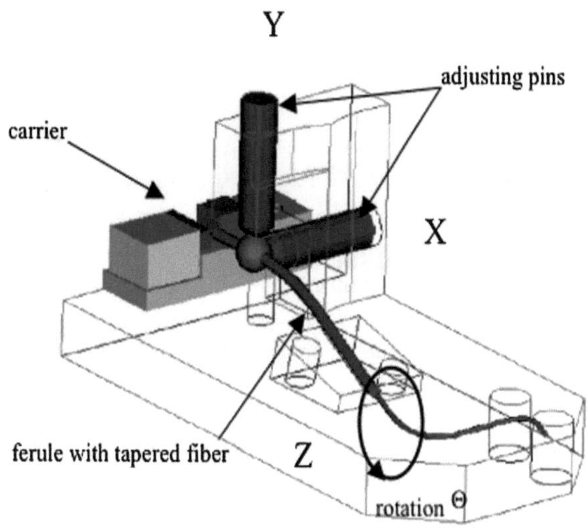

Fig. 11.10 Miniature optical coupling bench

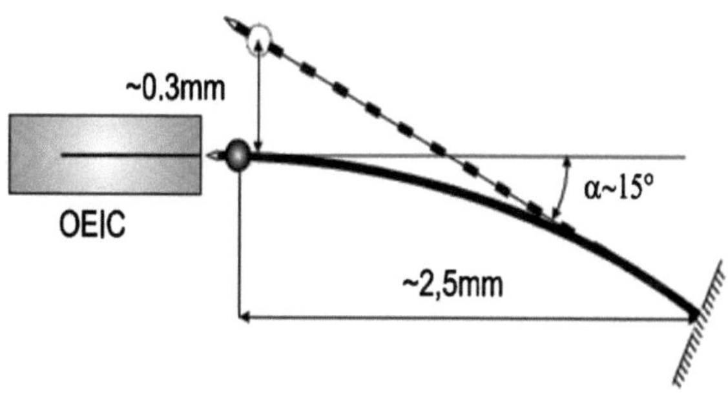

Fig. 11.11 Parallel adjustment of the fiber tip to the waveguide

is realized by shifting the fiber in the metal hollow needle. When the best optical power is launched into the fiber, a screw or laser welding secures the adjusting pins. If polarization-dependent fibers are used, rotating the cannula around the z-axis can optimize the polarization direction. The tuning is performed in a manipulator tool that allows tuning all directions and the angle θ.

11.3.1 Module Setup

The whole DIL module consists of a micro-optical bench, which includes the photonics device, a tapered fiber, and the adjusting needles. The bench is fixed on top of a Peltier element to stabilize the OEIC thermally. Via DIL pins and a thick-film circuit, the bias, the Peltier current, and the temperature sensor are fed. A semi-rigid cable realizes RF connection, which connects the OEIC directly. A good RF response up to 50 GHz was performed (Rohde et al. 1997). One module type was

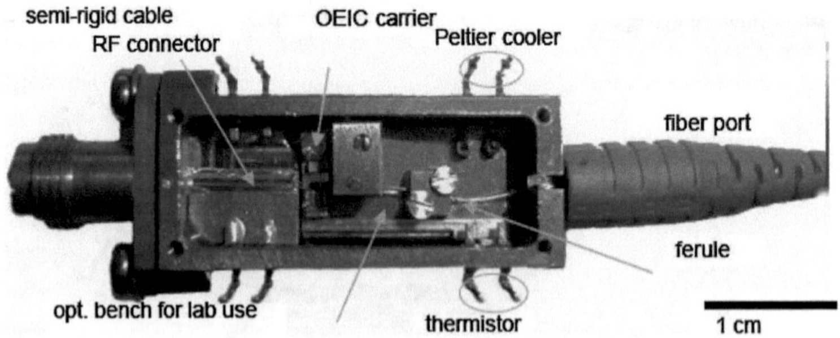

Fig. 11.12 Temperature-controlled DIL module for optical input with 50-GHz modulation bandwidth

realized by fixing the adjusting pins using laser welding, while in a second type fixing by screws (Fig. 11.12) was realized to reuse the module.

To use the DIL module type for waveguide-integrated photodiodes with 50 GHz RF response (Umbach et al. 1995), the temperature control and the Peltier cooler can be removed.

In order to avoid the mechanical bending of the optical bench, all materials are made of stainless steel. One has to keep in mind to compensate all the different extension coefficients: photonic device, fiber, needle, carrier heat sink, and glue. The used Peltier cooler has a heat pump performance of 2.2 W at 20 °C and the heat sink is made of copper. The Peltier is glued at the bottom side of the lid to prevent heat circulation inside the DIL module.

11.3.2 Laser Welding Results

For welding a Nd, YAG laser with up to 50 W output power was used. The welding spot size was 500 μm. In Fig. 11.13 it can be shown that how displacement measurements of the adjusting pins are performed while welding. To estimate optically the displacement of the adjusting pin during welding, a distance sensor with resolution of 10 nm was used. Laser welding tests are performed by several welding setups: lap weld and lap–fillet weld. The exact description of these methods is shown in Sect. 8.2.1. The lap weld was tested primarily. It shows low displacement, because the used welding spots are placed highly symmetrical. The welding spots at the coupling bench realized by this method can be seen in Fig. 11.14. The lap–fillet weld was tested additionally. It needs lower welding laser energy than the lap weld. The material connection shows insufficient adhesive power and high displacement (Fischer 2000).

The cumulative yield of the laser weld tests are shown in Fig. 11.15. A good performance of more than 96 % of the welding spots with dislocations less than 1 μm are depicted for welding connections without additional finishing. The best score was realized using stainless steel with a very low temperature coefficient, while the connection power after welding was better than 50 N.

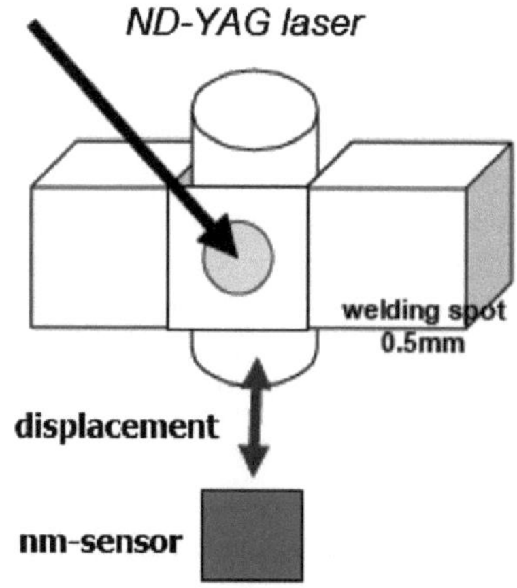

Fig. 11.13 Test setup for measuring the welding

Fig. 11.14 Welded adjusting pins

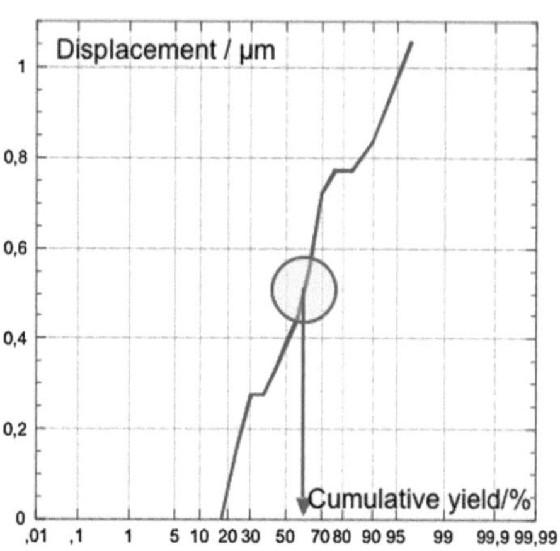

Fig. 11.15 Cumulative yield of displacements of the performed laser welds

11.3.3 RF Results

The high-frequency characteristics of the DIL module were collected using a network analyzer with a range of 45 MHz–50 GHz. The overall distance between V-connector and photonic device was 11 mm. We measured the electrical transmission parameter S_{21} that is shown in Fig. 11.16. The attenuation is strongly induced by the loss of the used RF cable of 50 dB/m. The curvature shows only low variations. At 50 GHz, a loss of less than 6 dB can be noticed. This kind of RF setup is indeed highly suitable for frequencies of even higher than 50 GHz.

11.3.4 Environmental Tests

Several modules were proved in long-term temperature tests between +15 and +40 °C and between −20 and +70 °C. The coupling stability of the modules was tested using waveguided photodiodes and a stabilized laser source.

Fortunately, the output signal of the photodiode is not temperature-dependent and exhibits only low polarization sensitivity of maximum 1 dB. It was found that the temperature behavior of the module (Fig. 11.17) has a maximum output variation of ±0.1 dB. The tests were repeated 50 times, and no significant degradation of the coupling efficiency was detected.

11.4 Receiver Modules

This section describes some commercially available receiver modules as well as some examples of prototypes of research developments. First, a commercially available photodiode module manufactured by ATT is described (Fig. 11.18).

Fig. 11.16 Electrical transmission S_{21} of DIL module

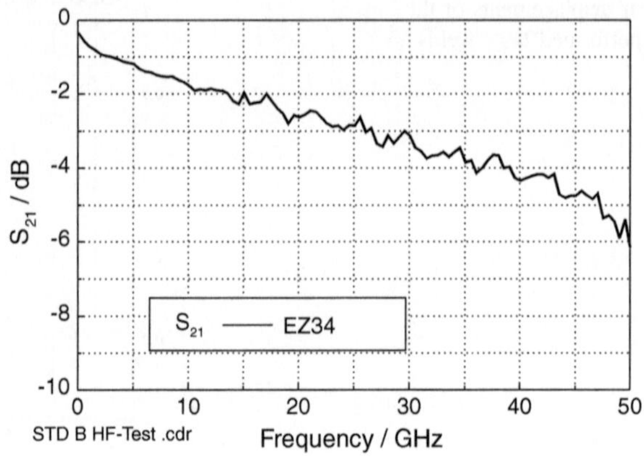

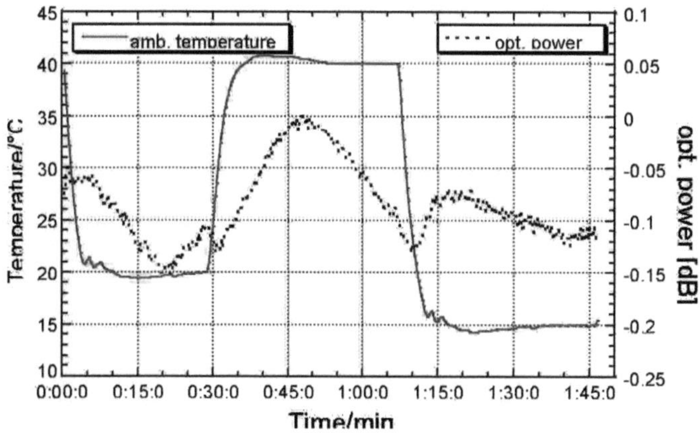

Fig. 11.17 Temperature cycling test of DIL module

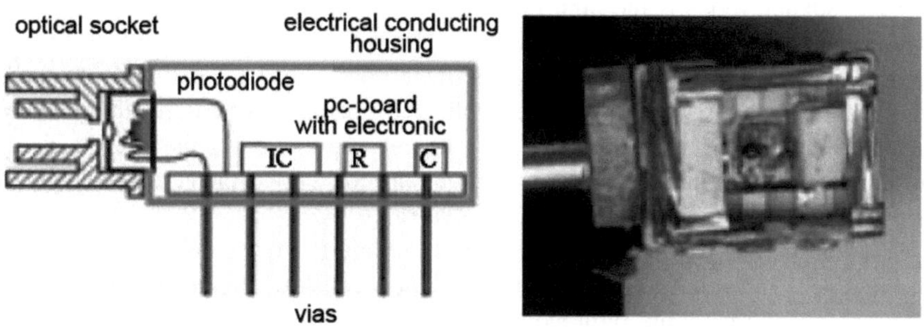

Fig. 11.18 DIL receiver module with lens coupling (ATT)

The fiber is inserted from the left side into the housing and a lens onto the photodiode images the light. It can be seen that a ceramic feed for the high-frequency connection has been bonded onto the diode to the connector. The housing has dimensions of 1 cm × 0.5 cm × 0.3 cm.

The optical–microwave converter (Rohde et al. 1997), which is shown in Figs. 11.19 and 11.20 will now be further discussed. It serves as a receiver module substantially for generating microwave signals, using the optical heterodyne method.

Two optical laser transmitters are tuned to each other in the optical frequency range of 60 GHz, and fed into a glass fiber. The signals on a single fiber can then be combined using an optical coupler. At the end of the fiber, the light is irradiated on a photodiode, on which then the differential signal of 60 GHz is mixed between the transmitter laser frequencies. The major advantage of this mixed method compared to conventional microwave generation is its high efficiency in the millimeter wave range. Since an electric TWA (traveling wave) pre-amplifier is monolithically integrated together with the photodiode on the chip, the RF efficiency of the converter was effective greater than one.

Fig. 11.19 65-GHz receiver module with lens coupling

Also advantageous is the high flexibility in the choice of the center frequency, which can be easily adjusted by shifting the laser center frequency and may range from 55 to 65 GHz. The shift is generally accomplished by detuning the laser temperature (see Sect. 3.1.4.1). The applicable optical wavelength range can be freely selected between 1300 and 1600 nm. As applications, the optical signal generation, the terahertz wave generation, spectroscopy, radar, mobile phone applications, or sensor systems are conceivable.

The properties of another type of a prototype receiver module shown in Fig. 11.21 are briefly described here. A one-sided fiber connection for tapered fibers is inserted from the right side of the module.

On the opposite side, the high-frequency connection in the form of a V-type plug was implemented to get a valuable frequency range up to 40 GHz. The receiver chip on which a waveguide photodiode and an electrical wideband amplifier were monolithically jointly integrated is located directly between the fiber and RF connection. The RF cable between the chip and RF connector is designed as a coplanar structure.

An RF electrical network for the bias supply with a polarity reversal protection is seen in the lower part of the module in which five bias lines are led out of the housing. The module does not require temperature control. Its application range is between 15 and 40 °C, where the optical power is changed not more than ±0.5 dB.

Fig. 11.20 Photodiode and TWA amplifier

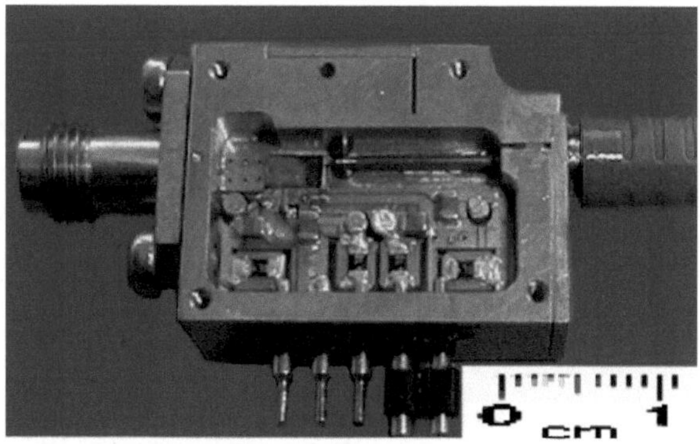

Fig. 11.21 45-GHz receiver module with fiber taper coupling and coplanar waveguides

Receiver modules with photodiodes containing integrated spot-size expanders can use easy Butt-coupling from single-mode glass fiber to the photodiode chip. Here, the use of this technique within a photodiode package with an RF connection with modulation bandwidth beyond the 40 Gbit/s domain will be shown (Eckhardt et al. 2000). The structure and the fabrication process of the used integrated spot-size transformers are described in Sect. 4.8.

The pre-alignment was made by manually adjusting the fiber in a line with the waveguide of the waveguide-fed photodiode using a microscope and simple micrometer stages. The output of the photodiode was monitored with an electrical power meter and send into a computer. The computer is equipped with a controller for the adjustment of the micrometer stages and can adjust automatically the optical coupling efficiency in two lateral dimensions. In combination with an AGILENT Precision Reflectometer (8504B), the back reflection of the optical connection was examined with a spatial resolution of better than 20 μm. Finally, the adhesive was inserted into the gap between the SMF and the photodiode chip and hardened by UV-light.

To adapt the effective index of SMF ($n = 1.47$) and the InP wave-guide ($n = 3.2$), the end facet of the OEIC was coated by several layers of SiO_2 and TiO_2. To adapt the refractive index of the air gap between OEIC and SMF, we used an AR-coating that adapts the step between OEIC and air: ($n = 3.2$ OEIC $\rightarrow n = 1$ air). Using the epoxy, the AR-coating was adjusted to the index step between OEIC and glue: ($n = 3.2$ OEIC $\rightarrow n = 1.501$ glue). For cost reduction reasons, the SMF was not AR-coated. Therefore, additional reflections at the fiber end are expected, especially for the air gap version (Table 11.2).

With UV-Epoxy Vitralit 1507 applied, the return loss can be extended right away to 28.0 dB as an acceptable value for communication system issues.

The overall loss of the "fiber → tapered waveguide → photodiode" coupling was measured at the best to less than 1.5 dB. This results in a coupling efficiency of

Table 11.2 Measured optical reflections in air and with glue

Air gap	UV-Epoxy	
	Harded	Not harded
RL = 14.70 dB	RL = 28.00 dB	RL = 26.93 dB

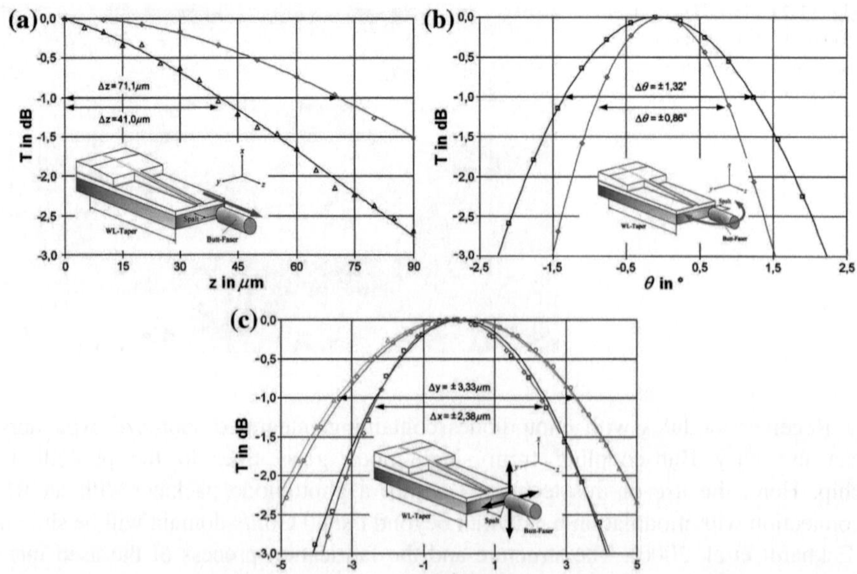

Fig. 11.22 Fiber–chip coupling loss with butt-ended fiber and optical adhesive. **a** Lateral displacement, **b** longitudinal displacement, **c** angular displacement

more than 50–70 %. The behavior of the coupling efficiency with lateral, longitudinal, and angular displacements is depicted in detail in Fig. 11.22. The loss in both lateral directions (*x*, *y*) shows no difference between air gap and glue. In Table 11.3, the values for an optical loss of 1 dB of the glued connection in comparison with an air gap are listed.

There is a small degradation in the permitted angle deviation between the two waveguides using glue (1.34° → 0.88°@ 1 dB) that can be seen in detail in Fig. 11.22. In practice, a gap of 15 μm between SMF and chip gave the best results in long-term stability.

The module consists of gold-plated milled brass. It includes one optical input and one RF output and several DC-inputs with RF blocking features (Fig. 11.23). For the operation of waveguide-integrated photodiodes with an RF response of more than 50 GHz, the RF response up to 45 GHz of the modules had been published from Rohde et al. (1997).

Table 11.3 Measured values for an additional optical loss of 1 dB in air and with glue

	z-direction	Angled shift	*x*-direction	*y*-direction
Air gap	Δ*z* 1 dB = 37.0 μm	ΔΘ 1 dB = 1.34°	Δ*x* 1 dB = 2.42 μm	Δ*y* 1 dB = 3.23 μm
UV-Epoxy	Δ*z* 1 dB = 56.0 μm	ΔΘ 1 dB = 0.88°	Δ*x* 1 dB = 2.34 μm	Δ*y* 1 dB = 3.32 μm

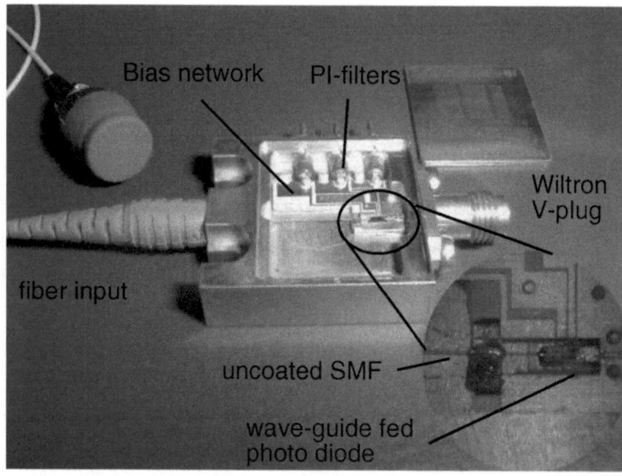

Fig. 11.23 Complete photoreceiver module

For application in optical networks, receiver modules must be stable with respect to temperature changes and mechanical stresses. Therefore, the stability results of the photodiode modules are investigated with reference to the Telcordia requirements (see Chap. 13).

As shown in Fig. 11.24, a maximum output variation of ±14 % between −10 and +40 °C is to be noticed. At higher temperatures than 40 °C, the glue is swelling too much that the temperature-dependent loss is unacceptable high. Nevertheless, after several temperature and vibration test runs, no significant degradation (<5 %) of the coupling efficiency was detected (Fig. 11.24).

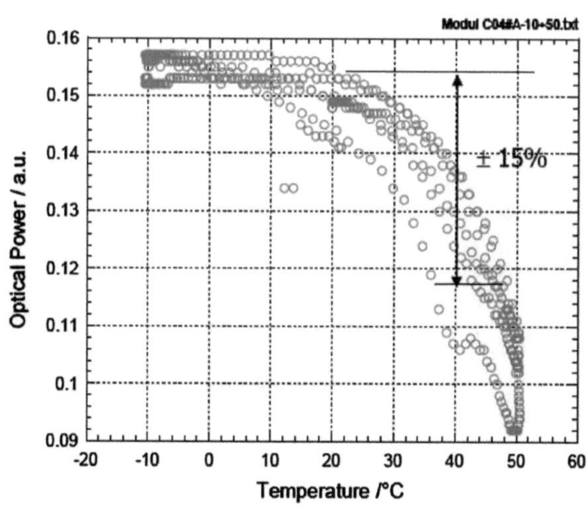

Fig. 11.24 Temperature behavior of photoreceiver module

11.5 Transceiver Modules

Transceivers are combined transmitter and receiver modules. The here-presented modules send with 1.3 μm wavelength and receive by 1.55 μm. This dual function has the advantage of taking advantage of the bandwidth of the fiber using simple dual wavelength-division multiplexing. In Fig. 11.25, the construction of a module from Siemens is shown. The mechanical construction is carried out using conventional micro-optics in a simple metal housing (a). The transmitting and receiving units are located in separately soldered TO packages (b).

The beam and wavelength selection components are located with two spherical lenses in a plastic injection-molded part, which takes over the automatic alignment of the optical elements (c). It has been successfully tried to achieve a cost reduction by utilizing a partially passive adjustment technique.

A functionally equal module, but in monolithic design manner, is shown in Fig. 11.26. All the optical elements were monolithically integrated on one InP chip (Hamacher et al. 2000). The cost of such a component is in mass production significantly below the cost of the micro-optical variant.

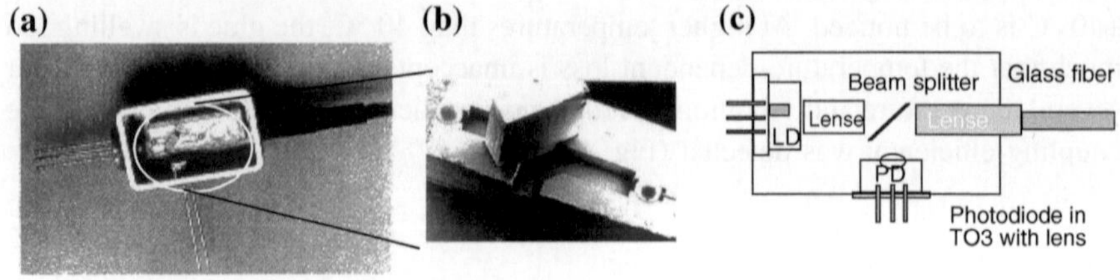

Fig. 11.25 Micro-optical transceiver

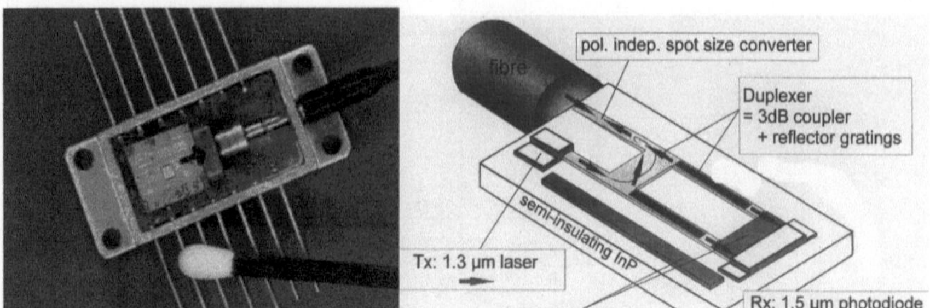

Fig. 11.26 Monolithic integrated transceiver

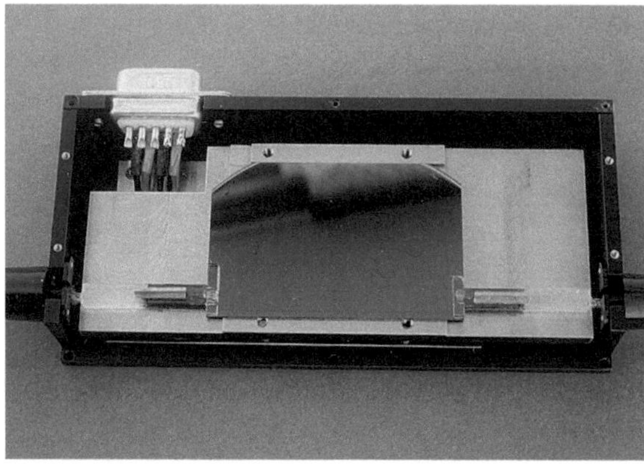

Fig. 11.27 Fiber array module with 8 fiber inputs and Peltier cooling

11.6 Multi-fiber Modules with Butt-Ended Fibers

Multi-fiber modules, which contain typically silicon oxide photonic devices, require only electrical current supplies. There is no need for RF modulation. The photonic devices are only temperature-controlled by a Peltier cooler as it can be seen in Fig. 11.27. On the other hand, many photonic chips are applicable without temperature control. These devices can be enclosed in a very inexpensive variant of plastic injection-molded enclosure (Fig. 11.28). More description of the possible housings can be found in Chap. 12. In this chapter, only an overview of the module variants with their basic properties will be given.

The arrayed waveguide grating chip (AWG) works as an optical spectrometer in planar waveguide technique. To get a good resolution of the spectrometer, it works with a high order of the grating of 50–250. To use the device in optical communication systems, a precise wavelength control and long-term wavelength stability is a prerequisite. A standard arrayed waveguide grating chip typically has a strong temperature drift of its center wavelength. One method to overcome this drift is to enclose the AWG in a housing combined with a temperature controller in order to detune and hold the required central wavelength.

In this chapter, a description of the optical and mechanical setup and the results on the wavelength controllability and environmental stability of modules with 8×8

Fig. 11.28 Array module in plastic housing

and 16 × 16 optical in/output ports is reviewed. Additionally, a comprehension of the measured optical losses and the reflections at the interface between the chip end faces and the coupled fiber arrays is given.

11.6.1 Device Characteristics

A drawing of the used arrayed waveguide grating chips is shown in Fig. 11.29. The DUT has 8 × 8 of input/output waveguides, two focusing slab waveguides, and a grating in with more than 100 optical paths of definitely lengths to realize the phase difference.

11.6.2 Optical Coupling Setup

The photonic chips are made of a Si/SiO_2 wafer fabricated by flame hydrolysis deposition. Reactive ion etching (RIE) is typically used to perform the waveguides. Normal chip sizes are 40 × 30 mm^2 or rather 80 × 30 mm^2. The extensions of the waveguides are 3 × 3 μm^2 up to 6 × 6 μm^2. The difference of refractive index between core (n_{co}) and cladding (n_{cl}) is 0.7–3 % by high confinement applications. A core index of $n = 1.454$ from doped SiO_2 is typically. The numerical aperture of the light point at the end of the waveguide can be calculated as follows:

$$NA = \sqrt{n^2 - n^2{}_{cl}} = \sin\alpha = 0.17 \tag{11.1}$$

$\alpha = 9.8°$ (angle of aperture)

A numerical aperture angle of the standard monomode fiber (SMF) is calculated using the standardized parameters of the ITU (ITU-T G652 2000) as 6.5°. To get a high efficient overlap of optical energy from the SMF and the chip waveguide, the electro-magnetic wave should “overlap” as large as possible (see Chap. 4). The approximation after Ladany (1993) computes the mode-field diameter mismatch R of the single-mode components and the loss $L(R)$:

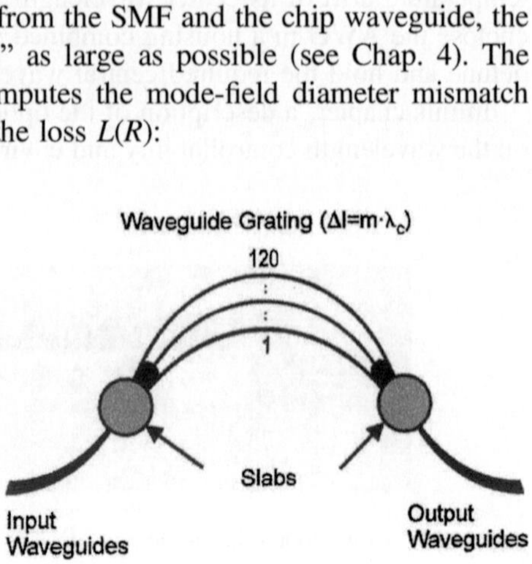

Fig. 11.29 Schematic sketch of AWG chip outline

$$R = \frac{2(\omega_{\mathrm{SMF}} - \omega_{\mathrm{AWG}})}{\omega_{\mathrm{SMF}} + \omega_{\mathrm{AWG}}} \tag{11.2}$$

ω_{SMF} mode-field diameter of SMF
ω_{AWG} mode-field diameter of AWG

$$L(R) \cong 4.343\, R^2 \tag{11.3}$$

The diameter is a function of wavelength (1550 nm) and aperture angle:

$$\omega = \frac{\lambda}{\pi n \tan \alpha} \tag{11.4}$$

λ wavelength
n refractive index of ambient medium

The mode-field diameters of the chip are 2.85 and 4.3 μm for the single-mode fiber, respectively. In Eq. (11.3), the calculation of the loss $L(R)$ between the two waveguides is realized. This gives an expected value of 0.75 ± 0.5 dB. The direct measure of the coupling attenuation could not be measured directly, because the chip has no optical active device in its structure. Values for the coupling losses are average values of 0.935 dB for the 8 × 8 AWG (Fig. 11.30a) and 0.91 dB and for the 16 × 16 AWG are shown in Fig. 11.30b.

The additional losses of 0.185 and 0.16 dB, respectively, are caused by imperfections in the alignment and fixing process of the fiber ribbon at the chip interface. It is not possible to adjust all fibers of the array optimal to the SiO_2 waveguides. A misfit in angle of 1° results in 0.9 dB additional loss and an offset of 1 μm in the lateral position gives an additional loss of 0.6 dB. The fiber array is fixed by UV-hardening glue with an optimal adaptation of the refractive index of $n = 1482$ to the fiber and the SiO_2 waveguide.

In this case, no anti-reflection coating of the end surfaces of the waveguide or the SMF have been applied which are straight cut. Expected values for the reflections of the fiber–chip interfaces are -30 dB.

For the 16 × 16 device, a mean value of −32.5 dB and for the 8 × 8 AWG, an even lower value of −38 dB was measured. The reason for this small value is the tilted fixation of the fiber array to the chip waveguide. Typically, additional tilting reduces the reflection values but increases the coupling loss, which is higher than expected mode-mismatch loss (Fig. 11.31).

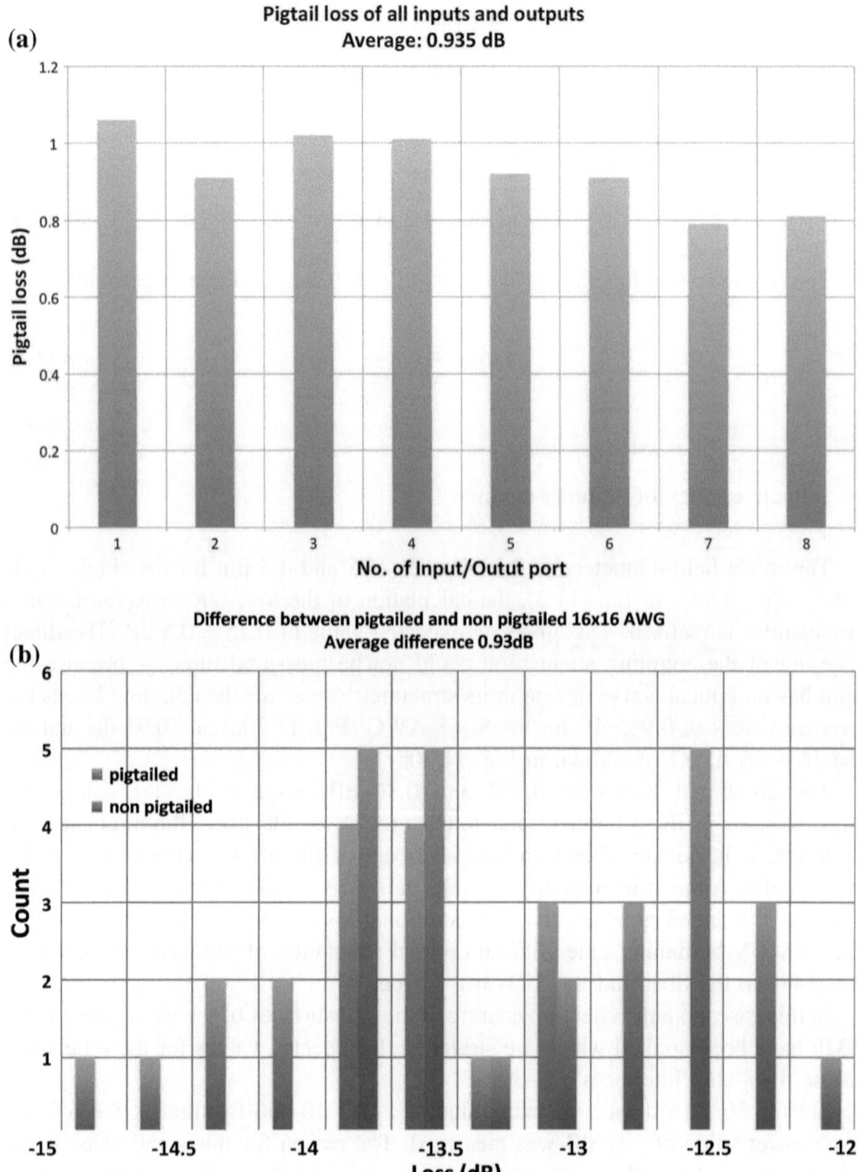

Fig. 11.30 **a** Loss of fiber-chip connection (8*8). **b** Loss difference between pigtailed and direct coupled 16*16 AWG

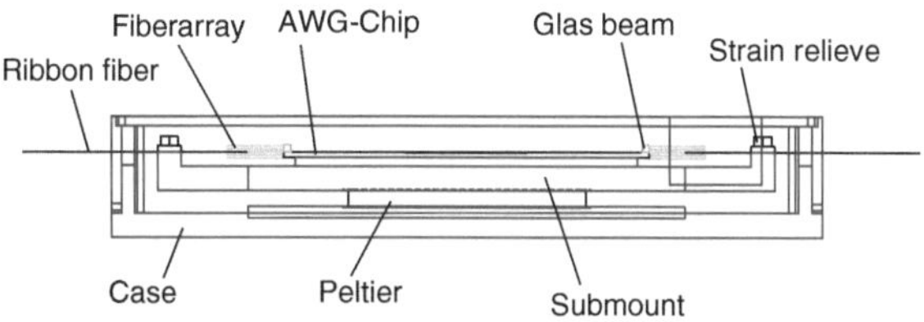

Fig. 11.31 Cross-sectional view of AWG housing with ribbon fibers

11.6.3 Module Packaging

Figures 11.33a and 11.34b show the photographs of the AWG module and the fiber-chip connection. The module consists of the AWG device, a pair of optical fiber arrays, and a chip carrier in a case. Additionally, an electrical plug for the cooling of the device is implemented in the case. The chip was assembled in the following way: first, the chip with the fiber arrays was fixed to a heat spreader by using a stress-free gluing technique. This realizes a high efficient adjustment of the extension coefficients of the chip with $\alpha_{Si} = 7 \times 10^{-6}$/°C, $\alpha_{SiO2} = 0.7 \times 10^{-6}$/°C to that of the heat sink, consisting of stainless steel: Covar $\alpha_{Co} = 4.8 \times 10^{-6}$/°C (Fig. 11.31).

If the length of the chip is $L_{chip} = 78$ mm and the temperature difference is $\Delta T = 2$ °C between heat sink and chip, there is a small difference in length of $\Delta l_{Kovar} - \Delta l_{Si} = 0.3$ μm. For the mounting of chip to the heat sink, a low Young's modulus adhesive is used. For monitoring of the chip temperature, thermal NTC sensor was embedded in the chip carrier. At the end, the heat sink was mounted on the Peltier cooler, which has been formerly thermally good connected to the case. The case itself was realized of aluminum to act as a global heat sink.

After finishing the electrical and mechanical structure, the chip and the fiber arrays were adjusted and bonded together with an UV-curable glue which is explained in more detail in Chap. 9. To perform the chip–fiber coupling procedure, the chip, the module and the sub-mount were installed (see Fig. 11.32) onto a three-axes stage (x, y, z).

Additionally, the fiber array was mounted on a second stage with three linear and two angular stages. Using a tunable laser source, the light was fed by a $1 \times N$ ($N = 8.16$) optical splitter to all fibers of the array. The pre-alignment was made by manual adjusting. To start the coupling procedure, the fiber array was manually adjusted parallel to the waveguides of the chip. The output at the opposite side of the chip was monitored using a lens with a split-field optic, and a camera system.

In the beginning, the optical outputs near the center of the AWG were aligned. After that, the outputs of the fiber array were aligned more precisely to the chip waveguides. Here, the laser must be tuned to the center wavelength of the filter curve of the observed output fiber. If the light gets a maximum value, the adhesive

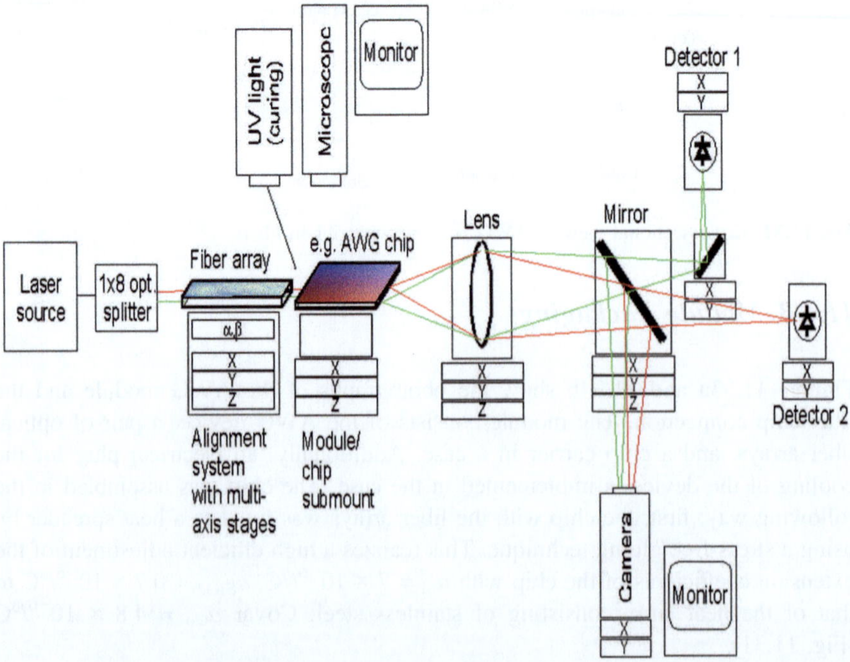

Fig. 11.32 Experimental setup for fiber array-chip coupling

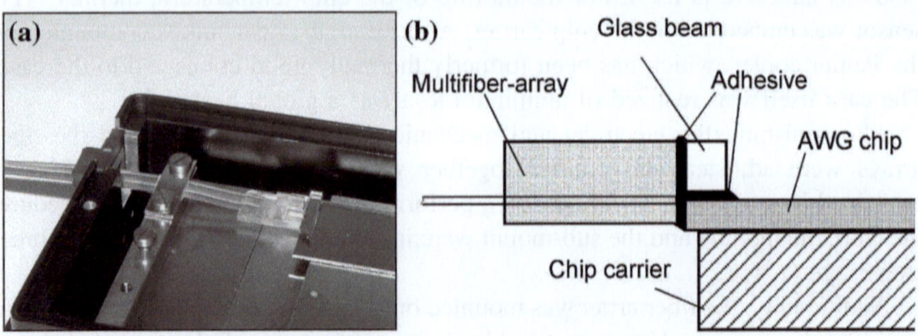

Fig. 11.33 **a** Photograph, **b** cross section of the fiber array-chip connection

was inserted into the gap between the array and the AWG chip. After that, the glue was hardened by UV-light.

To get a steady optical connection, a glass block was positioned on top of the chip. This was introduced as a mechanical stabilization for the array and fixed there as shown in Fig. 11.33a, b. After curing, on the opposite side of the chip, the same coupling procedure was performed, using the already fixed fibers at the opposite side for launching the input light.

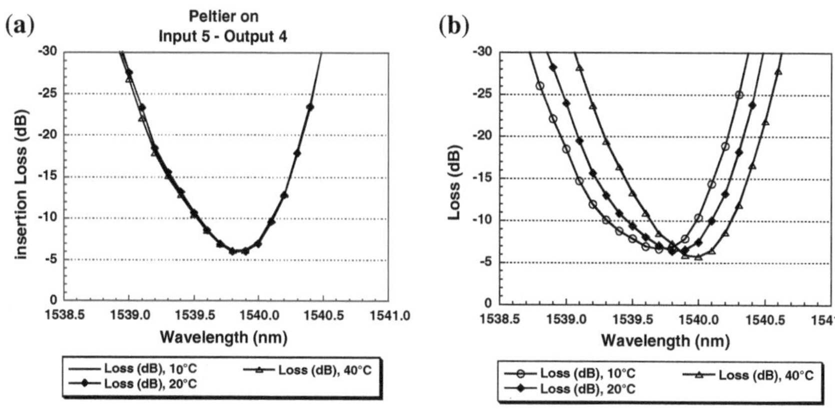

Fig. 11.34 **a** Controlled spectral response of T_{chip} = 23 °C at $T_{ambient}$ = 10, 20, and 40 °C. **b** Spectral response of center wavelength at $T_{ambient}$ = 10, 20, and 40 °C

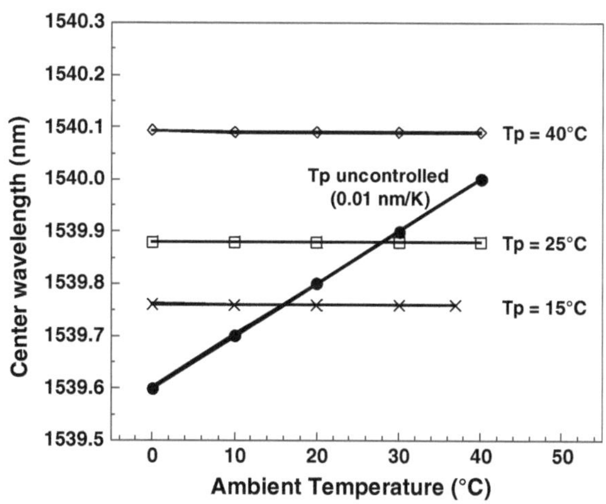

Fig. 11.35 Relationship between ambient temperature and center wavelength

11.6.4 Environmental Stability

To get long-term stable modules with respect to temperature changes and mechanical stresses, several kinds of environmental tests, including a vibration test as described in the following have been performed. Insertion losses, center wavelength, and return losses were measured for each fiber at a chip temperature of 23 °C. To prove the temperature controllability of the complete module, the changes in center wavelength are measured. It was measured with and without the Peltier cooler at different climate chamber temperatures.

For this, the module is fixed into a humidity-controlled environmental test chamber. To check the wavelength shift of the AWG, a tunable laser in combination with a polarization controller was implemented. The temperature behavior of the

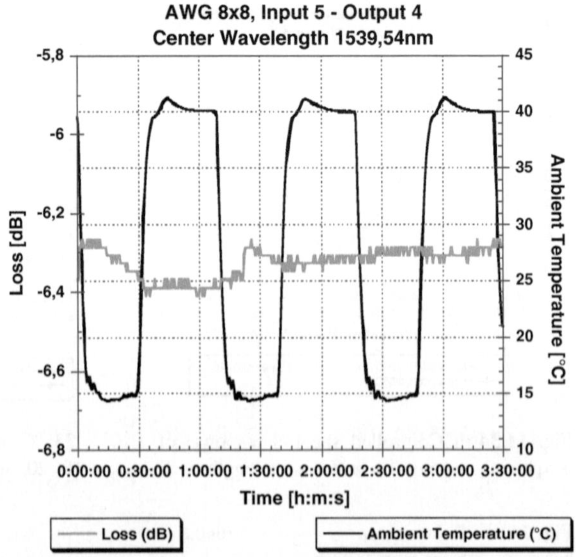

Fig. 11.36 Temperature cycling of (T_{chip} = 23 °C) 8 × 8 module

center wavelength as a function of ambient temperature is shown in Fig. 11.34. The chip temperature was held constant (T = 23 °C) and the ambient temperature was tuned from 10 to 40 °C shown in Fig. 11.34a.

A stability of the temperature control circuit of better than 0.1 °C was measured. No significant deviation between the measured curves can be observed. Without temperature control, the center wavelengths change with the formerly calculated drift of 0.01 nm (1.25 GHz)/°C. The results of the center wavelength drift are summarized in Fig. 11.35.

To confirm the temperature stability of the setup, many cycles between 15 and 40 °C were driven. The variation of the optical coupling loss is less than ±0.1 dB, which is an acceptable good value for the use of this setup in optical transmission systems shown in Fig. 11.36.

11.7 Multi-fiber Modules with Lensed Fibers

In optical communications systems, integrated interferometers with integrated semiconductor optical amplifiers (SOA-MZI) are highly attractive chips because of their small sizes (Tekin et al. 2000; Ehlers et al. 2001). These monolithically integrated photonic chips which were used in the field and in laboratory test beds require a special module that takes important system requirements such as high RF data rate and low optical insertion loss into consideration. These chips have several optical output ports with a distance of 250 μm between the optical waveguides. Several concepts for fiber–chip coupling with single tapered fibers had been shown in Chaps. 8 and 9. Here, multiple fibers with tapered ends must be handled to realize a long-term stable connection.

Coupling of several tapered fibers to an OEIC is still under development. Here, a coupling scheme is described and it connects an array of tapered fibers by fixing the array indirectly to the OEIC by using two-side holders.

11.7.1 Coupling Concept

The fiber tapers that are used in this package are performed by grinding the tip to a conical shape ($\sim 140°$). Using this type of tapered fibers, a coupling efficiency of up to −3 dB can be reached (see Chap. 4). In the case of multiple fiber tapers arranged in an array of V-grooves in a silicon-based plate, additional tolerances due to manufacturing of the arrays must be taken in account. In this module, four fibers are used. The variation of the center-to-center spacing between the fibers is ±0.5 μm. The additional lateral losses due to the tolerances will be in the range −1 dB for the 4-fiber array and the overall attenuation will be ca. 4 dB. To fix the fiber array to the SOA-MZI, the device is positioned in the center of a micro-optical bench on a sub-mount, that is fixed by a screw (Fig. 11.37).

The optical fiber taper array is mounted in front of the chip, and all optical fibers are adjusted to the waveguides of the chip to maximum coupling efficiency. The typical distance of 10–15 μm from chip facet to the OEIC must be kept constant to get the best coupling efficiency. Both sides of the fiber taper array are placed sliding on the adhesive and held by adhesion.

The array is glued to the chip sub-mount, which also acts as heat sink. Shrinking forces of the glue during curing act symmetrically on the array and force only a negligible lateral displacement.

In z-direction, the shrinking has only small effect on the coupling efficiency. The coupling is a factor of 10 less sensitive to longitudinal separation. In Fig. 11.38, a

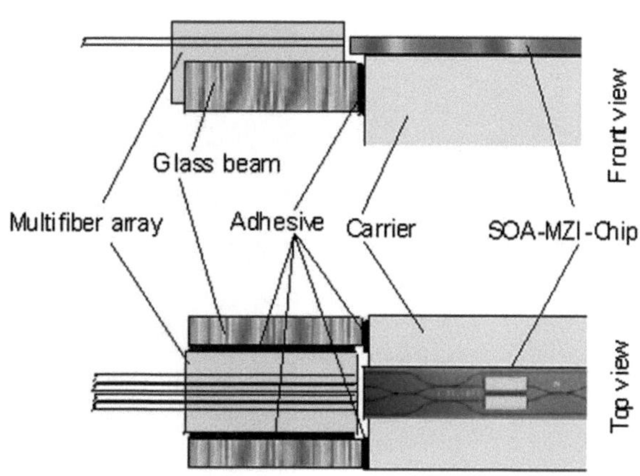

Fig. 11.37 Fixation setup

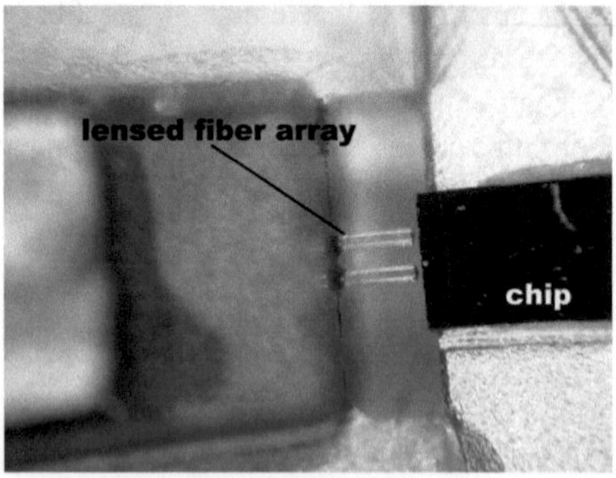

Fig. 11.38 Photograph of the fixed fiber array

photograph of the complete fixation part is shown. Here, the chip is located on the right with the two fiber tapers in front of its output facet at a distance of 15 μm and the fiber's pitch is 250 μm.

11.7.2 Module Setup

The module setup of the SOA-MZI module will be described in this section; a photograph of the SOA-MZI module shows (Fig. 11.39) a pair of optical fiber arrays and with the chip arranged on a carrier on top of a micro-optical bench. The setup is mounted in a housing with an electrical plug for the Peltier element and the RF connectors for the electrical switching.

The chip was fixed to a heat sink by using thermal glue with a low Young's modulus adhesive. This will be a stress-free mounting because the extension coefficient of the InP-substrate was adjusted for minimum mechanical stress due to

Fig. 11.39 Photograph of the SOA-MZI module

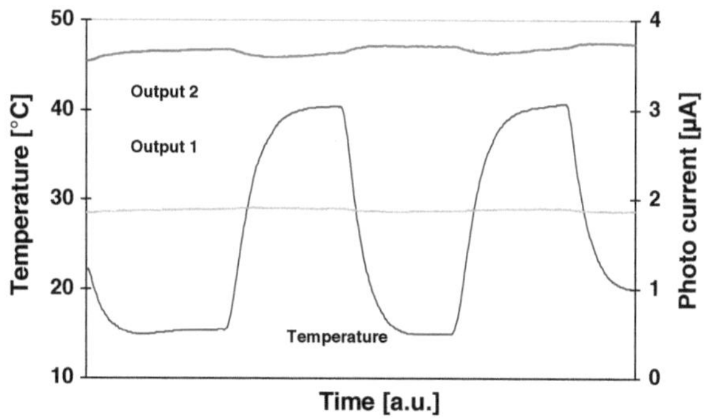

Fig. 11.40 Temperature test of SOA-MZI module between 15 and 40 °C

temperature changes of the sub-mount. It used a special heat conducting glue, which also connects the OEIC to electrical ground. A thermal NTC sensor was embedded on the chip carrier nearby the OEIC for monitoring the chip temperature. The optical sub-mount is fixed on top of Peltier coolers for the thermal stabilization of the photonic chip.

To get a good heat conduction, the case was made of brass. The temperature sensor connections and the electrical connection of the Peltier are fed via a special connector to the device. The bias currents of the amplifier region are connected by SMA electrical sockets and a thick-film circuit. The chip carrier is fixed by a screw to be removed from the module quickly and exchanged for other photonic chips. The tapered fiber array has a lens radius of 10 µm.

11.7.3 Coupling Procedure

After finishing the electromechanical packaging, the chip and the fiber arrays were adjusted. For chip–fiber coupling, the module with the chip sub-mount was fixed onto a three-axes stage (x, y, z). The fiber array was also fixed on a stage with three linear and three angular adjustments. The pre-alignment was made by manual adjusting the fiber array in a line with the waveguides of the SOA-MZI using a microscope.

Two photodiodes are used to check the optical power launch at the alignment process of taper to chip. Finally, glass cuboids were placed at both sides of the array and fixed with glue. After hardening, in the same way as described before the opposite chip side was aligned and connected.

To check the thermal module stability, temperature tests are carried out with many modules between +15 and +40 °C and also at stronger temperature changes between −20 and +70 °C. A typical temperature behavior of the modules is depicted in Fig. 11.40. One can see a maximum output variation of ± 0.2 dB. Also vibration tests are run through with accelerations of 15 g in all Cartesian directions. The stability of the modules was excellent because no significant degradation of the coupling efficiency was detected after many temperature and vibrational test runs.

References

Eckhardt, T., Fischer, U.H.P., Ziegler, R.: DIL-size reusable modules with up to 50 GHz modulation bandwidth for optical communications systems. In: Faulkner, D.W., Harmer, A.L. (eds.) WDM and Photonic Networks, pp. 181–184. IOS Press, Netherlands (2000)

Ehlers, H., Schlak, M., Fischer, U.H.P.: Multi-fiber-chip-coupling modules for monolithically integrated Mach-Zehnder interferometers for TDM/WDM communication systems. In: Conference on Optical Fiber Communication. Technical Digest Series, pp. WDD66/1–WDD66/3, Anaheim, CA (2001)

Fischer, U.H.P., Zech, S., Peters, K.: Transmitter modules with reusable fiber-chip coupling method for optical communications systems. http://www.eetimes.com/design/communications-design/4017993/A-Reusable-Fiber-Chip-Coupling-Method-for-Optical-Communication-Transmitter-Modules (2001)

Fischer, U.H.P.: Packaging of OEICs with tapered fibers for optical communications systems with up to 45 GHz modulation bandwidth. In: Proceedings of the European Conference on Networks and Optical Communications: Broadband Access and Technology, pp. 296–300 (1999)

Hamacher, M., et al.: Monolithic integration of lasers, photodiodes, waveguides and spot size converters on GaInAsP/InP for photonic applications. In: InP and Related Materials Conference (IPRM 2000), paper MA1.3, pp. 21–24, Williamsburg USA (2000)

ITU-T G652: Transmission systems and media. Digital systems and networks, Transmission media characteristics—optical fibre cables (2000)

Ladany, I.: Laser to single-mode fiber coupling in the laboratory. Appl. Opt. **32**, 3233–3236 (1993)

Peters, K.: DE 195 36 185.7 (1994a)

Peters, K.: DE 195 36 173.3 (1994b)

Fischer, UHP: Laser Micro Welding for Fiber-chip-coupling Modules with Lensed Fiber Ends for Photonic Communication Systems. 14th International Scientific Conference Mittweida, 8–11 November 2000

Rohde, D., et al.: Optic/millimeter-wave converter for 60 GHz radio-over-fiber systems. In: MIOP ′97, Conference Proceedings, pp. 311–315 (1997)

Rosin, T., Bornholdt, C., Hoffmann, D., Burghardt, R., Bornhold, C.: Opto-electronic packaging for broadband high speed (40 Gb/s) optical demultiplexer chip. In: LEOS 98, pp. 123–126, Orlando, Florida (1998)

Tekin, T., Schlak, M., Brinker, W., Maul, B., Molt, R.: Monolithically integrated MZI comprising band gap shifted SOAs: a new switching scheme for generic all-optical signal processing. In: European Conference on Optical Communication (ECOC), pp. 123–124, München (2000)

Umbach, A., Trommer, D., Siefke, A., Unterbörsch, G.: 50 GHz operation of waveguide integrated photodiode at 1.55 μm. In: Proceedings of the 21st ECOC, vol. 17, p. 1075, Brussels, Belgien, 21 Sept 1995

Chapter 12
From Chip Design to the Optimum Package

Abstract In this chapter, the construction way of mechanical and optical design of photonic modules is described in detail. First, an overview of the general requirements of photonic packaging techniques is given followed by an exact example of the mechanical design of a laser module, which is shown in Chap. 11. Afterward, several optical simulation tools are described, which are important to apply for the fiber–chip coupling and the design of the waveguides on an optoelectronic integrated chip (OEIC).

12.1 General Requirements

The application of the optoelectronic integrated chips (OEICs) in communication system determines the system construction of the housing, which can be very inexpensive in one case to be used in future television tuners, for example. But it can also be very costly, if only some specific functions are used in wide area networks. Here, only a small number of components are applied, but need to handle high data rates, e.g., in optical X-connect nodes.

To develop an OEIC packaging, very different cross-sectional technologies are necessary, which are listed here as follows:

- High-frequency technology
- Classical optics and wave optics
- Precision-engineered design and CAD Design
- Wire bonding and flip-chip technique
- Cooling technology and thermal management
- Communications Engineering
- Solid-state physics
- Etching of silicon substrates,
- Mask techniques
- Thick film and thin film technology
- Gluing, soldering, and welding

© Springer-Verlag Berlin Heidelberg 2015

U.H.P. Fischer-Hirchert, *Photonic Packaging Sourcebook*,
DOI 10.1007/978-3-642-25376-8_12

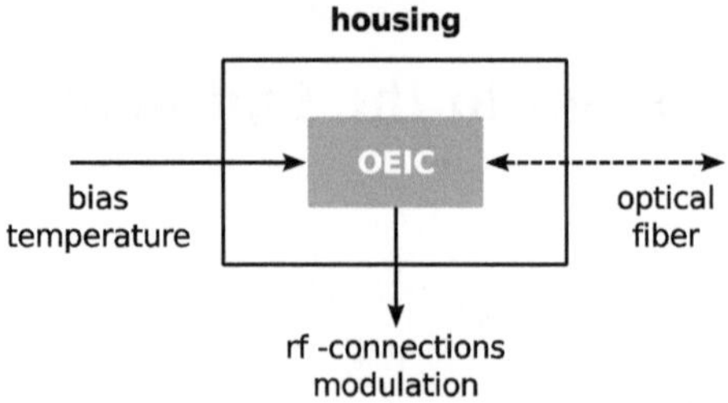

Fig. 12.1 Module structure in the optical assembly and interconnection technology

It is necessary to match the chip manufacturer with the construction specialists in planning a new chip. In the beginning, the following significant milestones must be fixed:

- Chip dimensions
- Number of waveguide outputs
- Distance between the waveguide outputs
- Requirements for the dissipation of energy losses
- RF leads and their contact
- Bond pad sizes and location

Such benchmarks should be determined in order to realize the best but also economical solution with the least effort.

The interrelationships of all elements are shown in a diagram (Fig. 12.1) for better understanding. Simple electrical leads for the temperature control and high-frequency connections are typically implemented for modulation. Supplying the optical signals is performed by glass fiber, which must be secure and long-term stable and connected to the OEIC. As a practical example, I would like to present the development of laser modules in detail below and thus give the reader a deeper insight into the problems of optical packaging in this area.

12.2 Design Optimization with the Finite Element Method

In this section, it is not possible to give a complete introduction to ANSYS (2015). By means of a practical example, it will be shown how the design process of an optoelectronic component can be supported by the results of a finite elements analysis based on the general-purpose software ANSYS.

In order to capture complex relationships, a common way of solving the problem is the idealization by splitting the problem into a finite number of simple sub-problems. This is the basic idea behind the finite element analysis FEA. A basic approach is as follows:

- Decomposition of the body to be considered in a finite number of geometrically recordable structure elements
- Formulation of the equilibrium conditions or equations of motion which control the idealized system
- Solution of these equations
- Evaluation and illustration of the results

Thus, converting a continuous problem into a discrete problem delivers an approximate solution. With an increased discretization, the approximate solution approaches to the exact solution. A finite element calculation model consists of finite elements which are connected to each other only at certain points (nodes). In general, FEA is a suitable way to simulate the behavior of components where analytical solutions are impossible due to the complex conditions. Not only structural aspects but also thermal and electromagnetic problems as well as fluid and electrostatic problems can be solved.

Prior to a FEA, some general questions should be answered, for example:

- What are the objectives of this analysis?
- Should the entire physical system be modeled, or is it possible to reduce, e.g., a 3D problem to a 2D dimensionality or can the user take advantage of symmetric conditions?
- Which level of abstraction regarding the geometrical modeling has to be chosen? Should all details be modeled?
- Is a fine element mesh of the whole structure required or is a selective mesh refinement in the appropriate regions sufficient?

12.3 Procedure of Finite Element Analysis and Software Packages

Generally, a finite element solution may be arranged in the following three steps:

Preprocessing

- Define the types of elements (beams, shells, and volume elements)
- Define the material properties (properties, material behavior, etc.)
- Create or import the geometry (points, lines, surfaces, and volumes)
- Discretization of the model (cross-linking of geometry)

- Syntax:

```
/PREP7
...
FINISH
```

Solution

Applying the loads and boundary conditions on the FE model (forces, supports, temperatures, accelerations, etc.)

- Setting the solver options (static, transient, harmonic, etc., analysis)
- Solving the equations

- Syntax: /SOLUTION
  ...
  FINISH

Postprocessing

Show the simulation results

- displacements, reaction forces, temperature distribution, deflection plots, stress contour diagrams)

- Syntax: /POST1
  ...
  FINISH

In Fig. 12.2, the software structure and the main ANSYS Programming Design Language (APDL) commands to switch between the different parts are shown. In general, ANSYS can be used in several ways: interactive or in batch, serial or parallel, or with a command line.

You may start with the ANSYS module "Product Launcher" where you have to choose between the most common interactive and the batch mode. The difference is that the interactive mode involves the so-called graphical user interface (GUI), with dialog boxes, menus, push buttons, and so on. You can see the result of your commands immediately—that is an advantage—but you have to wait until a computationally intensive calculation is completed, before you may apply the next command—that is the drawback. The batch mode consists of a APDL source code file which will be processed gradually while the results are written to a RESULT

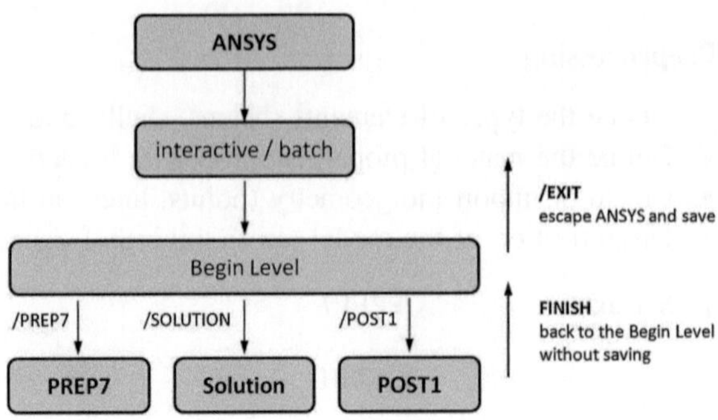

Fig. 12.2 ANSYS programming levels

file. With the ANSYS module "Display", the results can then be analyzed separately. From the Begin Level, you may switch between the preprocessor, the solver, and the postprocessor by means of the commands /PREP7, /SOLUTION and /POST1.

The preprocessing stage where the problem has to be defined is the most time-consuming part of a finite element analysis. This includes the modeling of the problem where keypoints, lines, areas, and volumes have to be defined depending on a two- or three-dimensional modeling.

Figure 12.3 shows how to generate the model from bottom to top (bottom-up): Keypoints, representing the basic points of the model, are the entities of lowest order. When you start with keypoints and then use these to generate higher order entities (such as lines, surfaces, and volumes), this is called a bottom-up modeling.

To generate the model from top to bottom (top-down), you can apply geometric primitives that consist of pre-defined lines, areas, or volumes. The program then automatically generates the entities of lower order. This allows you to model very complex structures by means of Boolean operations, e.g., adding or subtracting volumes. As shown in the APDL programming example below, you can choose the "bottom-up" and the "top-down" technology in combination if necessary.

Furthermore, the user has to choose appropriate element types and material properties before the meshing process is initiated which is one of the most critical procedures of the whole design process because the mesh influences the accuracy and speed of the solution. A high number of elements will result in long solver runs, whereas too few produce inaccurate results. ANSYS provides several options for controlling the mesh, particularly in terms of element size and shape. The meshing procedure assigns material properties and any additional information to the element types.

Meshing Types

Basically, two different types of meshing are provided:

Free Meshing (unstructured)

- Any geometry can be meshed unstructured
- Mixed use of triangles and squares

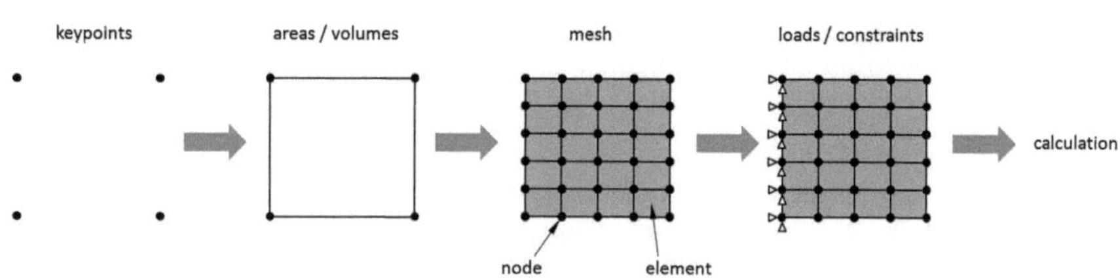

Fig. 12.3 Bottom-up modeling starting with the lowest order entities

Mapped meshing

- Uses only four-sided surface elements or hexahedral-shaped volumes
- Only applicable for geometries that meet the following criteria:
  - Surfaces must be bounded by 3 or 4 lines
  - When the surface is bounded by three lines, the lines must have an even number of subdivisions
  - Opposite sides should have the same number of subdivision
  - Volumes must be bounded by 4, 5, or 6 areas (tetrahedron, triangular prism, or hexahedron)

The following step upon completion of the "preprocessing" is the so-called solution phase where loads and boundary conditions have to be applied and the problem is solved. These include forces, forced deformations, temperatures, accelerations, supports, etc., applied to, e.g., nodes or elements (e.g., pressure loads). The default setting is a (linear) static analysis. If another solution type is required, you can choose for example static and transient analysis, modal analysis, etc.

In the postprocessor, the calculation results such as displacements, stresses, strains, etc., are read and displayed. This is achieved in the form of graphically treated plots (Plot Results) or of numerical values (Results List) for a further use of data performed in other software tools. Graphical representations are probably the most effective way to verify results. In the POST1 general postprocessor, the following modes are possible:

- Plots of the deformed structure
- Plots of the "forces of reaction"
- Contour line plots
- Vector plots

In a final step, the Results Viewer provides an animation of the results (e.g., modal analysis). Additional postprocessors, e.g., a time-dependent postprocessor to represent time-dependent results, are also implemented in ANSYS.

Among a variety of available FEM software which generally solves, e.g., either structural (deformation, stress, and strain fields) or thermal (temperature fields and heat flux) problems, the ANSYS software has the advantage to combine both results in a so-called coupled-field simulation in the multi-physics environment where multiple types of coupled physics interact. This means that, e.g., a particular temperature distribution within a component will lead to a corresponding stress distribution depending on the mechanical constraints. The physics mentioned above can be coupled together and solved in either a sequential or a direct-coupled analysis.

In a sequential coupled-field calculation, results from an analysis usually can be applied as loads for a second analysis. Thus, for example, the nodal temperatures of a thermal analysis can be used as volume loads in a structural analysis (for the calculation of strains). This is a typical thermal/mechanical coupled-field problem.

In the same way, even magnetic forces, which have been determined in a magnetic field calculation, are further processed as nodal forces in a structure analysis (e.g., in electrical machines). Other fields of application are as follows:

- The piezoelectric transducer (electromechanical)
- Electromagnetic problems (e.g., field displacement)
- Electrothermal dependencies (e.g., change in resistance due to heating)

Among static and dynamic solutions, the general-purpose finite element modeling package ANSYS (ANalysis SYStem) also provides linear/nonlinear solutions, respectively. Many physical real-world phenomena exhibit nonlinear behavior that means, e.g., nonlinear magnetic saturation effects or a nonlinear heat conduction coefficient can be taken into account. In the example below, we will restrict to linear conditions.

12.4 ANSYS Classic Introduction

12.4.1 Getting Help

ANSYS has an excellent online help consisting of the following basic manuals, for example:

(1) Theory manual,
(2) Analysis guides,
(3) Commands Manual,
(4) Element Reference Manual
(5) Verification Manual with many examples, including the source code of the input and the scientific reference
(6) Structural Analysis Guide and finally
(7) Thermal Analysis Guide

Furthermore, there are many tutorials and literature available:

- Examples of ANSYS Workbench and from courses at Pennsylvania State University
- ANSYS Tutorials from the University of Alberta
- ANSYS Tutorials provided by the manufacturer and installed with the software, respectively
- Literature in German: FEM for practitioners in four volumes where basics as well as examples are explained
  - Volume 1: Principles
  - Volume 2: Structural Dynamics
  - Volume 3: Temperature Fields
  - Volume 4: Electromagnetic Fields

12.4.2 Command Input

ANSYS uses commands that are entered directly or created in the background in the ANSYS command language APDL. In the example below, we will chose the so-called batch mode which means that ANSYS uses an input file with the source code, but the GUI as well as the input from the command line is also possible. The batch mode has the advantage that it runs faster than GUI and you have a permanent source file, which you can edit and copy. The GUI on the other hand provides an instantaneous graphical feedback. Experience has shown that it might be advantageous to run ANSYS interactively, which means that you might edit the source code with a well known and then successively copy the source code sequences into the command line.

If commands are entered directly from the command line or in an input file, it is important to know that the input is case-insensitive which means there is no distinction between upper and lower case. However, this can be well used for the sake of clarity, e.g., all commands are large and all variables are small.

- Dots and commas have a certain meaning. Dots are used to separate *integer digits* from *decimal digits*. Commas are used in commands to separate the parts.
- A "!" indicates a line as a comment
- It is not necessary to fully write out the commands. ANSYS accepts as much letters as needed to identify a certain command. This is often very practical, since the input is shorter, but it carries the risk that the value is set to a preset value unintentionally.

12.4.3 System of Units

In the classic ANSYS application, no units are defined so it is necessary to use consistent units. If for example the input is in [N] and [mm], ANSYS provides results in [N/mm^2]. Table 12.1 shows examples of consistent systems of units.

12.4.4 ANSYS Selection Techniques

The selection techniques allow selecting a subset of nodes, elements, keypoints, lines, etc., to work with a small amount of sizes, without destroying the remaining

Table 12.1 Examples of consistent unit systems

Mass	Length	Time	Force	Stress	Density	Young's modulus	Frequency
t	mm	s	N = t mm/s^2	MPa = N/mm^2	t/mm^3	MPa = N/mm^2	Hz = 1/s
kg	m	s	N = kg m/s^2	Pa = N/m^2	kg/m^3	N/m^2	Hz = 1/s
lbm	in.	s	lbf	psi = lbf/in.^{2}	lbm/in.^{3}	lbf/in.^{2}	Hz = 1/s

data. Selecting a subset is useful to define bearing loads or to check results in a particular area, volume, etc. If you, e.g., want to fix some parts of your model, you have to select the corresponding nodes either by selecting the area first and then select the nodes within this area or alternatively you may select the nodes by their coordinates.

12.4.5 Working Plane

The working plane is a plane anywhere in space, which serves as a reference system for the definition of the primitives. Unless otherwise defined, the working plane corresponds to the global coordinate system by default. If required, you can define a new working plane by the APDL command WPLANE or you can rotate it to a new orientation by the command WPROTA or move it by WPOFFS.

12.4.6 CAD Interfaces

In principle, ANSYS supports a broad range of CAD interfaces, for example, SolidWorks, CATIA, Pro/Engineer, and Autodesk Inventor. This allows you to use your existing CAD geometry directly, without translation to a common file format. Although this is a very useful and time-saving feature, the focus here shall be the solid modeling within the ANSYS environment by means of geometrical primitives (top-down modeling) or basically by keypoints (bottom-up modeling). The advantage of this method is that you are able to change the dimensions very easily just by changing some values of the pre-defined variables.

12.4.7 ANSYS Files

During an ANSYS session, several files are created, each with the same file name plus an extension, which shows the content of the file. The file name corresponds to the previously specified job name in the ANSYS Product Launcher. If no job name is specified, this will default to FILE. However, it can also be changed later within the user interface. The list below shows only the most important files and their contents:

.DB file with all model data such as geometry, boundary conditions, and any solutions (database file)

- .DBB backup of the database (not always available)
- .LOG command log file (LOG file). Here is the profile of the inputs of the current session logs

- .OUT output of all operations during an ANSYS session (ASCII)
- .ERR file with error messages and warnings (Error File)
- .RST file with results of a structural analysis (Results of Structural Analysis)
- .RTH file with the results of thermal analysis (Results of Thermal Analysis)

Depending on the purpose, the data are stored in various forms. Some files may be important for the user, such as the LOG file and the error file, so they are written in text format (ASCII). The others only exist in binary form.

12.5 Programming Example

12.5.1 Introduction

The following example deals with the design of a structure which is used to align optical fibers in modules for fiber–chip coupling (coupling modules). Such interface modules are components which are prepared for the application and for the operation of OEICs. Through these modules, OEICs and optical fibers have to be connected in such a way that the lowest possible radiation attenuation in the OEIC and the fiber occurs. OEICs can be semiconductor lasers, photodiodes, or similar active components. The method used for the insertion of the laser radiation requires the adjustment of the relative position of the OEIC to the end region of the optical fiber by a corresponding fine-adjusting device.

Figure 12.4 shows a sketch of the modeled structure with the main dimensions and the global coordinate system. The red line defines the surface with a constant temperature load, and the yellow line shows the path for temperature in Fig. 12.10. In order to give a better imagination of the module, Fig. 12.5 shows a three-dimensional view.

A detailed description of the functionality of temperature control of the modules is given in Chap. 11.

Within the operation of the module, the temperature will vary depending on the required laser frequency, so when the structure is heated or cooled, it will be deformed in a characteristic manner. If the deformation is restricted or the heat transfer to the environment is somehow asymmetric or if different materials are connected together, thermal stresses or structural deformations will occur.

In order to optimize the design, a FEA provides a simulation of the behavior, for example, with regard to the attenuation of the laser beam. In the following, a simplified model (Fig. 12.6) will be analyzed. It consists of a receptacle for the optical waveguide, a laser diode (OEIC) applied on a ceramic substrate, and steel structure with a certain area mounted on a Peltier element. The Peltier element itself is not modeled, but only the contact surface, whose temperature can be set explicitly in the source code.

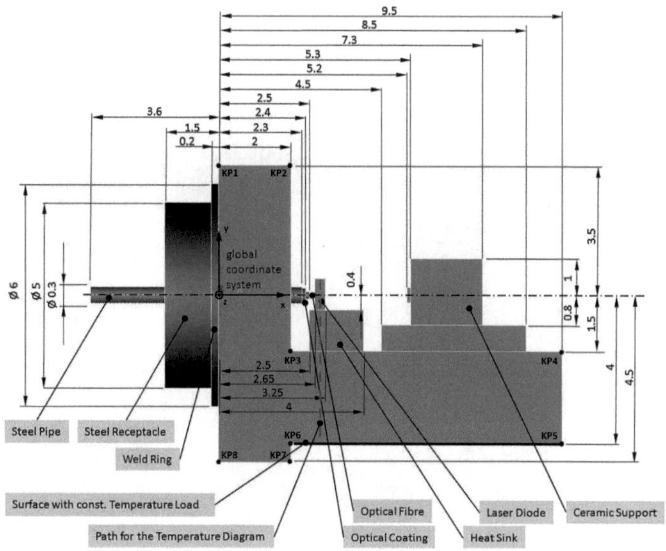

Fig. 12.4 Main dimensions of the modeled structure

Fig. 12.5 Three-dimensional view of the optical module

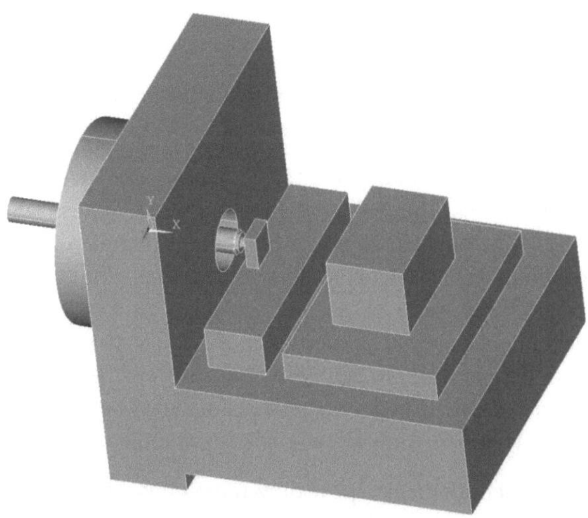

12.5.2 Procedure of the Sequential Thermal Stress Analysis

1. Begin with a steady-state (or transient if required) thermal analysis

- Model with thermal element types
- Define thermal material properties (thermal conductivity, specific heat, heat transfer coefficient)

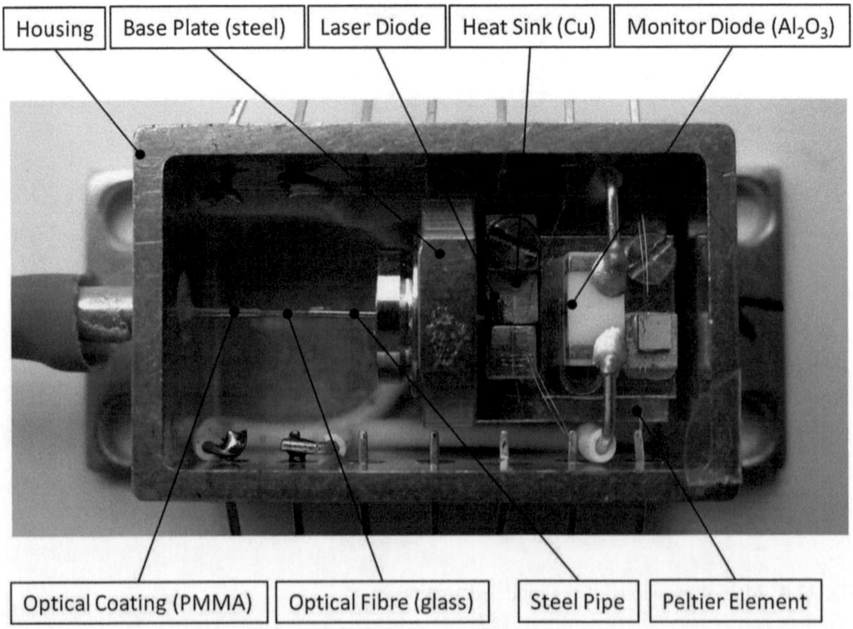

Fig. 12.6 View of the real housing

- Apply thermal loads (heat generation, heat transfer conditions)
- Solve and verify results

Static temperature field calculations are performed for the determination of temperatures, thermal gradients, and heat fluxes, which does not vary with time. The types of loads that can be applied include convection loads, heat fluxes, heat generation rates, and predetermined temperatures. A static thermal analysis can either be linear with constant material properties or nonlinear with temperature-dependent material properties. For most materials, the thermal characteristics vary with temperature, so generally, the analysis has to be nonlinear. However, to keep it simple, we will focus on a linear material behavior here.

2. Then, perform a static structural analysis:

- Switch element types (they must be compatible)
- Define structural material properties (Young's modulus, thermal expansion coefficient, density ...)
- Apply structural loads (degree of freedom (DOF) constraints, temperatures from thermal analysis ...)
- Solve and review results (Fig. 12.7)

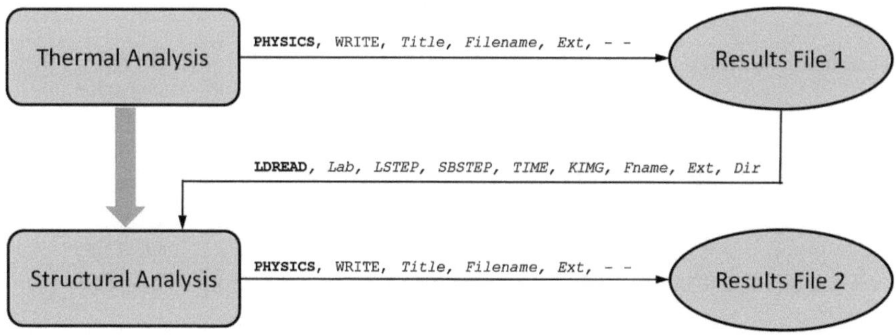

Fig. 12.7 ANSYS sequential coupled-field analysis

12.5.3 Element Types

ANSYS provides a great variety of element types such as beam, shell, and link elements as well as planes and solids. All of them have a historical background, and they are still available so it is up to the user to choose the correct element type for his calculation. ANSYS supports the user with many information manuals especially the "ANSYS Elements Reference" which for example is available online.

Each element is defined by an element name and a set of nodes in the corners (lower order/first order) or additionally nodes between the corners (higher order/ second order). Generally, the elements with midside nodes have quadratic base functions instead of linear functions and allow a more precise approach to the real conditions. Further, element items are defined in element input tables (in the elements reference) and usually contain

- The available degrees of freedom
- Real constants (additional element properties which sometimes may be required),
- Material properties,
- Surface loads,
- Body loads, and so-called
- Key options (which can be considered to be a kind of switch used to turn various element options on or off)

12.5.3.1 ANSYS Element Compatibility

In a sequential analysis, where results from a prior calculation are applied as loads for a second analysis, we have to ensure that the elements used are compatible which means that they must maintain a consistent base geometry. For example, a SOLID45 (10-node tetrahedral structural solid) is compatible with SOLID70 (a 10-node tetrahedral thermal solid).

Table 12.2 Compatible element types across physics environments

Structural	Thermal	Magnetic	Electrostatic	Fluid	Electric conduction
SOLID45	SOLID70	SOLID97, SOLID117	SOLID122	SOLID142	SOLID5, SOLID69
SOLID92	SOLID87	SOLID98, HF119	SOLID123	–	SOLID98
SOLID95	SOLID90	SOLID117, HF120	SOLID122	–	SOLID5, SOLID69
PLANE42	PLANE55	PLANE13, PLANE53	PLANE121	FLUID141	PLANE67
PLANE2	PLANE35	–	–	–	–
PLANE82	PLANE77	PLANE53	PLANE121	–	PLANE67
SHELL63	SHELL57	–	–	–	SHELL157
LINK1	LINK32	–	–	–	LINK68
LINK8	LINK33	–	–	–	–

You can ensure this in two ways:

1. Select the corresponding element type as listed in Table 12.2 or
2. Use the APDL command etchg which automatically changes element types to their corresponding types

12.5.3.2 SOLID90–3D 20-Node Thermal Solid

SOLID90 has three-dimensional heat conduction properties. It corresponds to the 3D 8-node element-type SOLID70, but has shape functions of higher order. The element has 20 nodes with one degree of freedom at each node. The element can be used in stationary or transient temperature field analysis. The element is defined by the specification of 20 corner nodes and the material properties.

12.5.3.3 SOLID95–3D 20-Node Structural Solid

The SOLID95 element is a three-dimensional element with 20 nodes similar to SOLID90. Each node has three degrees of freedom and the displacement in the *x*-, *y*-, and *z*-direction. The load can be applied on the element faces as surface load (force per unit area). Thermal body loads are also possible of course (Fig. 12.8).

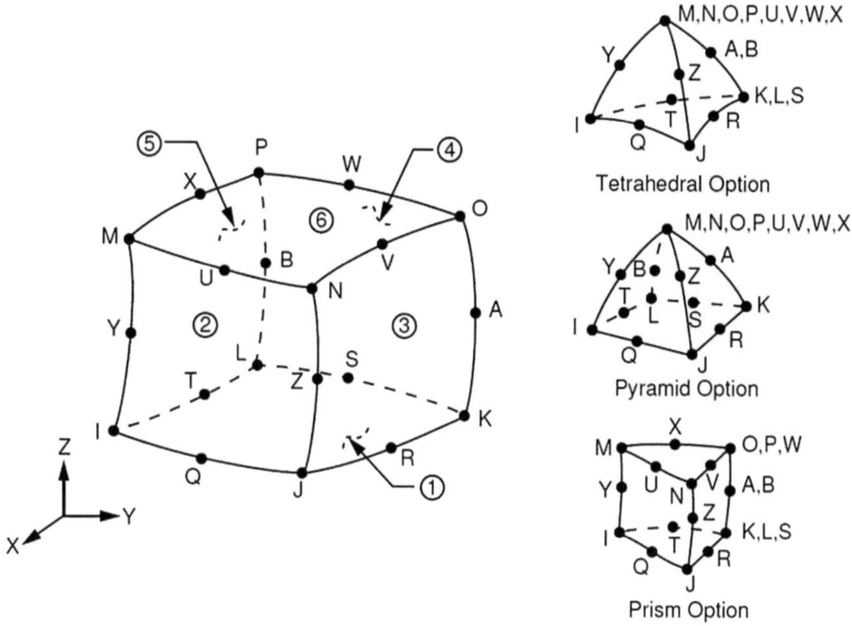

Fig. 12.8 Element shape of SOLID90 and SOLID95 with nodes and faces

SOLID90 input summary
Nodes
I, J, K, L, M, N, O, P, Q, R, S, T, U, V, W, X, Y, Z, A, B
Degrees of freedom
TEMP
Material properties
Thermal conductivities (*x*-, *y*-, and *z*-direction, KXX, KYY, KZZ),
Density (DENS),
Specific heat (C),
Enthalpy (ENTH)
Surface loads
Convection (conv) or Heat flux (but not both) and radiation
Body loads
Heat generations

(continued)

SOLID95 input summary
Nodes
I, J, K, L, M, N, O, P, Q, R, S, T, U, V, W, X, Y, Z, A, B
Degrees of freedom
Displacements (*x*-, *y*-, and *z*-direction)
Material properties
Young's moduli (EX, EX, EZ)
Coefficient of thermal expansions (ALPX, ALPX, ALPZ)
Poisson's ratio (NUXY, NUYZ, NUXZ)
Density (DENS)
Shear moduli (GXY, GYZ, GXZ)
Stiffness matrix multiplier for damping (DAMP)
Surface loads
Pressures
Body loads
Temperatures

12.5.4 ANSYS Command Listing

The following flowchart gives an impression of the general structure of a finite element analysis. You will find the detailed code together with some comments as well as results in the programming code appendix of this chapter.

PREPROCESSOR
• Define variables
• Define thermal material properties
• Define the geometry by keypoints and generate an area as an example of the bottom-up modeling
• Chose an appropriate element type for thermal problems
• Define the other geometry by top-down method
• Mesh the structure by mapping and free meshing
• Apply thermal loads and boundary conditions
• Write thermal conditions to the physics file
• Choose an appropriate element type for structural problems which is compatible with the thermal element
• Define structural material properties

(continued)

• Apply boundary conditions
• Write the structural physics file
SOLVER (Thermal solution)
• Read thermal data from the physics file
• Solve the thermal problem
POSTPROCESSOR (Thermal solution)
• Plot the temperature distribution
• Define a path through an interesting part along which a temperature diagram has to be plotted
• Define a cutting plane in order to get a look inside the model
SOLVER (Structural solution)
• Read structural data from the physics file
• Read thermal results from the .rth file
• Solve the structural problem
POSTPROCESSOR (Structural solution)
• Plot the displaced structure with a certain scaling factor
• Plot the von Mises equivalent stresses
• Plot the color-coded displacements
• Get the nodes of the opposed areas of the laser diode and the fiber in order to calculate the relative dislocation

12.5.5 Programming Code Listing

```
/batch

!*******************************************************************************
! File Name:                    FIBER.INP
! Execution environment:        ANSYS 12.1
! Language:                             APDL
! File type:                    ANSYS command input file
! Unit system:                  SI
! Created by:                   Guenter Buehler
!
! Keywords: Multi Physics Coupled Field Simulation (thermal, structural)
!           Static, linear solution
!
! Most of the comments were taken from the ANSYS Online Commands Reference
!*******************************************************

fini                            ! finish all previous jobs ...
/clear                          ! ... and delete the database
```

```
*****************************************************
!P R E P R O C E S S O R
!*******************************************************
/prep7                  ! Launch the preprocessor
/view,, 1, 1, 5
/show, FIBER, GRPH      ! Name of the output file 'FIBER.GRPH' in case of BATCH-Mode
                        ! used for the batch mode

l1_Laser     = 0.8e-3 ! Laserdiode Width in meter
l2_Laser     = 0.04e-3! Laserdiode Height in meter
l3_Laser     = 0.3e-3 ! Laserdiode Depth in meter
Vol_Laser    = l1_Laser*l2_Laser*l3_Laser       ! Volume of the Laser Diode
P_Laser      = 0.05                             ! Laser Thermal Losses in [W]
Heat_Gen     = P_Laser/Vol_Laser                ! Heat Generation in [W/m^3]

!------------------------------------------------------------------------------
mp, kxx  , 1, 50      ! Material Property: Thermal Conductivity Steel [W/mK]
mp, dens , 1, 7860    ! Material Property: Density Steel [kg/m^3]
mp, c    , 1, 490     ! Material Property: specific Heat Steel [J/kgK]

mp, kxx  , 2, 25      ! Material Property: Thermal Conductivity Al2O3 [W/mK]
mp, dens , 2, 3690    ! Material Property: Density Al2O3 [kg/m^3]
mp, c    , 2, 880     ! Material Property: specific Heat Al2O3 [J/kgK]

mp, kxx  , 3, 0.19    ! Material Property: Thermal Conductivity PMMA [W/mK]
mp, dens , 3, 1180    ! Material Property: Density PMMA [kg/m^3]
mp, c    , 3, 1470    ! Material Property: specific Heat PMMA [J/kgK]

mp, kxx  , 4, 398     ! Material Property: Thermal Conductivity Copper [W/mK]
mp, dens , 4, 8920    ! Material Property: Density Copper [kg/m^3]
mp, c    , 4, 395     ! Material Property: specific Heat Copper [J/kgK]

mp, kxx  , 5, 150     ! Material Property: Thermal Conductivity Silicon [W/mK]
mp, dens , 5, 2336    ! Material Property: Density Silicon [kg/m^3]
mp, c    , 5, 741     ! Material Property: specific Heat Silicon [J/kgK]

!*******************************************************
! Example of the Bottom Up Modeling of the basic Structure
!*******************************************************
k, 1,      0,  3.5e-3, -4e-3    ! Keypoint 1 (x=  0mm, y= 3.5mm, z=-4mm)
k, 2,   2e-3,  3.5e-3, -4e-3    ! Keypoint 2 (x=  2mm, y= 3.5mm, z=-4mm)
k, 3,   2e-3, -1.5e-3, -4e-3    ! Keypoint 3 (x=  2mm, y=-1.5mm, z=-4mm)
k, 4,9.5e-3, -1.5e-3, -4e-3     ! Keypoint 4 (x=9.5mm, y=-1.5mm, z=-4mm)
k, 5,9.5e-3, -4.0e-3, -4e-3     ! Keypoint 5 (x=9.5mm, y=  -4mm, z=-4mm)
k, 6,   2e-3, -4.0e-3, -4e-3    ! Keypoint 6 (x=  2mm, y=  -4mm, z=-4mm)
k, 7,   2e-3, -4.5e-3, -4e-3    ! Keypoint 7 (x=  2mm, y=-4.5mm, z=-4mm)
k, 8,      0, -4.5e-3, -4e-3    ! Keypoint 8 (x=  0mm, y=-4.5mm, z=-4mm)
a, 1, 2, 3, 4, 5, 6, 7, 8, 1    ! Defines an area by connecting keypoints
```

```
!*****************************************************
! Example of the Top Down Modeling with Primitives for all other volumes
!*****************************************************
et, 1, solid90                ! 20-Node Element (Element Type No. 1 in this simulation)
esize, 0, 20                  ! Sytax: ESIZE, SIZE, NDIV
                              ! Specifies the default number of line divisions
                              !  - SIZE: Default element edge length on surface boundaries (i.e., lines)
                              !         If SIZE is zero (or blank), use NDIV
                              !  - NDIV: Default number of element divisions along region boundary lines
vext, all,,, 0, 0, 8e-3       ! Extrusion in z-direction about 8 mm

!---------------------------------------------------------------------------
wprota, 0, 0, -90             ! Rotates the working plane
                              ! This is required because the length of a cylinder modeled with the
                              ! cylind-command is defined to correspond with the z-coordinates
                              ! Syntax: CYLIND, RAD1, RAD2, Z1, Z2, THETA1, THETA2
cylind, 0, 0.75e-3, 0, -2e-3, 0, 360 ! Massive cylinder with 0.75 mm radius and a length of 2 mm
vsbv, 1, 2                    ! Subtracts volumes from volumes. Here especially subtract the modeled
                              ! cylinder from the basic structure. The result is a hole

!---------------------------------------------------------------------------
                              ! Steel receptacle
cylind, 0.3e-3, 2.5e-3, 0.2e-3, 1.5e-3, -90, 90 ! Left half of the cylindrical receptacle
cylind, 0.3e-3, 2.5e-3, 0.2e-3, 1.5e-3,  90,270 ! Right half of the cylindrical receptacle
vglue, 1, 2                   ! Glue both halves together
                              ! This special modeling procedure is required for mapped meshing
numcmp, all                   ! Compresses the numbering of defined items (nodes, areas, lines ...)

vsel, s, volu,, 1             ! Select a new set of volumes (the left half no. 1) and ...
vsel, a, volu,, 3             ! ... additionally select the right half no. 3
esize, 0.4e-3                 ! Define the element size [mm]
type, 1                       ! Chose element type no. 1 (=solid90)
mat, 1                        ! Chose material no. 1 (=steel)
mshkey, 1                     ! Specifies whether free meshing or mapped meshing should be used to mesh
a model

mshape, 0, 3d                 ! For elements that support multiple shapes,
                              ! specifies the element shape to be used for meshing
                              ! Syntax: MSHAPE, KEY, Dimension
                              !   KEY: Key indicating the element shape to be used:
                              !    0  - Mesh with quadrilateral-shaped elements when Dimension=2D
                              !         Mesh with hexahedral-shaped elements when Dimension=3D
                              !    1  - Mesh with triangle-shaped elements when Dimension=2D
                              !         Mesh with tetrahedral-shaped elements when Dimension=3D
                              !   Dimension: Specifies the dimension of the model to be meshed:
                              !    2D - 2-D model (area mesh)
                              !    3D - 3-D model (volume mesh)
vmesh, all                    ! mesh all selected volumes (here no. 1 and no. 3)
allsel                        ! Selects all entities with a single command
!---------------------------------------------------------------------------
cylind, 2e-3    , 3e-3    , 0     , 0.2e-3 ,-90, 90    ! Weld ring left part
cylind, 2e-3    , 3e-3    , 0     , 0.2e-3 , 90, 270   ! Weld ring right part
cylind, 0.25e-3 , 0.3e-3  , 3.6e-3, -2.3e-3,  0, 360   ! Steel pipe
cylind, 0.125e-3, 0.25e-3 , 3.6e-3, -2.4e-3,  0, 360   ! Coating of the Optical Fibre
cylind, 0       , 0.125e-3, 3.6e-3, -2.5e-3,  0, 360   ! Optical Fibre
wpstyl, defa                  ! Return working plane to default setting
!---------------------------------------------------------------------------
                              ! Syntax: BLOCK, X1, X2, Y1, Y2, Z1, Z2
```

```
                                ! Creates a block volume based on working plane coordinates
                                !   X1, X2 - Working plane X coordinates of the block
                                !   Y1, Y2 - Working plane Y coordinates of the block
                                !   Z1, Z2 - Working plane Z coordinates of the block
block, 5.3e-3, 7.3e-3, -0.8e-3,  1.0e-3, -1.6e-3, 1.6e-3! Al2O3 Support Monitor Diode
block, 4.5e-3, 8.5e-3, -1.5e-3, -0.8e-3, -3e-3  ,   3e-3! Heat Sink Monitor Diode
block, 5.2e-3, 5.3e-3, -0.2e-3,  0.2e-3, -0.2e-3, 0.2e-3! Monitor Diode
block, 2.5e-3, 4.0e-3, -1.5e-3, -0.4e-3, -3e-3  ,   3e-3! Heat Sink Laser Diode
block, 2.65e-3, 2.65e-3+l3_Laser, -0.4e-3, 0.4e-3+l2_Laser, -l1_Laser/2, l1_Laser/2 ! Laser Diode

allsel
vglue, all
numcmp, all

!*******************************************************
! Meshing of all Volumes
!*******************************************************
vsel, s, volu,, 3               ! Selects a subset of volumes (Photo Diode)
                                ! Syntax: VSEL, Type, Item, Comp, VMIN, VMAX, VINC, KSWP
                                !   Type: Label identifying the type of volume select:
                                !     S - Select a new set (default)
                                !     R - Reselect a set from the current set
                                !     A - Additionally select a set and extend the current set
                                !     U - Unselect a set from the current set
                                !     ALL  - Restore the full set
                                !     NONE - Unselect the full set
                                !     INVE - Invert the current set
                                !     STAT - Display the current select status
                                !   Item: Label identifying data. Defaults to VOLU
                                !   Comp: Component of the item (if required)
                                !   VMIN: Minimum value of item range. Ranges are volume numbers,
                                !         coordinate values, attribute numbers, etc.
                                !   VMAX: Maximum value of item range
                                !   VINC: Value increment within range. Used only with integer ranges
                                !         (such as for volume numbers)
                                !   KSWP: Specifies whether only volumes are to be selected:
                                !         0 - Select volumes only
                                !         1 - Select volumes, as well as keypoints, lines, areas, nodes,
                                !             and elements associated with selected volumes
esize, 0.1e-3           ! Element Size 0.1 mm
type, 1                         ! Element type no. 1 (=solid90)
mat, 5                          ! 1 = steel, 2 = Al2O3, 3 = PMMA, 4 = copper, 5 = silicon
mshkey, 1                       ! Specifies whether free meshing or mapped meshing should be used to mesh
a model
                                ! 0 = free meshing, 1 = mapped
mshape, 0, 3d           ! Specifies the element shape to be used for meshing
                                ! Syntax: MSHAPE, KEY, Dimension
                                !   KEY: Key indicating the element shape to be used:
                                !    0  - Mesh with quadrilateral-shaped elements when Dimension=2D
                                !         Mesh with hexahedral-shaped elements when Dimension=3D
                                !    1  - Mesh with triangle-shaped elements when Dimension=2D
                                !         Mesh with tetrahedral-shaped elements when Dimension=3D
                                !   Dimension: Specifies the dimension of the model to be meshed:
                                !    2D - 2-D model (area mesh)
                                !    3D - 3-D model (volume mesh)

vmesh, all                      ! mesh all selectes volumes
allsel
```

```
!---------------------------------------------------------------------------
vsel, s, volu,, 8              ! Photo Diode Support
esize, 0.3e-3
type, 1
mat, 2                         ! 1 = steel, 2 = Al2O3, 3 = PMMA, 4 = copper, 5 = silicon
mshkey, 0
mshape, 1, 3d
vmesh, all
allsel

!---------------------------------------------------------------------------
vsel, s, volu,, 9              ! Heat Sink Photo Diode (Material: Copper)
esize, 0.3e-3
type, 1
mat, 4                         ! 1 = steel, 2 = Al2O3, 3 = PMMA, 4 = copper, 5 = silicon
mshkey, 0
mshape, 1, 3d
vmesh, all
allsel
!---------------------------------------------------------------------------
vsel, s, volu,, 6,7    ! Weld Ring (Material: Steel)
esize, 0.2e-3
type, 1
mat, 1                         ! 1 = steel, 2 = Al2O3, 3 = PMMA, 4 = copper, 5 = silicon
vmesh, all
allsel
!---------------------------------------------------------------------------
vsel, s, volu,, 5              ! Optical Fibre
esize, 0.1e-3
type, 1
mat, 3                         ! 1 = steel, 2 = Al2O3, 3 = PMMA, 4 = copper, 5 = silicon
vmesh, all
allsel
!---------------------------------------------------------------------------
vsel, s, volu,, 13     ! Optical Coating
esize, 0.1e-3
type, 1
mat, 3                         ! 1 = steel, 2 = Al2O3, 3 = PMMA, 4 = copper, 5 = silicon
vmesh, all
allsel
allsel
!---------------------------------------------------------------------------
vsel, s, volu,, 14     ! Steel Pipe
esize, 0.1e-3
type, 1
mat, 1                         ! 1 = steel, 2 = Al2O3, 3 = PMMA, 4 = copper, 5 = silicon
vmesh, all
allsel
!---------------------------------------------------------------------------
```

```
vsel, s, volu,, 4               ! Laser Diode
esize, 0.05e-3
type, 1
mat, 2                          ! 1 = steel, 2 = Al2O3, 3 = PMMA, 4 = copper, 5 = silicon
vmesh, all
allsel
!------------------------------------------------------------------------------
vsel, s, volu,, 10     ! Heat Sink Laser Diode (Material: Copper)
esize, 0.15e-3
type, 1
mat, 4                          ! 1 = steel, 2 = Al2O3, 3 = PMMA, 4 = copper, 5 = silicon
vmesh, all
allsel
!------------------------------------------------------------------------------
vsel, s, volu,, 15     ! Steel Structure
esize, 0.4e-3
type, 1
mat, 1                          ! 1 = steel, 2 = Al2O3, 3 = PMMA, 4 = copper, 5 = silicon
vmesh, all


nsel, s, loc, x, 2.0e-3, 2.5e-3
nsel, r, loc, y,-4.0e-3,-4.0e-3
nrefine, all,,, 1, 1, clean

nsel, s, loc, x, 8.0e-3, 8.6e-3
nsel, r, loc, y,-4.0e-3,-4.0e-3
nrefine, all,,, 1, 1, clean

eplo
allsel
!******************************************************************************
```

We expect that the solution near the edges of the Peltier element has a large temperature gradient, so that we use a higher resolution here in order to avoid incorrect results. One method is to start the meshing process with a coarse resolution and then refine the interesting sectors in a second step by selecting the corresponding nodes. The meshing results are shown in Fig. 12.9.

Now, we have to define the loads (power losses with respect to heat generation) and some constraints (constant temperature and mechanical DOF constraints at the contact area to the Peltier element) and surface loads (convection at all exterior surfaces). We assume an identical heat transfer coefficient of 2 W/(m^2 K) for all surfaces which is not correct in the strict sense because the flow conditions of the ambient air are versatile.

```
!****************************************************
! Thermal Loads and Boundary Conditions
!****************************************************
vsel, s, volu,, 4          ! Laser Diode
nslv, s                    ! Select all nodes within the Laser Diode
nplot                            ! Plot these nodes (just to verify)
bf, all, hgen, Heat_Gen    ! heat generation of Laser Diode [W/m^3]
/pbf, hgen,, 2             ! Shows body force loads as contours on displays
allsel
!----------------------------------------------------
nsel, s, loc, y, -4e-3, -4e-3     ! Define the contact area between Steel Structure and
Peltier Element
nsel, r, loc, x, 2.4e-3, 8.4e-3
nplo
d, all, temp, 10           ! Defines DOF constraints at nodes
                           ! Wall temperature = 10 °C (peltier controlled)
                           ! Syntax: D, NODE, Lab, VALUE, VALUE2, NEND, NINC, Lab2, Lab3,
Lab4, Lab5, Lab6
                           !   NODE - Node at which constraint is to be specified.
                           !          If ALL, NEND and NINC are ignored and constraints are
applied to all
                           !          selected nodes [NSEL]
                           !    Lab   - Valid degree of freedom label. If ALL, use all
appropriate labels
                           !                  Structural labels: UX    , UY    , or UZ
(displacements)
                           !                           ROTX, ROTY, or ROTZ (rotations)
                           !          Thermal labels:    TEMP                (temperature)
                           !          ...
                           !   VALUE- Degree of freedom value
allsel
!----------------------------------------------------
asel, all                  ! select all areas...
asel, u, area,, 6          ! ... and unselct area no. 6 (connecting surface to Peltier
element)
nsla,, 1                   ! Select all nodes in this area
                           ! Syntax: NSLA, Type, NKEY
                           !  Type:
                           !    S - Select a new set (default)
                           !    R - Reselect a set from the current set
                           !    A - Additionally select a set and extend the current set
                           !    U - Unselect a set from the current set
                           !   NKEY (Specifies whether only interior area nodes are to be
selected)
                           !    0 - Select only nodes interior to selected areas
                           !    1 - Select all nodes (interior to area, interior to lines,
and at keypoints)
                           !       associated with the selected areas
sf, all, conv, 2, 40       ! Apply convection: heat-transfer coefficient Alpha=2 W/(m^2K),
T_ambient=40°C
                           ! Syntax: SF, Nlist, Lab, VALUE, VALUE2
                           !    Nlist - Nodes defining the surface upon which the load is
to be applied
                           !     Lab    - Valid surface load label, PRES (pressure), CONV
(convection)
                           !           HFLUX (heat flux), RAD (radiation) ...
                           !     VALUE - Surface load value or table name reference for
specifying tabular
                           !           boundary conditions
```

```
                              !          If Lab=CONV, VALUE is typically the filr
coefficient and VALUE2 (below)
                              !          is typically the bulk temperature. If Lab=CONV anc
VALUE=-N,
                              !          the film coefficient may be a function o:
temperature and is determined
                              !          from the HF property table for material N [MP].
                              !          The temperature used to evaluate the filr
coefficient is usually the
                              !          average between the bulk and wall temperatures, bui
may be user-defined
                              !          for some elements
  allsel
  physics, write, thermal ! Write thermal conditions to the Physics-File
  physics, clear          ! Clear all bc's and options
  !*****************************************************
  ! Structural Simulation

  !et, 1, solid95             ! 20-Node Element (corresponding to thermal SOLID90) fo:
manually selection
  etchg, tts                  ! Converts the element types to the corresponding typ(
automatically
                              ! Valid labels are:
                              !       ETI - Explicit to Implicit
                              !       ITE - Implicit to Explicit
                              !       TTS - Thermal to Structural
                              !       STT - Structural to Thermal
                              !       MTT - Magnetic to Thermal
                              !       FTS - Fluid to Structural
                              !       ETS - Electrical to Structural

  mp, ex   , 1, 210e9 ! Material Property: Elastic Modulus Steel [Pa]
  mp, alpx , 1, 13e-6 ! Material Property: Coefficient of thermal expansion Steel [1/K]
  mp, nuxy , 1, 0.3         ! Material Property: Poissons Ratio

  mp, ex   , 2, 300e9 ! Material Property: Elastic Modulus Al2O3 [Pa]
  mp, alpx , 2, 8.2e-6      ! Material Property: Coefficient of thermal expansion Al2O:
[1/K]
  mp, nuxy , 2, 0.21  ! Material Property: Poissons Ratio Al2O3

  mp, ex   , 3, 3.3e9 ! Material Property: Elastic Modulus PMMA [Pa]
  mp, alpx , 3, 70e-6 ! Material Property: Coefficient of thermal expansion PMMA [1/K]
  mp, nuxy , 3, 0.39  ! Material Property: Poissons Ratio PMMA

  mp, ex   , 4, 110e9 ! Material Property: Elastic Modulus Copper [Pa]
  mp, alpx , 4, 16.5e-6     ! Material Property: Coefficient of thermal expansion Coppe:
[1/K]
  mp, nuxy , 4, 0.34  ! Material Property: Poissons Ratio Copper

  mp, ex   , 5, 150e9 ! Material Property: Elastic Modulus Silicon [Pa]
  mp, alpx , 5, 7.6e-6      ! Material Property: Coefficient of thermal expansion Silico:
[1/K]
  mp, nuxy , 5, 0.279 ! Material Property: Poissons Ratio Silicon
  !*****************************************************
```

The following command sequence selects the nodes of the area which is in contact with the Peltier element. Let us assume that those nodes are mechanically fixed, so we have to fix all degrees of freedom that means translational as well as rotational.

```
!*****************************************************
  nsel, s, loc, y, -4e-3, -4e-3    ! Select nodes on the contact surface to the Peltier
Element (y=-4mm)
  nsel, r, loc, x, 2.4e-3, 8.4e-3  ! from x = 2.4 mm to x = 8.4 mm
  d, all, all               ! Boundary conditions (bc's)
                            ! ALL: all selected nodes
                            ! ALL: fix all displacements and rotations
                            ! We assume that these nodes are fixed which might not be
realistic
  allsel

  physics, write, struct  ! Write the structural physics file
  physics, clear          ! Clear all bc's and options
!*****************************************************
```

All required input data are available, so the solver for the (static) thermal solution can be started.

```
!*****************************************************
! SOLVER - Thermal Solution
!*****************************************************
/solu
physics, read, thermal     ! Read thermal data from the physics file
antype, stat, new          ! Specifies the analysis type and restart status
                           ! Syntax: ANTYPE, Antype, Status
                           !   Antype: Analysis type
                           !     STATIC - Perform a static analysis
                           !     BUCKLE - Perform a buckling analysis
                           !     MODAL  - Perform a modal analysis
                           !     HARMIC - Perform a harmonic analysis
                           !     TRANS  - Perform a transient analysis
                           !     SUBSTR - Perform a substructure analysis
                           !     SPECTR - Perform a spectrum analysis
solve

finish
save,,,                    ! Saves all current database information

!*****************************************************
! POSTPROCESSOR - Thermal Solution
!*****************************************************
/post1                     ! Launch the postprocessor
wpstyle,,,,,,,,0           ! Delete working plane grid
/graphics, power

/contour,,10,10,,12 ! Specifies the uniform contour values on stress displays
                           ! Syntax: /CONTOUR, WN, NCONT, VMIN, VINC, VMAX
                           !    WN    - Window number (or ALL) to which command applies
(defaults to 1)
                           !   NCONT - Number of contour values. If VMAX is not specified,
NCONT defaults to 9
                           !   VMIN  - Minimum contour value
                           !   VINC  - Value increment (positive) between contour values
                           !   VMAX  - Maximum contour value. Ignored if both VMIN and VINC
are specified.
plnsol, temp        ! Displays (plots) results as continuous contours
                           ! Syntax: PLNSOL, Item, Comp, KUND, Fact
                           !      Item - Label identifying the item (here especially
temperature)

/cval,,10.1, 10.2, 10.4, 11, 12, 20, 50, 70
                           ! /CVAL, WN, V1, V2, V3, V4, V5, V6, V7, V8
                           ! Specifies non-uniform contour values on stress displays
/replo                     ! Replot command (which means 'plnsol, temp')
prnsol, temp        ! Prints the nodal solution results
                           ! Syntax: PRNSOL, Item, Comp
!*****************************************************
```

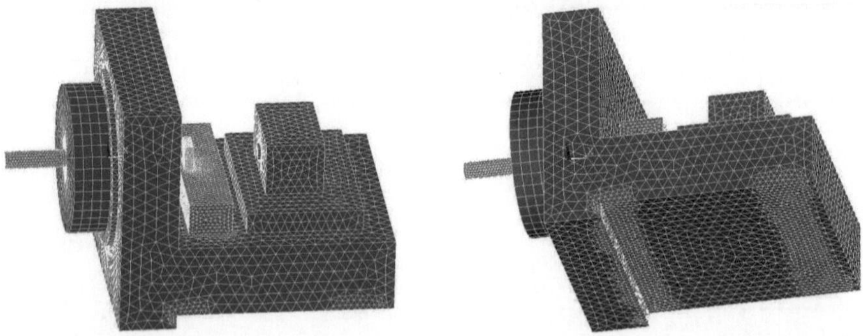

Fig. 12.9 Elements of the whole module

The thermal analysis results are shown in Fig. 12.9. On the left side, the temperature distribution is shown with uniform contour values, which means constant temperature steps between the colors. Due to the large gradient in the vicinity of the laser diode, wide areas are colored in gray because the temperature is above 12 °C. The /cval command enables you to define the steps individually.

If you want to show the temperature profile in a diagram (Fig. 12.10)—for example from the top of the laser diode downward through the heat sink (copper) and the steel structure—you may define a path with 2 or more coordinate locations:

```
!*******************************************************
path, my_path, 2,, 200      ! Syntax: PATH, NAME, Npts, nSets, nDiv
                            ! Defines a path name and establishes parameters for the path
                            !   NAME - Name for this path
                            !   nPts - Number of points used to define this path
                            !   nSets- Number of sets of data which you can map to this path
                            !   nDiv - Number of divisions between adjacent points

x1_path = (2.65e-3+2.65e-3+13_Laser)/2    ! x = 2.8 mm
y1_path = 0.4e-3 + 12_Laser               ! y = 0.44 mm
z1_path = 0                                       ! z = 0

x2_path = (2.65e-3 + 2.65e-3+13_Laser)/2  ! x = 2.8 mm
y2_path = -4e-3                                   ! y = -4 mm
z2_path = 0                                       ! z = 0

ppath, 1,, x1_path,  y1_path, z1_path     ! First coordinate location (upper point)
                                                  ! Syntax: PPATH, POINT, NODE, X, Y, Z, CS
                                                  ! Defines a path by entering specific
coordinate locations
ppath, 2,, x2_path,  y2_path, z2_path     ! Second coordinate location (lower point)

vsel, s, volu,, 1, 15       ! Select all defined volumes
eplo
pdef, Temp, temp            ! Interpolates an item onto a path
/pbc, path, 1               ! Shows boundary condition symbols and values on displays
                            ! Especially show the course of the defined path
/title, Temperature - vertical path through laser diode, heat sink and steel structure

/xrange, 0, 5e-3            ! x-range from 0 to 5 mm
/yrange, 0, 100             ! y-range from 0 °C to 100°C
/axlab, x, Path Length      ! Title x-axis
/axlab, y, Temperature      ! Title y-axis
plpath, Temp        ! Plot the temperature profile through the defined path
!*******************************************************
```

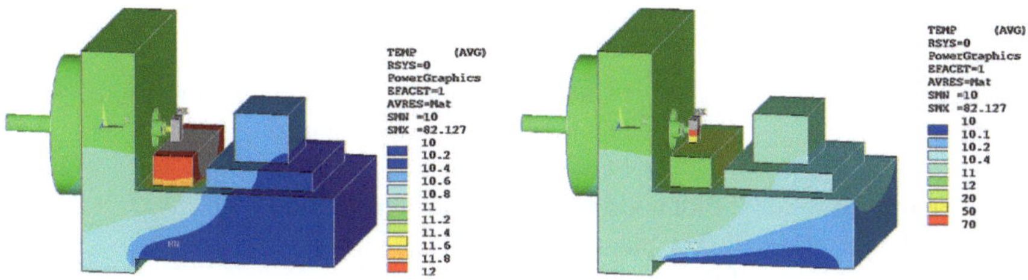

Fig. 12.10 Temperature distribution

Fig. 12.11 Temperature profile through a pre-defined path

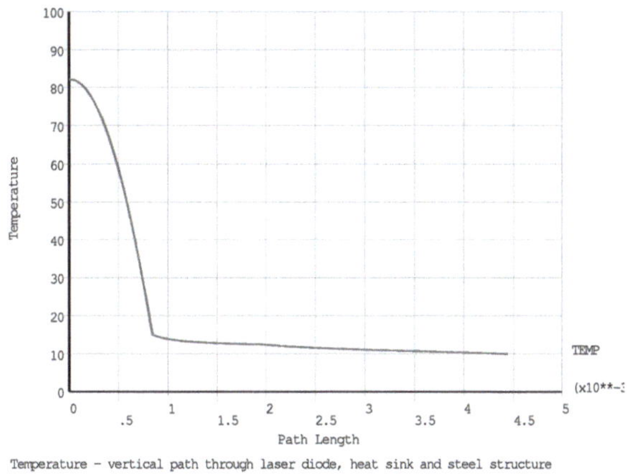

Figure 12.11 shows the image directly from ANSYS. If you, e.g., want to use EXCEL, you may add the command prpath, Temp, which means that you get the temperature values in ASCII. Now it is very easy to export them to an EXCEL table.

Another way to show the thermal results is to define a cutting plane with the APDL command /CPLANE. This allows you to look inside the model. A cutting plane is defined with one of the following methods:

1. It can be defined by a working plane WPLANE, KWPLAN, etc., or
2. The focus point or center of the window /FOCUS and the viewing direction for the display /VIEW, /ANGLE can define a cross-sectional plane, which is placed normal to the viewing direction and through the center of the image (focus point).

Here, we use the first method. By defining the variable z-plane, the position can be changed easily. In the example below, the cutting plane is identical with the mirror plane.

```
!*******************************************************************
/type,, 7
/cplane, 1                      ! Cutting plane relative to the working plane
z_plane = 0                     ! Cutting plane z-coordinate
                                ! your free choice
plnsol, temp             ! Plot node solution of the temperature field
wplane,, 0, 0, z_plane, 1, 0, z_plane, 0, 1, z_plane
wpstyl,,,,,,,,0          ! Do not show working plane
/replot                         ! ... which means 'plnsol, temp' again
/type,0                         ! Switch back to default settings for the display

finish
!*******************************************************************
```

Figure 12.12 shows the temperature profile within the structure by applying the cutting plane method. This is an example where the cutting plane is parallel to the axes of the coordinate system, but any rotation angle is also valid.

```
!*******************************************************
! SOLVER - Structural Solution
!*******************************************************
/solu
physics, read, struct
ldread, temp,,,,,, rth      ! Read the thermal results from the file 'Jobname.rth'
                            ! and apply loads derived from thermal environment
/pbf, temp,, 1              ! show temperatures in order to check the ldread command
antype, 0                   ! Analysis type 'static'
solve
finish

!*******************************************************
! POSTPROCESSOR - Mechanical Solution
!*******************************************************
/post1
/dscale, 1, 200             ! Scaling factor: 200 times
pldisp, 1                   ! Plot displacements (initial contour and displaced contour)
```

```
!----------------------------------------------------------------------------------------------------
/view,, 1, 1, 5
/contour                  ! Reset contour values
plesol, s, eqv         ! von Mises equivalent stress as element solution (not interpolated)
plnsol, s, eqv         ! von Mises equivalent stress as node solution (interpolated)
plnsol, u, sum         ! Plot the vector sum of the displacements (interpolated)
plnsol, u, x           ! plot only the displacement in x-direction
plnsol, u, y           ! plot only the displacement in y-direction
plnsol, u, z           ! plot only the displacement in z-direction
allsel
!----------------------------------------------------------------------------------------------------
```

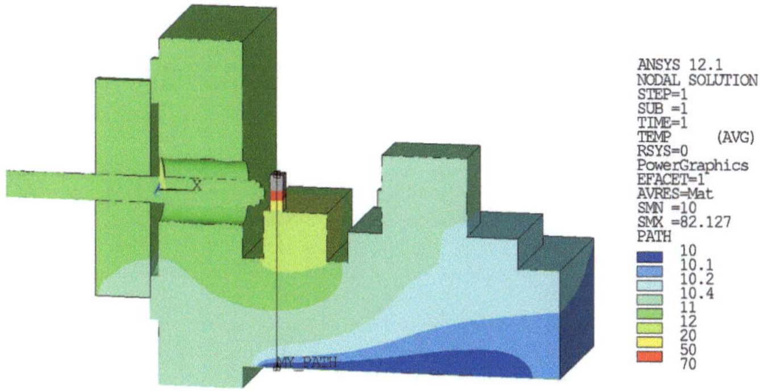

Fig. 12.12 Temperature distribution in a cutting plane at $z = 0$

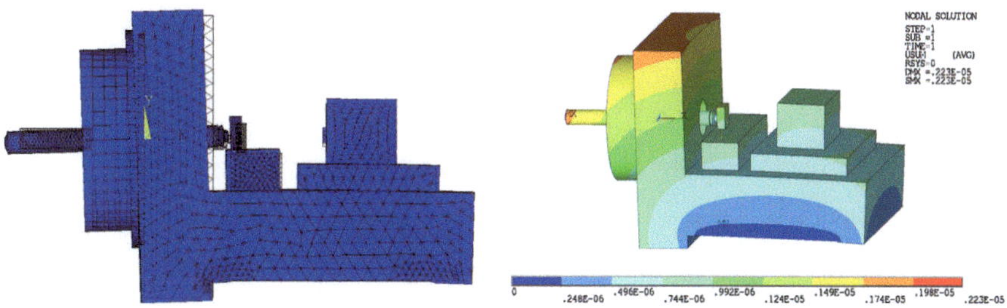

Fig. 12.13 Plot of the initial and displaced structure (*left*) and a color-coded illustration of the absolute value of the deflection vector u (*right*)

Figure 12.13 shows 2 examples of how you can plot the results. The left picture shows the superposition of the initial and the deflected shape with a scale factor of 200. The right plot uses colors to show the magnitude of the deflection vector. Of course it is also possible to plot the individual vector components separately as you can see in the source code above.

If you want to know the deflection of the fiber relative to the laser diode in order to check the quality of the optical coupling, you have to define the center nodes of the glass fiber and the center node of the light source. Since the node numbers vary depending on the element sizes, it is possible to get these two numbers automatically by the following APDL code. The NPLO command is not required, but a node plot helps you to check whether the selection is correct or not.

```
!--------------------------------------------------------

nsel, s, loc, x, 2.6e-3, 2.6e-3   ! Select all node with x-coordinate 2.6 mm (=inner end of the fibre)
nplo                              ! plot all nodes at the end area (to check the selection)
nsel, r, loc, y, -0.01e-3, 0.01e-3 ! Reselect those nodes from y=-0.01 mm to y=0.01 mm
nplo                              ! all nodes in a horizontal row
nsel, r, loc, z, -0.01e-3, 0.01e-3 ! Reselect those nodes from z=-0.01 mm to z=0.01 mm
/pnum, node, 1                    ! Show node numbers
nplo                   ! One node in the vicinity of the origin of the global coordinate system
*get, centernode_fibre, node, 0, num, min ! Save the node number in 'centernode_fibre'

/output, Fiber, vrt ! Syntax: /OUTPUT, Fname, Ext, Dir, Loc
                     ! Redirects text output to a file or to the screen
                     !   Fname - Filename to which text output will be redirected
                     !   Ext   - Filename extension
                     !   Dir   - Directory name
                     !   Loc   - Location within a file to which output will be written:
                     !          (blank) - Output is written starting at the top of the file
                     !          APPEND  - Output is appended to the existing file
                     ! Here file name: 'FIBER.VRT'

*get, Ux_Fibre, node, centernode_fibre, u, x ! Save the x-displacement of the center node in 'Ux_Fibre'
*get, Uy_Fibre, node, centernode_fibre, u, y ! Save the y-displacement of the center node in 'Uy_Fibre'
*get, Uz_Fibre, node, centernode_fibre, u, z ! Save the z-displacement of the center node in 'Uz_Fibre'
/output              ! Reset the output to avoid messages from being written to the file

allsel
!--------------------------------------------------------
nsel, s, loc, x,  2.65e-3, 2.65e-3 ! Select all nodes with x-coordinate 2.65 mm
                                   ! (=area of laser diode oppositely fibre)
nplo
nsel, r, loc, y, -0.1e-3, 0.1e-3
nplo
nsel, r, loc, z, -0.1e-3, 0.1e-3
nplo
*get, centernode_Laser, node, 0, num, min ! Save the node number in 'centernode_Laser'

/output, Fiber, vrt,, append
*get, Ux_Laser, node, centernode_Laser, u, x
*get, Uy_Laser, node, centernode_Laser, u, y
*get, Uz_Laser, node, centernode_Laser, u, z
/output
save, Fiber
*list, Fiber, vrt           ! Syntax: *LIST, Fname, Ext, Dir
                            ! Displays the contents of an external, coded file
finish
exit
!********************************************************
```

Here is the content of the file FIBER.VRT:

```
*GET  UX_FIBRE  FROM  NODE  26214  ITEM=U   X    VALUE=-0.214327872E-06
*GET  UY_FIBRE  FROM  NODE  26214  ITEM=U   Y    VALUE= 0.526976266E-06
*GET  UZ_FIBRE  FROM  NODE  26214  ITEM=U   Z    VALUE=-0.179158214E-08
*GET  UX_LASER  FROM  NODE  47084  ITEM=U   X    VALUE=-0.693208361E-06
*GET  UY_LASER  FROM  NODE  47084  ITEM=U   Y    VALUE= 0.808655301E-06
*GET  UZ_LASER  FROM  NODE  47084  ITEM=U   Z    VALUE= 0.452982095E-09
```

It lists all deflections in each coordinate axis and the corresponding node number. Now, you can calculate the difference by subtracting the numbers from each other for each direction.

The above-described sequence of commands performs a complete run of ANSYS. If you want to carry out further simulations with different parameters, such as other boundary conditions or loads, you do not have to run the whole process again. It will be sufficient to manipulate the so-called log file (file.log), which is created by ANSYS automatically when you start it.

12.6 Optical Simulation Programs

There are different simulation programs available on market. The basic underlying mathematical model corresponding to the propagation of light can classify them. This model is directly linked to the different models of light.

In general, on the one hand, light can be regarded as electromagnetic radiation, and on the other hand, it posses matter. But the simulation programs concentrate on the phenomena that are caused by the electromagnetic nature of light. The easiest model that can be used to describe the behavior of light is the paraxial image model. Light is described as rays, and only rays that are near the optical axis are considered. Solely refraction and reflection are examined to describe the path of light. This ideal behavior of light leads to very easy mathematical formulas. However, no aberrations are taken into account, so this model is far from being realistic.

The next step to get a realistic view of light is the so-called geometric model of light. The main difference to the paraxial image model is the consideration of light that is not only near the optical axis. This leads to aberrations and to mathematical formulas that are still analytically manageable. This is the basic model for ray tracing and is described in the next subsection.

The next step of complexity is a model based on Huygens' wave theory of light. In addition to previous models, this wave-optical model takes into account the phase of light and coherent phenomena like interference. It is the most complex model that is explained more in detail in the following subsections and is the basis for the beam propagation model (BPM).

Preferably, the simplest accurate model has to be used. Of course, the wave-optical model can be used every time, but that causes very complex mathematical models and that leads to very high calculating times. And in most cases, this is not necessary. It depends on the components that are placed in the path of light. If the dimensions of these components are large in comparison with the wavelength of light, easier models such as the geometric model of light should be used. If the dimensions of the components are in the same range as the wavelength of light, then wave-optical models should be used to calculate the distribution of light.

Diffractive optics such as gratings are a special case, the lens or mirror may be larger than the wavelength, but the smallest optical parts are in the dimension of the wavelength of light; therefore, the wave-optical model should be used. Some ray tracing simulation tools also offer special routines for that case, but they should be used with care. The mathematical model in the background could not handle this case normally, and hence, the formulas have to be modified. To get a deeper understanding of optical simulation programs, see Gross (2005a, b, 2007, 2008).

Rule of thumb:

- Dimensions of components in the optical systems are much larger than wavelength of light
  - Geometric model of light can be used
- Dimensions of components in the optical systems in same order as the wavelength of incident light
  - Wave-optical model of light should be used

12.6.1 Mathematical Model of Ray Tracing Programs

In 1930, T. Smith formulated the first unified mathematical model. However, the first application was in 1960 after suitably powerful computer became available, see also Mahajan (1998). This book summarizes the propagation of light rays which is fixed in the following mathematical expression, via optical surfaces as a linear equation. This can easily be done from the law of refraction to paraxial approximation. Therefore, the model of geometrical optics is applied:

$$n_1 \cdot \sin\alpha_1 = n_2 \cdot \sin\alpha_2 \quad \text{becomes } n_1 \cdot \alpha_1 = n_2 \cdot \alpha_2 \tag{12.1}$$

Figure 12.14 displays a random ray which hits a curved surface, thereby the height of the ray r in place x_1 can be described as the following equation:

$$r_1 = (x_1 - x_0)\alpha_0 + r_0 \tag{12.2}$$

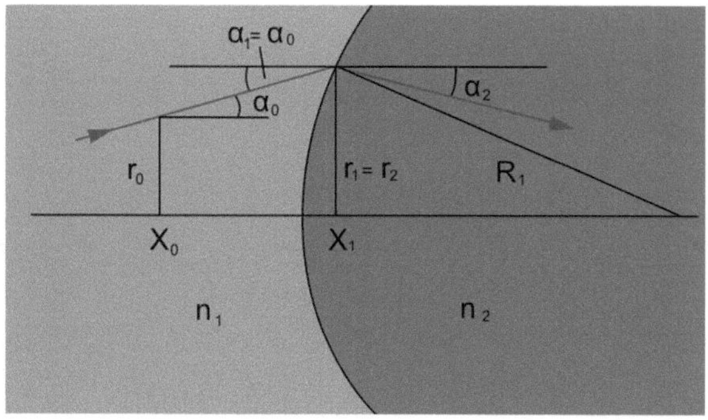

Fig. 12.14 Refraction of a light beam according to the matrix method

Moreover, a linear relationship at the point x_1 is given by the refraction itself:

$$n_2\alpha_2 = n_1\alpha_1 \frac{n_2 - n_1}{R_1} \cdot r_1 \tag{12.3}$$

Thereby $D = \frac{n_2 - n_1}{R_1}$ describes the refractive power of the spherical surface. It still applies $r_2 = r_1$ (see Fig. 12.14).

These two Eqs. 1.2 and 1.3 can be written in matrices as displayed in the following:

$$\begin{pmatrix} n_1\alpha_1 \\ r_1 \end{pmatrix} = \begin{pmatrix} 1 & 0 \\ \frac{x_1 - x_0}{n_1} & 1 \end{pmatrix} \cdot \begin{pmatrix} n_1\alpha_0 \\ r_0 \end{pmatrix} \tag{12.4}$$

$$\begin{pmatrix} n_2\alpha_2 \\ r_2 \end{pmatrix} = \begin{pmatrix} 1 & -\frac{n_2 - n_1}{R_1} r_1 \\ 0 & 1 \end{pmatrix} \cdot \begin{pmatrix} n_1\alpha_1 \\ r_1 \end{pmatrix} \tag{12.5}$$

This implies two 2 × 2 matrices. From Eq. 1.4 follows the so-called translation matrix T and from the Eq. 1.5 follows the refractive matrix B. This matrix notation allows the calculation of the propagation of light beams along optically effective surfaces by calculating the product of the corresponding matrices.

This method is particularly advantageous for numerical solving of lens systems and can therefore be calculated with the aid of computers. For more detailed explanations, see Mahajan (1998), Hecht (2001).

An example will illustrate this process:

The light strikes a medium 1 coming to a lens with a refractive index n_2 and then enters into a medium 3 as shown in Fig. 12.15.

The relevant parameters for the beginning are as follows: n_1, α_1, r_1. These are converted step by step into the final parameters n_3, α_3, r_3. The corresponding equation is as follows:

$$\begin{pmatrix} n_3\alpha_3 \\ r_3 \end{pmatrix} = B_2 \cdot T_{12} \cdot B_1 \cdot \begin{pmatrix} n_1\alpha_1 \\ r_1 \end{pmatrix} \tag{12.6}$$

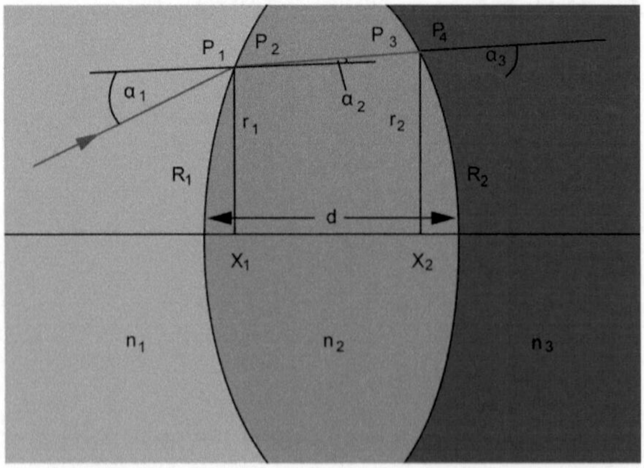

Fig. 12.15 Example for the matrix method

Therefore, initial and final parameters are directly linked to each other. This is a simple example in which the principle of the arrangement of matrices is clearly visible. The higher the number of lenses, the longer becomes the product of translational and refractive matrices.

But even for complex lens systems, the computing operations remain manageable. This basic principle also constitutes the basis for the calculation of the optical beam steering in simulation programs, also see Demtröder (2009).

However, it is further noted that these linear equations can only be applied for the Gaussian optics, i.e., using the paraxial approximation. This is the premise for the termination of the Taylor series for the sine function of Snell's law after the first element.

Furthermore, in this form, it is impossible to take the Seidel aberrations into account. As well for rays significantly beyond the axis, these linear approximations are not acceptable and the previous calculations need to be more complex.

However, to consider not only a few rays and imaging properties with this method, correction terms are used for computer calculations that are not explained in detail by the software developer. For some lens shapes, for example, gradient lenses or special aspherical lenses, easy paraxial beam paths are not sufficient and the imaging space is by so-called parabasal rays—rays that proceed diagonally through the paraxial imaging space—extended.

12.6.2 Basic Conventions of Optical Simulation Software

The construction of simulation programs with the presented working principle of ray tracing is based on the sequence of optically effective surfaces. Therefore, it is common to describe a lens through two areas. To achieve this, conventions are needed. Many sizes of technical optics are set out in DIN 1335 (2003). This includes, among other things, the definition of sign and axial directions.

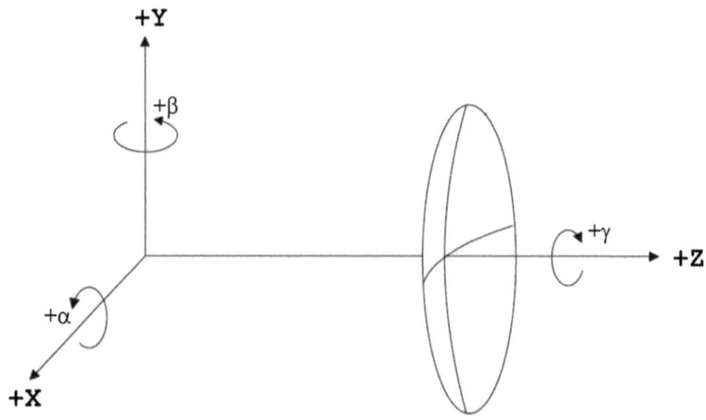

Fig. 12.16 Convention for the coordinate system in ray tracing programs

Usually, the positive direction of the z-axis coincides with the positive direction of the optical axis. The x- and the y-axis are located perpendicular to this. The angles around these axes are denoted by α, β, and γ.

A graphical representation is given in Fig. 12.16. If the focal point of a curved surface is in the negative direction along the z-axis, then the radius of curvature will be negative and positive if the focal point is to the right of the surface.

12.6.3 Strengths and Weaknesses of Ray Tracing

The optical designer and their experience are indispensable. The optical simulation programs can only be supportive. Furthermore, the critical assessment of the results is essential in this context. The limits of optics programs, which are based on the ray tracing principle, are introduced in the previous chapter. The calculation of beam steering is done by means of matrices which are based on the linearization of Snell's law of refraction. The focus of the programs is on the design of lenses and mirrors.

However, in general, an existing basic design is adapted to individual needs depending on the object and field of view. This basic fact has to be considered when not an image-generating optic is calculated, but rather when these simulation programs are used for elements of, e.g., optical telecommunications.

Due to the mathematical basic model, there are some peculiarities in the use of these programs of optical communications technology, which have to be considered:

- It is common for lenses that each optical surface is only traversed once, since they are arranged sequentially in a lens.
- In addition, reflective surfaces are implemented, but it is always to consider that the actual origins are breaking surfaces.

These two points can lead to very non-realistic beam path calculations. Therefore, each level has to be processed sequentially, which means in the order in which they are entered in the surface editor. After a single passage through the surface, this surface is not considered for further calculations.

Moreover, it may occur that light which passes the first time via a lens is calculated correctly in the context of the basic mathematical model. After that, it follows a beam deflection, and when the light passes through the lens a second time, it is possible that the lens has no further influence on the light rays.

Therefore, these rays proceed without deflection through this lens, which is undoubtedly false.

On the other hand, it may happen that a mirror does not reflect the rays but refracts the light beams when the next level to reach is positioned directly behind the mirror.

The optical designer must be aware of the limitations of these programs and enter a useful basic setup as origin into the surface editor from the start. This basic structure is generally created with paper and pencil, before the computer is used.

Another important point is the optimization: This is a powerful tool for proper adjustment of certain sizes. It is possible to modify variables for each surface such as size, radius of curvature, spacing, the angle of rotation, and others in the simulation program. Nevertheless, the right setup is crucial to achieve optimal results. If the basic structure is not well thought through, the optimization routine will not provide good results.

In the worst case, the structure can be optimized in the wrong direction and then has properties that are much worse. This situation is illustrated in Fig. 12.17. If the correct starting point is not chosen, the correct end point will not be reached.

Finally, one can say that these optical simulation programs are a great help for rapid calculation of optical systems and a good help for optimizing them. But it is only a limited aid for the optic designer, comparable to a good calculator. The designer must always check whether the supplied results are realistic or not and that can only be done with a lot of knowledge and experience.

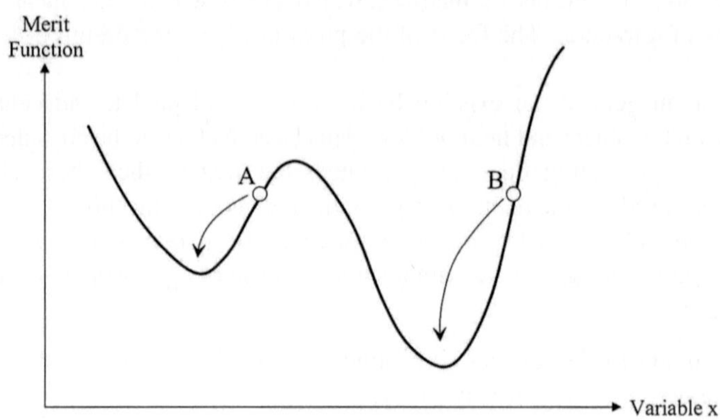

Fig. 12.17 Choosing the right starting point for optimization

12.7 Overview of Different Simulation Programs

In general, there are two different approaches for these simulation programs available on market. It depends on the setup calculated. For imaging optics, sequential ray tracing programs are normally used. These programs process the different parts in the optical path in a straight order. Every component is defined by its surface(s).

For that purpose, there are a lot of simulation programs available on market from inexpensive or even free to very expensive and of course powerful tools. It depends on the complexity of the setup, the analysis, and the optimization routines that are necessary to design and develop the optical part. For the beginning and for easy setups, cheap tools are adequate (Table 12.3).

The programs listed above offer partially more than classical ray tracing, most of them can also handle non-sequential and even physical optics. It is always a good idea to use the trial version (if available) to get familiar with the program.

These programs are even not limited to imaging optics and can also handle non-imaging optics, but are not specialist in that field. Non-imaging optics could be illumination purposes for LEDs or lasers or simulation of light propagation in different kinds of fibers. The user interface of the tools specialized in non-imaging simulations differs from the classical ray tracing programs. There are no surfaces defined normally; instead, complete parts are set up in the optical system. These programs work in the non-sequential mode and are not suitable to analyze and optimize classical ray tracing scenarios such as lens systems. But they still use ray tracing to analyze incoherent and even coherent light sources that works well for light phenomena that are not based on coherence (Table 12.4).

12.7.1 Simulation Programs Based on Wave Optics

As mentioned before, ray tracing is only accurate enough for setups that consist of optical components that are large in comparison with the wavelength of light used

Table 12.3 Sequential ray tracing programs (not complete)

Basic tools	Mid-range	High-class
OpTaliX (Optenso)	Zemax (Radiant)	Code V (Synopsys)
WinLens (Qioptiq)	Oslo (Lambda Research)	VirtualLab (LightTrans)

Table 12.4 Ray tracing programs for illumination (not complete)

Basic tools	Mid-range	High-class
OpTaliX (Optenso)	Zemax (Radiant)	LightTools (Synopsys)
	TracePro (Lambda Res)	VirtualLab (LightTrans)

in the system. In this case, wave optics must be applied. Based on wave optics, the BPM and the finite-difference time-domain (FDTD), which is more accurate and can be applied in more cases than BPM, can be used.

12.7.2 Beam Propagation Method

This method is based on the electromagnetic nature of light. The fundamental model uses the Helmholtz equation and is suitable for the slowly varying envelope approximation. This can be used for forward-traveling electromagnetic waves that vary slower in time and space according to the period or wavelength of light.

BPM can be applied for simulating the propagation of light in slowly varying optical waveguides, e.g., single-mode or multi-mode glass fibers. It was developed in the 1970s and it is much more complex in comparison with ray tracing and therefore needs much more computing time than ray tracing, which was introduced 40 years earlier.

BPM uses every transversal electric (TE) and transversal magnetic (TM) mode that can be propagated in the waveguide. But in longitudinal or z-direction, only slowly varying changes may occur. This fits really well for optical waveguides.

In this case, BPM is a very quick method to dissolve electromagnetic fields in integrated photonic components. Typically, it is used for calculating intensity and mode problems within core-cladding-structures with low numerical aperture, but is not limited to that. So it can handle bent, tapered, and terminated waveguides as well. These structures typically consist of isotropic optical materials such as fused silica, but this method has also been upgraded to be applicable to analyze the propagation of light in anisotropic materials such as liquid crystals. This extension offers the possibility to examine the polarization and rotation of light in anisotropic materials. Common examples are the tunability of a directional couplers based on liquid crystals or the light diffraction in LCD pixels.

But BPM is not the solution for every optical setup, it has limitations. As it is not suitable for the modeling of discretely or rapidly varying structures. Waveguides with large core diameters, such as polymer optical fibers, cannot be calculated accurately. Also large numerical apertures will lead to wrong results or in other words it is not suitable for light that propagates under a large range of angles. It is also not applicable for fibers with a high refractive index difference between the core and the cladding material; this is the case for polymer photonics. To get a more in depth view of BPM, the following literature is recommended (Okamoto 2006).

12.7.3 Finite-Difference Time-Domain Method

The FDTD method is a more rigorous method than BPM and can be applied for the cases where BPM is not suitable. But the calculations that are based on this method cannot longer be solved analytically; therefore, numerical approximations are used, which are only applicable for high computing power.

FDTD is also based on the time-dependent Maxwell's equations, and due to the fact it is a time-domain method, the solutions can cover a wide frequency range with a single simulation run and can include nonlinear effects and material properties. These equations, in partial differential form, are discretized using central-difference approximations to the space and time partial derivatives. The solution of these equations, the transmission modes, are solved for their electric field component in a volume of space, and at a given instant in time, then the magnetic field component is solved in the same environment at the next instant in time. This step-by-step process is repeated until the desired transient or steady-state electromagnetic field behavior is fully evolved.

For optical issues, two commercially available simulations programs are suitable, one is from Synopsys and belongs to the RSoft product family, e.g., BeamPROP (BPM) or FullWave (FDTD), and the other is OptiFDTD from Optiwave.

For a better understanding of the mathematical model of FDTD, Gedney (2011) is recommended.

References

ANSYS http://www.ansys.com (2015)

Demtröder, W.: Experimentalphysik 2 Elektrizität und Optik (2009)

DIN 1335:2003-12. DIN—Verlag Beuth, Germany (2003)

Gedney, S.D.: Introduction to the finite-difference time-domain (FDTD) method for electromagnetics. Synth. Lect. Comput. Electromagnet. **6**, 1–250 (2011)

Gross, H.: Handbook of optical systems. In: Gross, H. (ed.) Fundamentals of Technical Optics, vol. 1. Wiley-VCH, Weinheim (2005a)

Gross, H.: Handbook of optical systems. In: Gross, H. (ed.) Physical Image Formation, vol. 2. Wiley-VCH, Weinheim (2005b)

Gross, H.: Handbook of optical systems. In: Gross, H. (ed.) Aberration Theory and Correction of Optical Systems, vol. 3. Wiley-VCH, Weinheim (2007)

Gross, H.: Handbook of optical systems. In: Gross, H. (ed.) Survey of Optical Instruments, vol. 4. Wiley- VCH, Weinheim (2008)

Hecht, E.: Optik. Oldenbourg Verlag, München (2001)

Mahajan, V.N.: Ray Geometrical Optics. SPIE Optical Engineering Press, Bellingham (1998)

Okamoto, K.: Fundamentals of Optical Waveguides. Academic press, New York (2006)

Chapter 13
Reliability Tests

Abstract At the end of the development of a module, the question about the stability of the fiber-chip coupling and electrical and mechanical stability always stands. The following test runs are essential to modules, which must be tested for the datacom industry (DUT, device under test): (1) Temperature cycles. (2) Test at high dampness and high temperature. (3) Mechanical oscillations. (4) Mechanical shock. The environmental influences on products and their effects are numerous and from most different kind. The necessity for device and component manufacturers to bring long-lived products on the market requires suitable examinations under operating conditions. The goals of the environmental checks are the uncovering of cause-and-effects connections, the qualification of environmental conditions and the optimization of a long-term-stable product development. With aging and decomposition processes and with reliability studies, one mostly uses time lapse and artificial aging. The tests are accomplished in suitable environmental simulators, whereby commercial climate chambers are available from interior of some liters up to 100 m^2 large areas for automobile tests. A typical climatic chamber with an interior of 200 L and a usable temperature range between −40 and +280 °C is represented in Fig. 13.1. Basis of the audits provides international testing standards. In the specific case, the stress values must be adapted to the standards, which manifests itself, e.g., in the different procedures of the various standards for telecommunications goods.

13.1 Test Methods and Standards

If a reproduction of environmental condition is at all possible, the parameters must be well known. A classification arises, for example, as a result of collection of data of the air temperature and humidity to the world climatic map. Too many components that can be examined apply special operating conditions. These conditions must be evaluated and judged before a test series can start. The following test methods are currently used as follows:

© Springer-Verlag Berlin Heidelberg 2015
U.H.P. Fischer-Hirchert, *Photonic Packaging Sourcebook*,
DOI 10.1007/978-3-642-25376-8_13

- Testing under conditions of use
- Stricter conditions of use

The methods referred to in the first point, the operating conditions for the test pieces are exactly reproduced using the system. The second group of methods is always used when results must be available in a very short time, the so called accelerated testing. So it must be sought for a way with the help of tests to close with stricter conditions on the long-term behavior of samples. This can be achieved relatively easily by comparing the tests with natural stress and the actual stress, which was exposed to the sample. From this, the fast motion factor can be determined. The intensified stress can be achieved by in quick successive temperature changes or by elevated temperature exposure. A temperature cycle is defined by the lower temperature TA, an upper limit temperature TB, and the residence time in the achieved final temperatures (Figs. 13.1 and 13.2).

An accelerated testing program is typically cut into the following tasks:

(a) Define the objective and scope of the test program
(b) Identify the stress factors
(c) Determine volume of stress
(d) Conduct the accelerated test and analyze the collected data.
(e) Use models for examining the failure curve:
  - Arrhenius model
  - Temperature non-thermal model
  - Temperature–humidity model
  - Inverse power law model

In addition, the rate of change of temperature cycling is generally set to 1–3 °C per s. Other tests include very rapid temperature changes of up to 10 °C per s. Here, a two-chamber system is necessary. In addition to temperature, the relative humidity can be varied, which lies usually between 80 and 95 %.

Fig. 13.1 Climate chamber with control unit

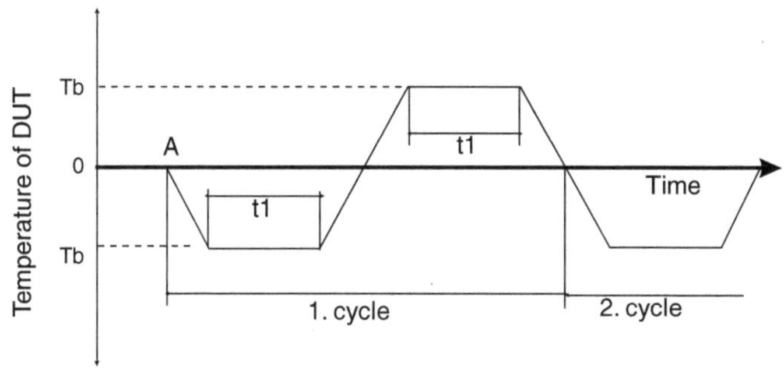

Fig. 13.2 Temperature change according to IEC 60068-2-14 cycle

There are several international and national standards for the examination of the long-term stability of modules for telecommunications use. Only the most important test standards are as follows: DIN (Deutsche Industrie Norm)

- MIL-STD (Military Standard/USA)
- IEC (International Engineering Committee)
- Telcordia (now Ericsson)

National standards, such as the German Industrial Standard (DIN), are supplemented by German military standards (e.g., VG). Similar to the German standard 40046 are in the UK the British Standard BS 2011 and the DEF STAN. The overall guidelines are summarized in the IEC 60068-X.

The most important American standards are MIL-STD 810 and MIL-STD 202 for equipment for components. These are taken into account in the framework of the IEC 60068-X.

13.2 Standards Reference Sources

- VDE VERLAG http://www.vde-verlag.de
- DIN http://www.din.de
- IEC-Shop http://www.iec-normen.de
- British Standard www.bsigroup.com/standards
- American MIL standards http://www.atec.army.mil

IEC 60068 contains various climatic conditions for measurements and tests. It establishes environmental testing procedures and their severity levels for electrotechnical products. These test methods are used to prove the resistance to expected environmental effects under conditions of use. Although intended primarily for electrical products, this standard is also applied to other technical products.

The origins come from the military regulations, such as MIL-STD 883E. More product-specific environmental assessments are specified in the relevant specification.

IEC 60068-X defines a set of uniform and reproducible environmental testing (especially climatic and mechanical) fixed with normal climates for measurements and tests. The IEC 60068-X can be referenced in specific provisions for electrical products. The tests are based on international experience. They provide insights into the behavior of electro-technical products, for example, through:

- Specified limits of temperature, air pressure, humidity, mechanical stress, and other environmental conditions that occur singly or in combination,
- Stress during storage and transport.

The tests allow comparing samples of work pieces together. To assess the goodness or the expected life of a product, the tests should be carried out according to a sampling plan and possibly supplemented by additional tests.

To customize the tests to the expected different environmental conditions, they sometimes contain multiple requirement levels that are determined according to the above-mentioned standards with fixed control parameters for humidity and temperature. The control times are strictly prescribed.

In addition, to test the seal of the module, salt solutions are introduced into the humidity. They speed up the aging process of the sample and thus allow a prediction of the expected service life.

The estimation of the failure rate later occurring during operation by the environmental assessments is a key objective in the evaluation of test results. The degradation of the fiber-chip coupling is the main cause of the reduced service life of optical modules. The reliability of connection is controlled by the choice of suitable connecting materials and holding mechanisms. Different connection systems for fiber-chip coupling have already come in the previous chapters of this book de-fitted for language. The failure rate of the optical coupling of these modules is typically specified in "FIT" (failure in time, failure per unit time): a FIT corresponds to a failure in 106 h or, in other words a million hours of operation without a single malfunction, which corresponds to at least a lifetime of about 1000 years (Nelson 1990).

The calculation models are based on a distribution curve for the life of a connection, showing an Arrhenius behavior, and thus can be calculated in analogy to the determination of the lifetime of a semiconductor device (Sect. 6.3). Results are, e.g., FIT rates of 100 (Hanafusa et al. 1998) for the optical coupling of AWG modules, which corresponds to a lifetime of 20 years, which is usually required for components for optical communication systems of the telecom companies. At this point, the issue will not be further deepened. It is recommended to read the above-mentioned literature. Table 13.1 provides an overview of the applicable test procedures for optical modules.

Table 13.1 Environmental tests

Influencing factors	Tests
Climatic	Cold
	Dust and sand
	Dry heat
	Vacuum
	Damp heat constant
	Humid heat cyclic
	Temperature changes
	Solar radiation
Mechanical	Case
	Accelerate
	Vibrations
Chemical and biological	Corrosive atmospheres
	Mold growth
Assembly and manufacturing	Solder
	Ultrasonic cleaning
	Mechanical Robustness of terminations
Other	Tightness

13.3 Stability Criteria of Telcordia Technologies

Internationally, the test series GR-486-CORE in the company Telcordia Technologies (1997) has prevailed for testing the optical modules of telecommunications. The following are the tests for laser modules, receiver modules, and AWG modules and will be presented shortly. The tests for at least eleven modules must be made as prototypes. Some tests even 20 modules are required. For accurate research, we refer to the previously mentioned literature.

13.3.1 Transmitter and Receiver Modules

Laser and receiver modules are tested in both the passive and the active state. Such tests include both mechanical tests and thermal cycling. In addition, long-term tests occur under high heat and humidity added, as in Table 13.2 illustrates in detail.

13.3.2 Arrayed Waveguide Grating Modules

AWG modules are tested analogously to the tests for transmitting and receiving modules. The test standard used here has the name of GR-1221-CORE:

Table 13.2 Stability criteria for transmission components and optical receiver for communication systems by Telcordia Technologies

Examination	Condition	Reference	Period of time	Failure criteria	Allowed failures
Mechanical shock	5-fach/axis	MIL-STD 883		$\Delta P_{out} < 0.5$ dB	0 of 11
	1500 g for 0.5 ms	Method 2002			
Vibration	20 g	MIL-STD 883	4 min/axis	$\Delta P_{out} < 0.5$ dB	0 of 11
	20–2000 Hz	Method 2007	4 cycles		
Mechanical shock	DT = 100 °C	MIL-STD 883	1 min	$\Delta P_{out} < 0.5$ dB	4 of 11
	0 °C till 100 °C	Method 1011	Transit time		
Fiber tensile	10 N, 3-fach 20 N, 3-times		5 s	$\Delta P_{out} < 0.5$ dB	0 of 11
Aging test I	70 and 85 °C, actively pursued in 0.1 mW (laser)	GR-453 5.18	1000 h	$\Delta P_{out} < 0.5$ dB	0 of 25
Aging test II	70 and 85 °C, actively pursued in 0.1 mW (laser)	GR-453 5.18	5000 h	$\Delta P_{out} < 0.5$ dB	0 of 10
Storage I	+85 °C		2000 h	$\Delta P_{out} < 0.5$ dB	0 of 11
Storage II	−40 °C		2000 h	$\Delta P_{out} < 0.5$ dB	0 of 11
Temperature changes	−40 °C bis +70 °C		100 cycles	$\Delta P_{out} < 0.5$ dB	0 of 11
	−40 °C bis +85 °C		500 cycles		
Steam test I	85 °C	MIL-STD 202	1000 h	$\Delta P_{out} < 0.5$ dB	2 of 11
	85 % moist	Method 1018			
		IEC 60068-2-3			
Steam test II	−10 °C bis +60 °C	MIL-STD 883	20 cycles	$\Delta P_{out} < 0.5$ dB	0 of 11
	95 % moist	Method 2007			
ESD		GR-453		$\Delta P_{out} < 0.5$ dB	0 of 6
		Section 5.18			

"Reliability Requirements for Optic Branching Components". Table 13.3 lists the corresponding test run. At the same time, the table contains a comparable certification process, in which the analog test methods are described.

At the same time, Table 13.3 contains standard comparable certification schemes, wherein analog test procedures are described.

Table 13.3 Stability criteria for AWG modules and other peripheral devices for optical communication systems by Telcordia Technologies

Examination	Condition	Reference	Period of time	Failure criteria	Allowed failures
Mechanical shock	8-fach/axis 1000 g for 0.5 ms	EIA/TIA-455-2A		$\Delta P_{out} < 0.5$ dB	0 of 11
		Section 6.2.1			
Vibration	20 g	EIA/TIA-455-11A	20 min/axis	$\Delta P_{out} < 0.5$ dB	0 of 11
	20–2000 Hz				
	12-fach/axis	Section 6.2.2			
Mechanical shock	DT = 100 °C	MIL-STD 883	1 min	$\Delta P_{out} < 0.5$ dB	4 of 11
	0 °C till 100 °C	Method 1011	Transit time 20 cycles		
Fiber stress	10 N, 3-fach		5 s	$\Delta P_{out} < 0.5$ dB	0 of 11
	20 N, 3-fach				
Storage I	+85 °C	EIA/TIA-455-5A	2500 h qualifying, 5000 h for information	$\Delta P_{out} < 0.5$ dB	0 of 11
		Section 6.2.5			
Storage II	−40 °C	EIA/TIA-455-4A	2500 h qualifying, 5000 h for information	$\Delta P_{out} < 0.5$ dB	0 of 11
		Section 6.2.6			
Temperature changes	−40 °C till +75 °C and −40 °C till +85 °C	EIA/TIA-455-3A	Holding time 15 min 500 h qualifying, 1000 h for information	$\Delta P_{out} < 0.5$ dB	0 of 11
Steam test I	75 °C	IEC 60068-2-3	2500 h qualifying, 5000 h for information	$\Delta P_{out} < 0.5$ dB	2 of 11
	90 % moist				
Steam test II	−40 °C to +75 °C	IEC 60068-2-38	5 cycles-holding time 4 h till 16 h, 2 °C/min transit time	$\Delta P_{out} < 0.5$ dB	0 of 11
	95 % moist 75 °C				
ESD		GR-453		$\Delta P_{out} < 0.5$ dB	0 of 6
		Section 5.18			
Water immission	43 °C, pH 5.5	EIA/TIA-455-12A	340 h	$\Delta P_{out} < 0.5$ dB	0 of 11
Salt spray	35 °C, 5 % NaCl-Solution	EIA/TIA-455-16A	168 h	$\Delta P_{out} < 0.5$ dB	0 of 11

13.4 Shaker Test Stations

In addition to the environmental tests, mechanical tests are carried out on the modules. An important part of this mechanical stress testing is the simulation of a case and the simulation of transportation by truck or plane. In the former case, high acceleration forces on the module will be applied, which can be up to 10,000 g at the start of a missile simulation. In most cases, however, the case is simulated from 1 m in height, which corresponds to an acceleration of about 1000 g.

In the transport simulation in the truck or plane, although smaller accelerations occur on, but in a very wide frequency range from 20 Hz to 3 kHz. This attempt is made to achieve by changing the excitation frequency is swept within a certain period of time between the two cutoff frequencies.

Apparatus for vibration testing consists of a vibration exciter, an associated electric drive amplifier, and a control unit with a computer package and application programs.

The vibration exciter is an electrodynamic exciter with peak forces of up to 45,000 N. The pathogen, which is depicted in Fig. 13.3, has an amplitude of 44 mm, which makes under sinusoidal excitation 20,000 N. For noise excitation between 20 Hz and 2 kHz, still a force of 22,800 N is applied to the DUT. Acceleration of 100 kg moving mass up to 350 g in one axis is possible as with this machine.

The associated amplifiers are mostly carried out in 19-inch racks, which are then placed together with the field excitation unit and the electrical distribution in a locked closet.

The associated control software allows control and monitoring of up to eight vibration systems. Test curves and gradient protocols can be detained arbitrarily and stored or backed up automatically according to certain specifications. In Fig. 13.4, a small plant with the corresponding test software is shown. This system allows for the following tests:

Fig. 13.3 Shaker test station for heavy DUT (RMS_GmbH 2014)

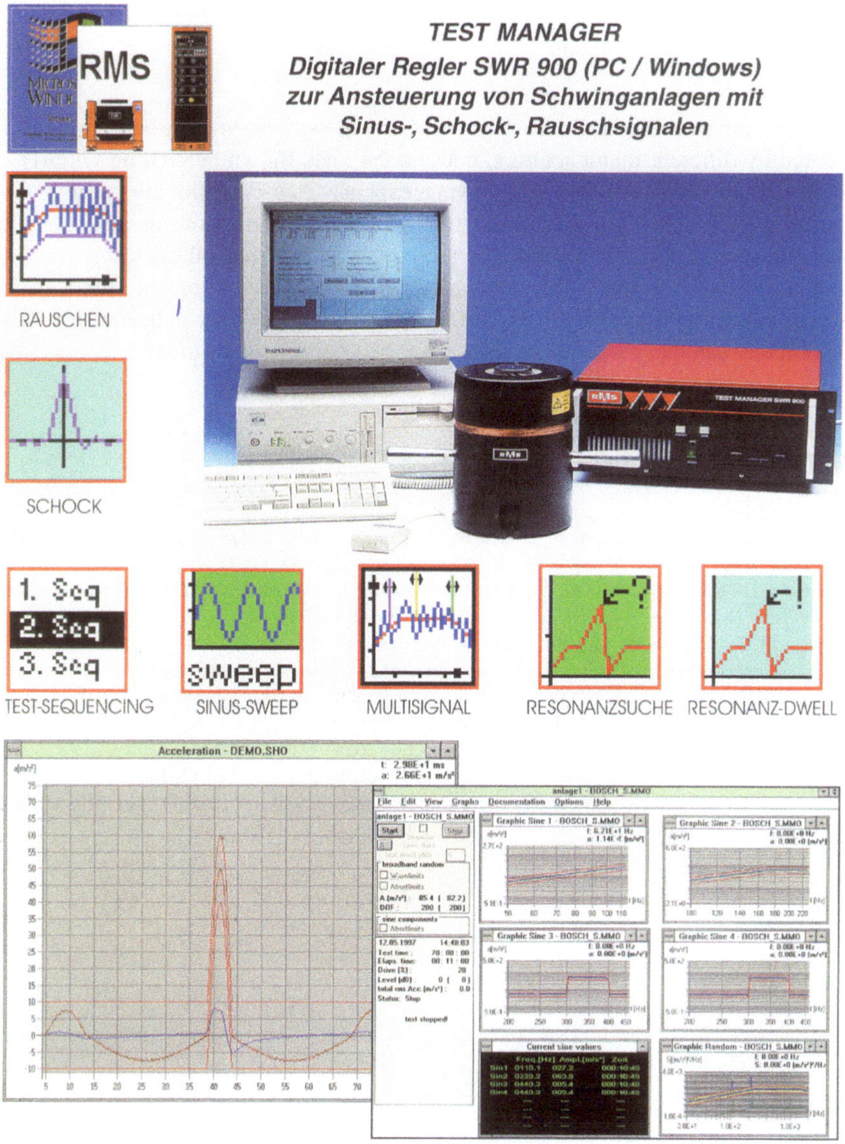

Fig. 13.4 Complete test station for vibration and shock tests

- Sine excitation
- Noise excitation
- Resonance search and stress
- Shock excitation
- Recording of the shock response spectrum

- Sine on noise excitation
- Noise on noise excitation

Thus, all possible test cases can be covered with such a system, which are offered by different manufacturers, not just the ones shown here (Ling GmbH).

For smaller companies worth buying expensive air-conditioning cabinets and shaker systems usually hardly, since the investments are still considerable. Another most inexpensive way to test their own products is to perform these test services of specialized institutions (IWM_ FHG 2014; Siemens 2014; Albrecht 1994), which are also certified according to ISO 9001 for this service. This indicates that these institutions can carry out all the tests with the appropriate accuracy.

References

Albrecht, H.: Photonic integrated device research at Siemens. LEOS summer topical meeting (1994)
Hanafusa, H., Sumida, S., Takato, N.: Long term reliability of silica-based planar lightwave circuit devices. Mater. Res. Soc. Symp. Proc. **531**, 349–358 (1998)
IWM_ FHG: Fraunhofer Institute for Mechanics of Materials IWM. http://www.en.iwm.fraunhofer.de/business-units/assessment-of-materials-lifetime-concepts/ (2014)
Ling GmbH: http://www.lds-group.com/Products/shakers-exciters/lds-vibration-test/shakers (2015)
Nelson: Accelerated Testing. Wiley, New York (1990)
RMS_GmbH: RMS. http://www.rms-testsystems.de (2014)
Siemens: Erprobung und Zulassung von LWL-Komponenten., München (2014)
Telcordia: http://telecom-info.telcordia.com/site-cgi/ido/docs.cgi?ID=111111111SEARCH&DOCUMENT=GR-468 (1997)

Index

© Springer-Verlag Berlin Heidelberg 2015
U.H.P. Fischer-Hirchert, *Photonic Packaging Sourcebook*,
DOI 10.1007/978-3-642-25376-8

MIX
Papier aus verantwortungsvollen Quellen
Paper from responsible sources
FSC® C105338

If you have any concerns about our products,
you can contact us on
ProductSafety@springernature.com

In case Publisher is established outside the EU,
the EU authorized representative is:
Springer Nature Customer Service Center GmbH
Europaplatz 3, 69115 Heidelberg, Germany

Printed by Libri Plureos GmbH
in Hamburg, Germany